Mathematics for Engineering

Mathematics for Engineering

Second edition

W. Bolton

Newnes

OXFORD AUCKLAND BOSTON JOHANNESBURG MELBOURNE NEW DELHI

Newnes
An imprint of Butterworth-Heinemann
Linacre House, Jordan Hill, Oxford OX2 8DP
225 Wildwood Avenue, Woburn, MA 01801 – 2041
A division of Reed Educational and Professional Publishing Ltd

℞ A member of the Reed Elsevier plc group

OXFORD   AUCKLAND   BOSTON
JOHANNESBURG   MELBOURNE   NEW DELHI

First published 1995
Second edition 2000

British Library Cataloguing in Publication Data
A catalogue record for this book is available from the British Library

ISBN 07506 49313

Printed and bound in Great Britain by Martins the Printers Ltd., Berwick-upon-Tweed

PLANT A TREE

BTCV
British Trust for
Conservation Volunteers

FOR EVERY TITLE THAT WE PUBLISH, BUTTERWORTH-HEINEMANN
WILL PAY FOR BTCV TO PLANT AND CARE FOR A TREE.

Contents

Preface   ix

Section A: Number and mensuration

1   Number and units   1.1   Introduction   2
                             1.2   Significant figures   2
                             1.3   Indices   5
                             1.4   SI units   8
                                    Problems   10

2   Mensuration   2.1   Introduction   12
                             2.2   Areas and perimeters   12
                             2.3   Volumes and surface areas   15
                             2.4   Irregular areas   21
                             2.5   Volumes of irregular solids   24
                                    Problems   25

Section B: Algebra

3   Algebraic equations 3.1   Introduction   30
                             3.2   Manipulating equations   30
                             3.3   Manipulation of formulae   37
                             3.4   Manipulation of units   40
                                    Problems   42

4   Linear equations   4.1   Introduction   45
                             4.2   Solving linear equations   45
                             4.3   Simultaneous equations   47
                                    Problems   55

5   Quadratic equations 5.1   Introduction   57
                             5.2   Factors   58
                             5.3   Completing the square   61
                             5.4   Solving by formula   64
                                    Problems   68

6   Equations in action 6.1   Introduction   71
                             6.2   Straight line motion   71
                             6.3   Electrical circuits   77
                                    Problems   87

Section C: Further algebra

7	Sequences and series	7.1	Introduction	92
		7.2	Sequences	92
		7.3	Series	95
		7.4	Binomial series	98
		7.5	Approximations	102
			Problems	104
8	Exponentials	8.1	Introduction	107
		8.2	Exponentials	107
		8.3	The exponential	108
		8.4	Manipulating exponentials	115
			Problems	117
9	Logarithms	9.1	Introduction	120
		9.2	Logarithms	120
		9.3	Properties of logarithms	123
		9.4	Solving exponential equations	125
			Problems	126

Section D: Trigonometry

10	Trigonometric ratios	10.1	Introduction	130
		10.2	Degrees and radians	130
		10.3	Trigonometric ratios	131
		10.4	Ratios for common angles	138
		10.5	The sine rule	145
		10.6	The cosine rule	148
		10.7	The areas of triangles	150
			Problems	153
11	Further trigonometry	11.1	Introduction	157
		11.2	Trigonometric relationships	157
		11.3	Trigonometric ratios of sums of angles	159
			Problems	161
12	Vectors	12.1	Introduction	162
		12.2	Addition of vectors	163
		12.3	Subtraction of vectors	169
		12.4	Resolution of vectors	171
			Problems	174

Section E: Graphs

13	Graphs	13.1	Introduction	178
		13.2	Cartesian graphs	178
		13.3	Graphs of equations	180
		13.4	Graphs of trigonometric functions	182

		13.5	Graphs of exponentials	186
		13.6	Adding and subtracting functions	186
		13.7	Solving equations graphically	187
		13.8	Gradients	190
		13.9	Areas under graphs	193
			Problems	195
14	Straight-line motion	14.1	Introduction	198
		14.2	Distance–time graphs	198
		14.3	Velocity–time graphs	200
			Problems	203
15	Linear graphs	15.1	Introduction	205
		15.2	Finding linear laws	205
		15.3	Reduction of laws to linear form	207
		15.4	Log graph paper	212
			Problems	214
16	Polar coordinates	16.1	Introduction	221
		16.2	Polar graphs	223
			Problems	225
17	Alternating waveforms	17.1	Introduction	227
		17.2	Average value	227
		17.3	Root-mean-square value	230
		17.4	Phasors	233
		17.5	Adding phasors	239
		17.6	Subtracting phasors	244
			Problems	246

Section F: Calculus

18	Differentiation	18.1	Introduction	252
		18.2	Gradients	252
		18.3	Derivative of $y = ax^n$	255
		18.4	Basic rules	258
		18.5	Derivatives of trigonometric functions	260
		18.6	Derivatives of exponential functions	263
		18.7	Derivatives of logarithmic functions	264
		18.8	The second derivative	265
		18.9	Derivatives	266
			Problems	266
19	Further differentiation	19.1	Introduction	269
		19.2	The product rule	269
		19.3	The quotient rule	270
		19.4	Function of a function	272
			Problems	273

20 Maxima and minima   20.1 Introduction   274
                              20.2 Turning points   274
                                    Problems   280

21 Integration   21.1 Introduction   282
                              21.2 Integration as the reverse of differentiation   282
                              21.3 Integration as the area under a graph   288
                                    Problems   292

22 Calculus in action   22.1 Introduction   294
                              22.2 Straight-line motion   294
                              22.3 Work   300
                              22.4 Root-mean square values   302
                              22.5 Current   303
                              22.6 Capacitors   304
                              22.7 Inductors   306
                                    Problems   308

                              Answers   312

                              Index   337

Preface

Aims

*   To provide an accessible, readable basic text in mathematics for students studying for General Vocational Qualifications (GNVQ) at the Advanced level in Engineering.

*   To develop the basic mathematical tools in number and mensuration, algebra, trigonometry, graphing and calculus needed by engineers and apply them in engineering

*   To provide a springboard for progression to further studies on Higher National Certificates, Higher National Diplomas or other higher education courses.

Audience

This book is aimed at engineering students studying the following GNVQ units in the Advanced level Engineering courses:

*   *Applied Mathematics in Engineering*

*   *Further Applied Mathematics*

*   *Further Mathematics for Engineering*

Content

The book is divided into a number of sections. At the beginning of each section are listed the chapters in the section, the aims of the section, and the skills required to tackle the various chapters.

*   **Section A: Number and mensuration**
    This covers the *Number and units* and *Mensuration* parts of the *Applied Mathematics in Engineering* unit.

*   **Section B: Algebra**
    This covers the *Algebra* part of the *Applied Mathematics in Engineering* unit.

*   **Section C: Further algebra**
    This covers the *Series* and *Indicial and exponential equations* parts of the *Further Applied Mathematics* unit and the *Algebra* part of the *Further Mathematics for Engineering* unit.

- **Section D: Trigonometry**

  This covers the *Trigonometry* parts of the *Applied Mathematics in Engineering* unit and the *Further Mathematics for Engineering* unit.

- **Section E: Graphs**

  This covers the *Functions and graphs* part of the *Applied Mathematics in Engineering* unit, the *Curve fitting* part of the *Further Applied Mathematics* unit and the *Graphical and Numerical Techniques* part of the *Further Mathematics for Engineering* unit.

- **Section F: Calculus**

  This covers the *Calculus* section of the *Further Applied Mathematics* unit and the *Differential and Integral Calculus* section of the *Further Mathematics for Engineering* unit.

Format

Each chapter has the basic format of: discussion of principles followed by worked examples and problems for revision of those principles. At the end of each chapter are further problems. Answers to all problems are supplied at the end of the book. While engineering applications are incorporated in each chapter, certain chapters are devoted entirely to showing how mathematics is applied in engineering applications.

Differences from the first edition

Since the first edition there have been changes in the GNVQ course units and the second edition has thus been expanded to enable all the changes to be covered. In addition there has been some restructuring of the text to enable it to be more easily segmented according to the three GNVQ units it is designed to cover.

W. Bolton

Section A
Number and mensuration

1   Number and units
2   Mensuration

The aims of this section are to enable the reader to:

*   Round calculations to the appropriate number of significant figures.
*   Use and calculate with indices and simplify combinations of indices.
*   Work with numbers in standard index form.
*   Express large or small values of a unit by using appropriate multiples and submultiples of the unit with their associated SI prefixes.
*   Calculate perimeter lengths and areas of basic and compound shapes involving rectangles, parallelograms, triangles, trapeziums and circles.
*   Calculate surface areas and volumes of basic and compound solids involving cylinders, cones, spheres, pyramids, prisms, frustum of cones, frustum of pyramids.
*   Estimate the areas of irregular shapes.
*   Estimate the volumes of irregular solids.

This section is about basic number skills and mensuration techniques required by engineers. It is is assumed that the reader has some skill in manipulating numbers and putting numbers in equations. If there are problems with equations, it might be necessary to preface Chapter 1 by the early parts of Chapter 3.

1 Number and units

1.1 Introduction

This chapter is a familiarisation with the manipulation techniques for numbers that are a necessary feature of engineering and includes a consideration of significant figures, rounding and approximation, the use of indices and the use of multiples and sub-multiples with SI units.

For example, if you measure the sides of a rectangular piece of metal as 120.3 mm and 83.3 mm and use a calculator to calculate the area as 10 010.99 mm^2, how certain are you that the actual area is this value and would it be better to quote that area as 10 011 mm^2? If you measure the voltage across a resistor as 4.3 V and the current through it as 0.33 A and use a calculator to determine the resistance, i.e. the value of voltage/current, as 13.03030303 Ω, how many decimal place figures should you be quoting?

With regard to units and indices, would you recognise that 0.0115 A, 11.5 mA and 11.5×10^{-3} A all represent the same value? Can you calculate the value of force/area when the force is given as 12 kN and the area as 120 mm^2 and recognise that the result can equally well be quoted as 100 000 000 N/m^2 or 0.1 kN/mm^2 or 100 MN/m^2?

1.2 Significant figures

When quoting values obtained from measurements, it is important to realise that when a measurement is quoted as being, say, 2.0 V that this implies that the actual value lies between 1.95 V and 2.04 V. The 0 in the 2.0 is assumed to have significance in that it implies we could have had a reading of perhaps 1.9 V or 2.1 V and that the value of 2.0 indicates that it is nearer to 2.0 than either of the possible readings of 1.9 or 2.1 on either side of it. Both the 2 and the 0 are significant in indicating the value of the measurement. Writing the value as 2.0 thus gives it to two *significant figures*.

Suppose, however, we had obtained a voltage reading to three significant figures of, say, 2.12 V. This indicates that the voltage value is nearer to 2.12 than possible readings of 2.11 and 2.13 on either side. When a value is quoted as 0.00145 A then the number of significant figures is the number of figures between the first non-zero figure and the last figure, i.e. between the 1 and the 5. Thus 0.00145 A is being quoted to three significant figures.

When quoting measurement values it is important that they should only be quoted to the number of significant figures with which the measurement can be made. Thus if a meter can only give a reading to one decimal place, e.g. 2.7 V, then it should not be quoted as 2.70 V since this would imply a greater accuracy than the measurement justifies.

Example

What are the number of significant figures in the distance measurements of (a) 3.1 m, (b) 12.50 mm, (c) 0.152 m, (d) 0.00189 m?

(a) 3.1 m has 2 significant figures.
(b) 12.50 mm has four significant figures.
(c) 0.152 m has three significant figures.
(d) 0.00189 m has three significant figures.

Example

State the following numbers to two significant figures: (a) 125, (b) 11.2, (c) 0.562, (d) 0.00159.

(a) Rounding to two significant figures gives 130.
(b) Rounding to two significant figures gives 11.
(c) Rounding to two significant figures gives 0.56.
(d) Rounding to two significant figures gives 0.0016.

1.2.1 Multiplying or dividing measurement values

When we calculate the value of quantity involving more than one measurement value then we might use a calculator and obtain a value with a large number of digits. For example, the value of a resistance might be being determined from measurements of the voltage across it and the current through it (resistance = voltage/current) and if the measured voltage was, say, 2.0 V and the measured current 0.13 A then my calculator gives the result of dividing these numbers as 15.384 615 38 Ω. However, not all these figures are significant. The voltage and the current were only measured to two significant figures and so it is pointless quoting all these numbers in the calculated resistance. Thus the number of significant figures for the voltage imply that it could have been 2.04 V and would still have been quoted as 2.0 V since this is nearer 2.0 then the next value of 2.1; the current might have been 0.132 A and still quoted as 0.13 A. These would give a resistance value of 15.454 545 45 Ω. The rule is adopted of *only quoting a result calculated from measurements to the same number of significant figures as the measurement with the least number of significant figures*. Thus we only quote the above resistance to two significant figures. To do this we *round* the calculated value to two significant figures. This involves looking at the third figure in the calculated value and deciding what clue it gives as to the possible number that should be quoted for the second figure. If the third figure is 5 or greater then the second figure is rounded up by 1; if the third figure is less than 5 then the second figure is left at its quoted value. Thus 15.384 615 38 Ω is rounded to 15 Ω. This then tells us that the resistance is likely to be somewhere between 14.5 Ω and 15.4 Ω but the accuracy of our measurements preclude us giving a more accurate value.

When multiplying or dividing measurement values, the result should be rounded to the same number of significant figures as the value which has the least number of significant figures.

Example

Determine the results of the following calculations involving measurement values: (a) 1.2×3.7, (b) 1.7×2.567, (c) $10.3/0.12$, (d) $0.123/0.045$.

(a) Each value is quoted to two significant figures and thus the calculated result should be rounded to two significant figures, i.e. 4.44 is rounded to 4.4.
(b) The 1.7 has the least number of significant figures and thus the result should be quoted to two significant figures, i.e. 4.3639 is rounded to 4.4.
(c) The 0.12 has the least number of significant figures and thus the calculated result should be rounded to two significant figures, i.e. 85.833 333 33 is rounded to 86.
(d) The 0.045 has the least number of significant figures and thus the calculated result should be rounded to two significant figures, i.e. 2.733 333 33 us rounded to 2.7.

1.2.2 Adding or subtracting measurement values

Suppose we measure two currents as 2.0 A and 0.6 A; note that one current is to two significant figures and the other to just one significant figure but both have the same number of decimal place values. What will be the value of the sum of the currents? The 2.0 A current lies somewhere between 1.95 A and 2.04 A. The 0.6 A current lies somewhere between 0.55 A and 0.64 A. Thus the sum must lie somewhere between $1.95 + 0.55 = 2.50$ A and $2.04 + 0.64 = 2.68$ A and quoting the result as 2.6 seems reasonable. Suppose we add the two measured values 0.10 and 0.014. The 0.10 value can lie between 0.095 and 0.104. Thus there is doubt as to the value of the third decimal place value. Hence, when we add the numbers to give 0.114 we need to round to just two decimal places and quote the result as 0.11.

When adding, or subtracting measurement values, the result should be rounded to the same number of decimal places as the number in the calculation with the least number of decimal places.

Example

Determine the values that should be quoted for the results of the following calculations involving measured values: (a) $10.0 + 0.52$, (b) $0.045 + 0.005$, (c) $5.43 - 0.102$.

(a) The number with the least number of decimal places is 10.0 and thus the result should be quoted as 10.5.
(b) Both the numbers have the same number of decimal places and so the result should be quoted as 0.050.
(c) The 5.43 has the least number of decimal places and so the result should be quoted as 5.33.

Revision

1   Determine the values that should be quoted for the following
    calculations if all the numbers quoted refer to measured values: (a) 2.0
    $\times$ 0.65, (b) 1.1 $\times$ 0.01, (c) 0.50/0.050, (d) 0.015 + 0.13, (e) 1.2 − 0.55.

1.3 Indices

If we have 2×2 then we can represent this as 2^2; likewise $2 \times 2 \times 2$ is 2^3. In
general we have, where a represents some number:

a^1 represents a,
a^2 represents $a \times a$,
a^3 represents $a \times a \times a$,
and so on.

The terms used for a^x are that x is termed the *index* or *power* to which the
base a is raised. Indices (the plural of index) can be integers, positive,
negative or fractional.

1.3.1 Rules of indices

Since $a \times a \times a$ is the same as $a \times (a \times a)$ then a^3 has the same value as $a^1 \times a^2$. In general:

$$a^{n+m} = a^n \times a^m \qquad \text{Rule 1}$$

You may wonder what the value a^0 is. Consider $a^0 \times a^5$. According to
the above rule, this must have the value a^{0+5} and so a^5. The only way this
can occur is if $a^0 = 1$.
   Suppose we have $(ab)^2$. We can write this as $ab \times ab$ and hence it can
be written as $a \times a \times b \times b$ and so $(ab)^2 = a^2b^2$. In general:

$$(ab)^n = a^n b^n \qquad \text{Rule 2}$$

We can write $(a^2)^3$ as $a^2 \times a^2 \times a^2 = a^6$; thus, in general:

$$(a^n)^m = a^{n \times m} \qquad \text{Rule 3}$$

Suppose we have a^5/a^2. We can write this as $(a \times a \times a \times a \times a)/(a \times a)$
$= (a \times a \times a) = a^3$. Thus, when we divide two quantities with the same base,
we subtract the indices, i.e. $a^5/a^2 = a^{5-2} = a^3$. In general:

$$\frac{a^n}{a^m} = a^{n-m} \qquad \text{Rule 4}$$

If we have $(a/b)^2$, we can write it as $(a/b) \times (a/b) = a^2/b^2$. In general:

$$\left(\frac{a}{b}\right)^n = \frac{a^n}{b^n} \qquad \text{Rule 5}$$

Example

Simplify: (a) $3^2 \times 3^4 \times 3^{-1}$, (b) $x^2 \times x^5$, (c) $(x^2y)^3$, (d) $2^5/2^3$, (e) x^{12}/x^5.

(a) Because the terms are multiplied we can add the powers to give 3^{2+4-1} or 3^5.

(b) Because the terms are multiplied we can add the powers to give x^{2+5} or x^7.

(c) Each of the terms in the brackets is cubed and so we can write this as $(x^2)^3 \times (y)^3$ and hence x^6y^3.

(d) Division of quantities with the same base mean subtracting the indices, hence $2^5/2^3 = 2^{5-3} = 2^2$.

(e) Division of quantities with the same base mean subtracting the indices, hence $x^{12}/x^5 = x^{12-5} = x^7$.

Revision

2 Simplify: (a) $2^3 \times 2^4$, (b) $a^2 \times a^5$, (c) $x^2 \times x^4 \times x^3$, (d) $(ab^2)^5$, (e) $(2x)^2 \times (2x)^4$, (f) $2^9/2^2$, (g) x^5/x^2, (h) $(x^5 \times x^6)/x^7$.

1.3.2 Negative powers

Powers are not restricted to just positive numbers, we can also have powers with negative numbers. Suppose we now have $\frac{1}{2} \times 2 \times 2 \times 2$. This has the value 4. What we have is $2^3/2^1$. Thus we effectively write 2^{-1} for $\frac{1}{2}$ and then we can use the rule for the multiplication of powers and have $2^{3-1} = 2^2$. In general we have:

$$\frac{1}{a^x} = a^{-x}$$

Example

Simplify the following by using negative powers for the denominators:

(a) $\frac{a^4}{a^2}$, (b) $\frac{a^4 \times a^2}{a^3}$, (c) $\frac{10^5 \times 10^2}{10^8}$

(a) $a^4 \times a^{-2} = a^{4-2} = a^2$
(b) $a^4 \times a^2 \times a^{-3} = a^3$
(c) $10^5 \times 10^2 \times 10^{-8} = 10^{-1}$

Revision

3 Simplify the following by using negative powers for the denominators:

(a) $\frac{2^4}{2^2}$, (b) $\frac{a^5}{a^2 \times a^4}$, (c) $\frac{(2x)^2 \times (3x)^3}{2x^4}$

1.3.3 Fractional powers

We can use the multiplication rule to write $a^1 = a^{1/2}a^{1/2}$. But the square root of a number is defined as the two identical numbers which have to be multiplied together to give the original number. Thus a can also be written as the product of square roots, i.e. $a = \sqrt{a} \times \sqrt{a}$. Thus $a^{1/2}$ is the same as the square root of a. We could have written $a^1 = a^{1/3} \times a^{1/3} \times a^{1/3}$ and a as the product of cube roots, i.e. $\sqrt[3]{3} \times \sqrt[3]{3} \times \sqrt[3]{3}$. Thus $a^{1/3}$ is the cube root of a. In general, $a^{1/n}$ is the nth root of a:

$$a^{1/n} = \sqrt[n]{a}$$

It follows that:

$$a^{m/n} = (a^{1/n})^m = \left(\sqrt[n]{a} \right)^m$$

Example

Simplify the following: (a) $2^{1/2} \times 2^2$, (b) $a^{-1/2} \times a^3$, (c) $\dfrac{a^2 \times \sqrt{a}}{a^3}$.

(a) $2^{1/2} \times 2^2 = 2^{1/2+2} = 2^{5/2}$.

(b) $a^{-1/2} \times a^3 = a^{-1/2+3} = a^{5/2}$.

(c) We can write the square root as $a^{1/2}$ and so we have

$$\frac{a^2 \times a^{1/2}}{a^3} = a^{5/2} \times a^{-3} = a^{-1/2} = \frac{1}{\sqrt{a}}.$$

Example

What is the cube root of 10^6?

$$(10^6)^{1/3} = 10^{6/3} = 10^2$$

Revision

4   Simplify the following:

(a) $a^3 \times a^{1/2}$, (b) $x\sqrt{x}$, (c) $\dfrac{x^2}{\sqrt{x^3}}$, (d) $2^2 \times \dfrac{\sqrt{2}}{2^3}$, (e) $a^{5/2}a^2a^{1/3}$.

1.3.4 Standard form

When writing large numbers, or small numbers, it is more convenient to write them in *standard index form*; this involving writing a number as a product of a number and base ten raised to some index. Thus we can write 2500 as 2.5×10^3 and 0.025 as 2.5×10^{-2}.

Example

The tensile stress acting on a rod is 4 500 000 Pa. Write this value in standard index form.

$4\ 500\ 000 = 4.5 \times 10^6$ Pa

Example

The expansion of a rod is 0.0051 m. Write this in standard index form.

$0.0051 = 5.1 \times 10^{-3}$ m

Revision

5 Write the following numbers in standard index form: (a) 120, (b) 5600, (c) 650 000 000, (d) 0.0045, (e) 0.000 000 12.

1.3.5 Using a calculator

You may have a calculator which can be easily used for the working out of powers. It will have a key labelled x^y. Thus, for example, to use this to obtain 2^4: the procedure could be: press the key for 2, then press the x^y key, then press 4 and finally the key for = to give the answer. In this case 16. To obtain 2^{-4}, the procedure could be: press the key for 2, then press the x^y key, then the key for 4 followed by the +/− key, and finally = to give the answer of 0.0625.

Example

Use a calculator to evaluate (a) 5^3, (b) $3^{5/2}$.

(a) The key sequences are: press 5, press x^y, press 3, press =. The result is 125.
(b) The key sequences are: press 3, press x^y, press 2.5, press =. The result is 15.6 (to one decimal place).

Revision

6 Use a calculator to evaluate: (a) 3^5, (b) 2^{-5}, (c) $5^{0.3}$, (d) $3^{1/3}$, (e) $2^{1.5}$, (f) $3.5^{1.2}$, (g) $1.2^{-0.67}$, (h) $2.7^{-0.5}$.

1.4 SI units

The International System (SI) of units has seven base units and two supplementary units, the radian and the steradian. Table 1.1 lists these units and their symbols.

Table 1.1 *SI units*

Quantity	Unit	Unit symbol
Length	metre	m
Mass	kilogram	kg
Time	second	s
Electric current	ampere	A
Temperature	kelvin	K
Luminous intensity	candela	cd
Amount of substance	mole	mol

The SI units for other physical quantities are formed from the base units via the equation defining the quantity concerned. Thus, for example, volume is defined by the equation volume = a length × a length × a length, thus the SI unit for volume is that of length cubed, i.e. m^3. Velocity is defined as distance/time and so has the SI unit of m/s or m s^{-1}. Acceleration is velocity/time and so has the SI unit of $(m/s)/s = m/s^2$ or m s^{-2}. Some of these derived units are given special names. Thus, for example, the unit of force is defined by the equation force = mass × acceleration and so has the SI unit of kg m s^{-2}, this being given the special name of newton (N); hence, $1 N = 1 kg m s^{-2}$.

Standard prefixes are used for multiples and submultiples of units, the SI preferred ones being multiples of 10^3. Table 1.2 shows the standard prefixes.

Table 1.2 *Standard prefixes*

Multiplication factor		Prefix	
1 000 000 000 000 000 000 000 000	= 10^{24}	yotta	Y
1 000 000 000 000 000 000 000	= 10^{21}	zetta	Z
1 000 000 000 000 000 000	= 10^{18}	exa	E
1 000 000 000 000 000	= 10^{15}	peta	P
1 000 000 000 000	= 10^{12}	tera	T
1 000 000 000	= 10^9	giga	G
1 000 000	= 10^6	mega	M
1 000	= 10^3	kilo	k
100	= 10^2	hecto	h
10	= 10	deca	da
0.1	= 10^{-1}	deci	d
0.01	= 10^{-2}	centi	c
0.001	= 10^{-3}	milli	m
0.000 001	= 10^{-6}	micro	μ
0.000 000 001	= 10^{-9}	nano	n
0.000 000 000 001	= 10^{-12}	pico	p
0.000 000 000 000 001	= 10^{-15}	femto	f
0.000 000 000 000 000 001	= 10^{-18}	atto	a
0.000 000 000 000 000 000 001	= 10^{-21}	zepto	z
0.000 000 000 000 000 000 000 001	= 10^{-24}	yocto	y

Thus, for example, 1000 N can be written as 1 kN, 4 000 000 N as 4 MN, 0.001 m as 1 mm, and 0.000 001 A as 1 μA.

When calculating quantities involving equations with physical quantities expressed in units with prefixes, it is generally easiest to first write the quantities in standard index form with units having no prefixes before carrying out the calculation. For example, in the calculation of resistance by means of Ohm's law we might have a current I of 120 mA through a resistor giving a potential difference V of 10 V across it. The resistance R is then:

$$R = \frac{V}{I} = \frac{10\ [\text{V}]}{120\ [\text{mA}]} = \frac{10\ [\text{V}]}{120 \times 10^{-3}\ [\text{A}]} = 83\ \Omega$$

Note that when writing quantities with their units in equations, it is customary to write the units in square brackets to indicate that they are not algebraic symbols but units.

Revision

7 Write the following physical quantities with units using standard prefixes: (a) 120 000 m, (b) 0.0051 m, (c) 0.120 A, (d) 1200 V, (e) 4 700 000 Ω.

8 Use the equation $V = IR$ to determine: (a) V when $I = 20$ mA and $R = 5$ kΩ, (b) I when $V = 5.0$ kV and $R = 2.0$ MΩ, (c) R when $V = 1.5$ V and $I = 360$ μA.

Problems

1 State the number of significant figures for each of the following measured quantities: (a) 3.5, (b) 0.014, (c) 0.0001, (d) 1230.

2 Round each of the following numbers to three significant figures: (a) 120.3, (b) 53.99, (c) 0.015 10, (d) 0.000 6925.

3 Determine the values that should be quoted for the results of the following calculations involving measured quantities: (a) 43 × 0.21, (b) 0.45 × 0.02, (c) 12.3 × 0.123, (d) 0.45/0.10, (e) 0.0025/0.012, (f) 0.120 + 0.045, (g) 0.015 + 0.12, (h) 120.3 − 5.55.

4 Simplify the following:

(a) $3^2 \times 3^6$, (b) $a^3 \times a^2$, (c) $a^5 \times a^{-2}$, (d) $\dfrac{a^4}{a^2}$, (e) $(2 \times 3)^2$, (f) $\left(\dfrac{1}{x^2}\right)^3$,

(g) $a^{1/2}a^2a^3$, (h) $\dfrac{x\sqrt{x}}{x^3}$, (i) $\dfrac{a^3}{\sqrt{a}}$, (j) $\sqrt{x^5}$, (k) $(-4)^{-2}$, (l) $2^5 2^{-2} 2^{1/2}$,

(m) $(a^{-1}b^{-2})^{-1}$, (n) $a^{-1} + a^{-2}$, (o) $\dfrac{3a^2 b^{-1}}{12 b^3 a}$.

5 Use a calculator to evaluate: (a) 2^7, (b) 2^{-4}, (c) $3^{4/5}$, (d) $5^{0.7}$.

6 Simplify the following and express your answer in standard index form:

(a) $\dfrac{1.2 \times 10^5}{0.6 \times 10^2}$, (b) $\dfrac{2 \times 5 \times 10^6 \times 8 \times 10^{-2}}{4 \times 10^3}$, (c) $\pi(1.2 \times 10^{-3})^2$.

7 Write the following physical quantities with units involving standard unit prefixes: (a) 5200 V, (b) 1500 W, (c) 0.160 A, (d) 0.000 120 A, (e) 0.010 m, (f) 120 000 000 m, (g) 4.0×10^{-3} m, (h) 1.5×10^6 Ω.

8 Use the equation $s = \frac{1}{2}at^2$ to determine: (a) s when $a = 9.8$ m/s^2 and $t = 120$ ms, (b) a when $s = 2.0$ km and $t = 400$ s, (c) t when $s = 500$ mm and $a = 2.0$ m/s^2.

9 Use the equation $F = ma$ to determine: (a) F when $m = 120$ kg and $a = 2.0$ m/s^2, (b) F when $m = 10$ g and $a = 0.2$ m/s^2, (c) m when $F = 5$ kN and $a = 2.0$ m/s^2, (d) a when $F = 2$ MN and $m = 1.5$ Mg.

2 Mensuration

2.1 Introduction

Engineers need to be able to solve problems involving the perimeters and areas of simple two-dimensional shapes such as rectangles, parallelograms, triangles, trapeziums, circles and compound shapes built up from the basic simple shapes. They need to be able to determine the surface areas and volumes of three-dimensional shapes such as cylinders, cones, spheres, pyramids, prisms and compound shapes built up from the basic simple shapes. In addition, they need to be able to determine irregular areas, such as those under some graphs, and irregular volumes. This chapter is about the techniques of determining the above quantities.

2.2 Areas and perimeters

The following are basic two-dimensional shapes:

Figure 2.1 *Basic shapes*

1   *Rectangle*
A rectangle of length l and breadth b (Figure 2.1(a)) has an area of:

$$\text{area} = l \times b$$

and a perimeter of length $2l + 2b$.

2   *Triangle*
A triangle with a base b and height h (Figure 2.1(b)) can be considered to be parts of two rectangles, with the diagonals of the rectangles forming the sloping sides of the triangle. The area of the triangle is half the sum of the areas of the two rectangles and thus has an area of:

$$\text{area} = \tfrac{1}{2}b \times h$$

Sometimes we are not given the vertical height of a triangle but the lengths of its sides and its internal angles. We can often use trigonometry to calculate the vertical height. Thus, for the triangle shown in Figure 2.1(b), we have $h/L = \sin \theta$.

3   *Parallelogram*
A parallelogram of length l and breadth b (Figure 2.1(c)) can be considered to be divided by a diagonal into two triangles and, since each triangle has an area of $\tfrac{1}{2}b \times h$, has an area of:

$$\text{area} = l \times b$$

4   *Trapezium*
A trapezium with parallel sides b and a separated by a perpendicular distance h (Figure 2.1(d)) can be considered to be divided by a diagonal into two triangles and, since they will have areas of $\tfrac{1}{2}ah$ and $\tfrac{1}{2}bh$, will have an area of:

Figure 2.2   *Example*

Figure 2.3   *Example*

$$\text{area} = \tfrac{1}{2}h(a + b)$$

5 *Circle*

A circle with radius r (Figure 2.1(e)), diameter $d = 2r$, has an area of:

$$\text{area} = \pi r^2 = \tfrac{1}{4}\pi d^2$$

and circumference of:

$$\text{circumference} = 2\pi r = \pi d$$

Example

Determine the area of the triangle shown in Figure 2.2.

$$\text{Area} = \tfrac{1}{2} \times 60 \times 50 = 1500 \text{ mm}^2$$

Example

Determine the area of the trapezium shown in Figure 2.3.

$$\text{Area} = \tfrac{1}{2} \times 50(60 + 15) = 1875 \text{ mm}^2$$

Revision

1   Determine the areas of the shapes shown in Figure 2.4.

Figure 2.4   *Revision problem 1*

2.2.1 Areas of compound shapes

Many more complex shapes can be considered has being built up from the basic shapes described above and their areas then determined by summing the areas of the constituent shapes. The following examples illustrate this.

Figure 2.5   *Example*

Example

Determine the area of the shaded part of Figure 2.5.

We can consider the shaded area as made up of two triangles; the line through the shaded area indicates one possible breakdown into the two triangles. The larger triangle has a base of 15 cm and a height of 10 cm;

it thus has an area of ½ × 15 × 10 = 75 cm². The small triangle has a base of 3 cm and a height of 9 cm and thus an area of ½ × 3 × 9 = 13.5 cm². The total area of the shaded area is thus 75 + 13.5 = 88.5 cm².

Example

A steel bar has a regular hexagonal section, with the flat lengths being 10 mm, and contains a central hole of diameter 5.0 mm. Determine the area of metal in the section.

Figure 2.6 shows the section. The hexagon can be considered to be made up of four triangles and a rectangle. Each triangle has a height given by $h/10 = \sin 30°$, i.e. $h = 5.0$ mm. Each triangle has a base given by $b/10 = \cos 30°$, i.e. $b = 8.7$ mm. Thus each triangle has an area of ½ × 8.7 × 5.0 = 21.75 mm². The height of the rectangle is $2H = 17.4$ mm and so its area is 10 × 17.4 = 174 mm². Hence the total area of the hexagon is 4 × 21.75 + 174 = 261 mm².

Alternatively we could consider the hexagon to be made up of six triangles. Each has a base of 10 mm and height $h = 10 \sin 60° = 8.7$ mm, hence area ½ % 10 % 8.7 = 43.5 mm². Thus the total area of the hexagon is 6 % 43.5 = 261 mm².

The area of the hole is $¼\pi × 10^2 = 78.5$ mm². Hence the area of the section is 261 – 78.5 = 182.5 mm², or 180 mm² when rounded to two significant figures.

Figure 2.6 *Example*

Example

Determine the area of the section shown in Figure 2.7. All dimensions are in centimetres.

We can regard this area as being the sum of the areas of two rectangles, namely 3 × 6 = 18 cm² and 12 × 4 = 48 cm². Hence the total area is 18 + 48 = 66 cm². Alternatively we can regard it as a rectangle 10 × 12 = 120 cm² with a rectangle 6 × 9 = 54 cm² cut out. Hence the total area is 120 – 54 = 66 cm².

Figure 2.7 *Example*

Revision

2   Determine the shaded areas of the sections shown in Figure 2.8.

Figure 2.8   *Revision Problem 2*

2.3 Volumes and surface areas

(a)

(b)

(c)

(d)

(e)

Figure 2.9 *Prisms*

Figure 2.10 *Forming an oblique prism from a right prism*

2πr

Figure 2.11 *Cylinder*

A *sphere* is a solid such that every point on its surface is the same distance, the radius r, from a fixed point, the centre. The volume V of a sphere is:

$$V = \tfrac{4}{3}\pi r^3$$

and is surface area A is:

$$A = 4\pi r^2$$

2.3.1 Prisms

A solid with two end parallel to each other and the same plane figure is called a *prism*. The term *right prism* is used if the side faces are perpendicular to the ends, the term *oblique* when this is not the case. Figure 2.9 shows some examples; Figures 2.9(a), (b), (c) and (d) are right prisms; Figure 2.9(e) is an oblique prism. The volume of a right prism is:

volume = area of base × height

and the surface area can be found by adding together the face areas. If we regard the right prism of Figure 2.9(d) as made up of a number of layers, then we can turn it into the oblique prism of Figure 2.9(e) by just sliding the layers over each other (Figure 2.10). Since the volume is unchanged by pushing the layers from a right prism into an oblique prism, the volume of an oblique prism is calculated using the same equation but using the perpendicular height.

A *cylinder* is a particular form of prism (Figure 2.9(a)), having a circular cross-section. Thus the volume of a cylinder is:

$$V = \pi r^2 h$$

The area A of the curved surface of a cylinder is the perimeter of the base multiplied by the height (think of a sheet of paper wrapped round the cylinder and then being unrolled to give a rectangular area of height h and length $2\pi r$ (Figure 2.11), i.e.

$$A = 2\pi rh$$

The area of each end face is πr^2 and so the total surface area is:

surface area = $2\pi rh + 2\pi r^2$

For the prism shown in Figure 2.9(b), if the triangle has equal length sides with each internal angle 60°, then the height of the apex of the triangle above its base is $L \sin 60°$ and so the volume is:

$$V = Ah = \tfrac{1}{2} \times L \times L \sin 60° \times h$$

The area of each side face is hL and so the total area of the side faces is $3hL$. The area of the each end faces is $½ \times L \times L \sin 60°$. Thus:

$$\text{total surface area} = 3hL + 2 \times ½ \times L \times L \sin 60°$$

For an oblique prism, such as that in Figure 2.9(e), the volume is:

$$\text{volume} = \text{base area} \times \text{perpendicular distance between parallel faces}$$

and thus for Figure 2.9(e), volume $= Ah$.

Example

Determine the volume and surface area of a prism with an equilateral triangle of side 3.0 m as the base and a height 9.0 m.

Figure 2.12 *Example*

The prism has the shape shown in Figure 2.9(b). The equilateral triangle base has each side of length 3.0 m and internal angles each 60° (Figure 2.12). For such a triangle, trigonometry gives $L/3.0 = \sin 60°$ and hence $L = 3.0 \times \sin 60° = 2.6$ m. Hence the area of the base of the prism is $½ \times 3.0 \times 2.6 = 2.0$ m^2. The prism has a height of 9 m and so the volume is $2.0 \times 9.0 = 18$ m^3.

Each of the side faces of the prism has an area of $3.0 \times 9.0 = 27$ m^2. The base and top surfaces each have an area of 2.0 m^2. Hence the total surface area is $3 \times 27 + 2 \times 2.0 = 85$ m^2.

Example

Determine the volume of a steel bar which has a regular hexagonal section with each hexagonal side being 10 mm, a length of 15 mm, and a circular hole of diameter 10 mm drilled centrally through its length.

Figure 2.13 *Example*

Figure 2.13 shows the shape, it being a right prism. The area of the end face of the steel bar was determined in the Example on page 14 as 180 mm^2. Hence the volume of the prism is $180 \times 15 = 2700$ mm^3.

Revision

3 Determine the volume of a right hexagonal prism with each of the base sides of the hexagonal being 2.0 m and the distance between the parallel hexagonal faces being 6.0 m.

4 Determine the total surface area of a cylinder of length 6.0 mm and diameter 4.0 mm.

5 A pipe has an internal diameter of 40 mm, an external diameter of 50 mm and a length of 6.0 m. Determine the volume of metal required for the pipe.

Figure 2.14   *Frustum of a cylinder*

Figure 2.15   *Surface area of frustrum*

Figure 2.16   *Revision problem 7*

6   Determine the volume of a prism when (a) the base is 4.0 cm square and its height is 6.0 cm, (b) the base is a a rectangle 2.5 cm by 2.0 cm and its height is 5.0 cm, (c) the base is circular of diameter 4.0 cm and its height is 8.5 cm.

2.3.2 Frusta of prisms

If a prism is cut by two planes inclined to one another, the solid contained between them is called a *frustum of the prism*. For example, Figure 2.14 shows a frustum of a cylinder. The average height of the cylinder is $\frac{1}{2}(h_1 + h_2)$ and so its volume is:

$$\text{volume} = \frac{1}{2}(h_1 + h_2) \times \pi r^2$$

The surface area of the curved surface is the circumference of the transverse section multiplied by the mean height (Figure 2.15), i.e.

$$\text{area} = 2\pi r \times \frac{1}{2}(h_1 + h_2)$$

The volume of the frustum of any triangular prism is:

$$\text{volume} = \text{average length of lateral sides} \times \text{area of transverse section}$$

Example

A frustum of mean length 1.20 m is cut from a cylindrical bar of diameter 0.20 m. Determine its volume and surface area of the curved surface.

Volume $= 1.20 \times \frac{1}{4}\pi \times 0.20^2 = 0.038$ m^2

Surface area of curved surface $= 1.20 \times \pi \times 0.20 = 0.75$ m^2

Revision

7   Determine the volume of the frustum of the cylinder shown in Figure 2.16. All dimensions are in millimetres.

2.3.3 Pyramids

A solid with a plane end and straight sides meeting in a point is called a *pyramid*. The shape of the plane end may have any plane shape, thus it may be rectangular, triangular, hexagonal, circular, etc. Figure 2.17 shows some examples. When the plane end is circular, as in Figure 2.17(c), the term *cone* is used.

Figure 2.17 *Pyramids*

Figure 2.18 *Right and oblique pyramids*

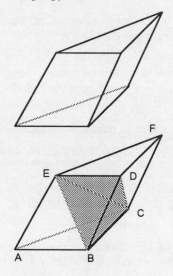

Figure 2.19 *A prism divided into pyramids*

Figure 2.20 *Area of curved surface*

The term *right pyramid* is used when the apex is vertically above the centre of the base, otherwise the term *oblique pyramid* is used. If we think of a right pyramid as built up from a number of layers, then sliding successive layers over another generates an oblique pyramid will the same volume (Figure 2.18). A right pyramid and an oblique pyramid will have the same volumes if their base areas are the same and their vertical heights equal.

The volume V of any pyramid is:

$$V = \tfrac{1}{3} \times \text{area of base} \times \text{perrpendicular height}$$

This can be demonstrated by first considering a triangular-based prism (Figure 2.19). Diagonal planes can be used to divide the prism into three equal volume triangular-based pyramids ABCE, DEFC and BCDE. Thus, since the volume of the prism is the base area multiplied by the perpendicular height, the volume of the triangular prism is one third the base area multiplied by the perpendicular height. This argument applies to other prisms.

The total surface area of a pyramid is the sum of the areas of the triangles forming the sides plus the area of the base.

A cone with a base of radius r and vertical height h (Figure 2.17(c)) has a base area of πr^2 and thus the volume V of a cone is:

$$V = \tfrac{1}{3}\pi r^2 h$$

The area of the curved surface of a cone may be obtained by imagining the cone surface to be a sheet of paper which is cut along the line OA and then unrolled onto a flat surface (Figure 2.20). The sheet is a segment of a circle with a radius equal to the slant height l of the cone and segment having a circumference equal to that of the cone, i.e. $2\pi r$. The area of a circle of radius l is πl^2 and its circumference is $2\pi l$; thus the area of a segment with an arc length $2\pi r$ is $(2\pi r/2\pi l) \times \pi l^2 = \pi r l$. Thus:

$$\text{Surface area of cone} = \pi r l$$

Example

A right cone has a base radius of 100 mm and a perpendicular height of 150 mm. Determine the volume of the cone and the area of its curved surface.

Figure 2.21 *Example*

$$\text{Volume} = \tfrac{1}{3}\pi r^2 h = \tfrac{1}{3}\pi \times 100^2 \times 150 = 1.57 \times 10^6 \text{ mm}^3$$

The slant height l of the cone (Figure 2.21) is given by the Pythagoras theorem as:

$$l = \sqrt{100^2 + 150^2} = 180 \text{ mm}$$

$$\text{Surface area} = \pi r l = \pi \times 100 \times 180 = 5.65 \times 10^4 \text{ mm}^2$$

Revision

8 Determine the volume and total surface area of a pyramid with (a) a square base of side 60 mm and a perpendicular height of 100 mm, (b) a rectangular base 7.0 m by 4.0 m and height 10.0 m.

2.3.4 Frusta of pyramids

Figure 2.22 *Frustum of a pyramid*

The *frustum of pyramid* is the portion cut off between the base and a plane parallel to it; it is sometimes called a *truncated pyramid*. Figure 2.22 illustrates this for a square-based pyramid with the frustum cutting off an upper pyramid from the original pyramid of vertical height H to leave a height h. Let A be the base area of the original pyramid and a the area of the cut surface. The areas are equal to the squares of the lengths of the corresponding sides, hence:

$$\frac{a}{A} = \frac{(ab)^2}{(AB)^2}$$

Since the sides are formed by the intersection of parallel planes with the triangular sides of the original pyramid, we have $(ab)/(AB) = (H - h)/H$. Thus:

$$\frac{a}{A} = \frac{(H-h)^2}{H^2}$$

and we can write:

$$\frac{a}{(H-h)^2} = \frac{A}{H^2} = k$$

where k is a constant. The volume V of the frustum is the volume of the whole pyramid minus the volume of the part removed. Thus:

$$V = \tfrac{1}{3}AH - \tfrac{1}{3}a(H-h)$$

$$= \tfrac{1}{3}kH^3 - \tfrac{1}{3}k(H-h)^3 = \tfrac{1}{3}k\big[H^3 - (H-h)^3\big]$$

$$= \tfrac{1}{3}k\big[H - (H-h)\big]\big[H^2 + H(H-h) + (H-h)^2\big]$$

Radius
r

x

h

l

Radius R

Figure 2.23   *Cone*

$$= \tfrac{1}{3}\left[H - (H-h)\right]\left[kH^2 + kH(H-h) + k(H-h)^2\right]$$

$$= \tfrac{1}{3}h\left[A + \sqrt{Aa} + a\right]$$

The surface area of the frustum of a pyramid is the sum of the areas of the trapeziums forming the sides plus the areas of the top and the base.

The volume V of the frustum of a cone (Figure 2.23) can be determined using the above equation with $A = \pi R^2$ and $a = \pi r^2$, where R is the radius of the base and r the radius of the cut surface, as:

$$V = \tfrac{1}{3}h\left[\pi R^2 + \sqrt{\pi R^2 \pi r^2} + \pi r^2\right] = \tfrac{1}{3}\pi h(R^2 + Rr + r^2)$$

The area of the curved surface of the cone is the area of the curved surface of the original cone minus the curved surface of the cut-off cone:

area of curved surface $= \pi R(l + x) - \pi r x = \pi R L + \pi(R - r)x$

Since $r/x = R/(l + x)$ then $rl + rx = Rx$ and so $x = rl/(R - r)$. Hence:

area of curved surface $= \pi R l + \pi(R - r)\dfrac{rl}{R - r}$

$$= \pi l(R + r)$$

The total surface area of the frustum of the cone is thus the area of the curved surface plus the areas of the two end faces:

total surface area $= \pi l(R + r) + \pi R^2 + \pi r^2$

Example

Determine the volume and area of the curved surface of the frustum of a cone if the diameter of the ends are 160 mm and 70 mm and its perpendicular height is 30 mm.

Volume $= \tfrac{1}{3}\pi h(R^2 + Rr + r^2) = \tfrac{1}{3}\pi \times 35(80^2 + 80 \times 35 + 35^2)$

$$= 3.8 \times 10^5 \text{ mm}^3$$

Area of curved surface $= \pi l(R + r)$

Since, using the Pythagoras theorem, $l^2 = h^2 + (R - r)^2 = 30^2 + 15^2$ then $l = 33.5$ mm and so:

area of curved surface $= \pi \times 33.5(80 + 35) = 1.2 \times 10^4 \text{ cm}^2$

Example

A hopper is in the shape of a frustum of a square-based pyramid. Determine its volume if the ends of the frustum are squares of sides 6.0 m and 4.2 m respectively and the perpendicular height is 3.5 m.

The hopper has the shape shown in Figure 2.22. Its volume V is:

$$V = \tfrac{1}{3}h\left[A + \sqrt{Aa} + a\right] = \tfrac{1}{3} \times 3.5\left[6.0^2 + \sqrt{6.0 \times 4.2} + 4.2^2\right]$$

$$= 68.4 \text{ m}^3$$

Revision

9 Determine the volume of the frustum of a prism which has a rectangular base 80 mm by 120 mm, an upper rectangular surface 30 mm by 45 mm and a vertical height of 80 mm.

10 A bucket is to be made from sheet metal and have a bottom with a diameter of 270 mm and a top with diameter 360 mm. If the vertical height is to be 360 mm, what area of sheet metal will be required?

2.4 Irregular areas

Engineers often have to estimate the areas of irregular shapes, e.g. the areas under irregular graphs. The following are three methods that can be used.

2.4.1 Counting squares

A simple method that can be used with graphs is to count the squares under the graph line and then multiply the number of squares by the value corresponding to their area.

Example

Estimate the area under the graph in Figure 2.24 between the ordinates $x = 0$ and $x = 3$.

Each square on the graph as an area of $0.5 \times 0.5 = 0.25$ square units. Between the ordinates $x = 0$ and $x = 3$, I estimate there are 16 whole squares and part square amounting to about 4 squares, i.e. a total of 20 squares. The area is thus 20×0.25 square units or about 5 square units.

Revision

11 Estimate the area under the graph in Figure 2.25 between the ordinates $x = 1$ and $x = 6$.

Figure 2.24 *Example*

Figure 2.25 *Revision problem 11*

Figure 2.26 *Mid-ordinate value*

Figure 2.27 *Example*

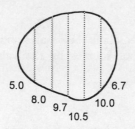

5.0   6.7
8.0   10.0
9.7
10.5

Figure 2.28 *Example*

Figure 2.29 *Trapezoidal rule*

2.4.2 Mid-ordinate rule

With the *mid-ordinate rule*, the area under a curve is divided into a number of equal width vertical strips. The area of the strips is then determined by multiplying the mid-ordinate value (Figure 2.26), i.e. the value of a line drawn through the middle of the strip, by the width of the strip and repeating this for each of the strips under the curve. This is the same as multiplying the width of the strips by the sum of the mid-ordinate values:

area = width of strips × sum of mid-ordinate values

The more strips that are crammed into the space under the curve, the more accurate will be the estimated area.

Example

Use the mid-ordinate rule to determine the area under the graph in Figure 2.27 between the ordinates $x = 0$ and $x = 3$.

The area has been divided into three strips, each of width 1.0 units. The values at the mid-ordinates are estimated as 0.4, 1.8 and 3.1. Thus the estimate for the area is $1.0 \times (0.4 + 1.8 + 3.1) = 5.3$ square units.

Example

Use the mid-ordinate rule to determine the area of the shape shown in Figure 2.28. The shape has been divided into 10 mm strips and the mid-ordinate values measured and quoted on the figure in millimetres.

The estimate for the area is $10 \% (5.0 + 8.0 + 9.7 + 10.5 + 10.0 + 6.7) = 499$ mm^2

Revision

12 Use the mid-ordinate rule to estimate the area under the graph in Figure 2.25 between the ordinates $x = 1$ and $x = 6$. Use strips of width 1 unit.

2.4.3 Trapezoidal rule

With the *trapezoidal rule* the area under the curve is divided into equal width strips. Each strip is then assumed to be a trapezium with an area of the base width multiplied by half the sum of the two parallel sides of the trapezium, i.e. half the sum of the ordinates forming its sides (Figure 2.29). Hence the area under the curve in Figure 2.29 is:

$$\text{area} = b \times \tfrac{1}{2}(y_0 + y_1) + b \times \tfrac{1}{2}(y_1 + y_2) + b \times \tfrac{1}{2}(y_2 + y_3)$$
$$+ b \times \tfrac{1}{2}(y_3 + y_4) + b \times \tfrac{1}{2}(y_4 + y_5) + b \times \tfrac{1}{2}(y_5 + y_6)$$

$$= b[\tfrac{1}{2}(y_0 + y_6) + y_1 + y_2 + y_3 + y_4 + y_5]$$

Thus, in general, we can write the trapezoidal rule as:

area = width of strip × [½(sum of first and last ordinates)
+ (sum of the remaining ordinates)]

As with the mid-ordinate rule, the greater the number of strips used the greater the accuracy of the estimated area.

Example

Use the trapezoidal rule to determine the area under the graph in Figure 2.30 between the ordinates $x = 0$ and $x = 3$.

Figure 2.30   *Example*

The area has been divided into three strips of width 1.0 unit. The values of the ordinates are $y_0 = 0$, $y_1 = 1.0$, $y_2 = 2.6$, $y_3 = 3.4$. Thus the area is:

area = $1.0 \times [½(0 + 3.4) + (1.0 + 2.6)] = 5.3$ square units

Revision

13 Use the trapezoidal rule to estimate the area under the graph in Figure 2.25 between the ordinates $x = 1$ and $x = 6$. Use strips of width 1 unit.

2.4.4 Velocity–time graphs

The area under a velocity–time graph between two time ordinates is the distance covered in that time interval.

Example

Determine the distance travelled during the interval $t = 0$ to $t = 6.0$ s for the movement giving the following distance–time data:

Velocity m/s	0	10	20	30	36	41	44
Time s	0	1.0	2.0	3.0	4.0	5.0	6.0

Using the trapezoidal rule with strips of width 1.0 s:

area = $1.0 \times [½(0 + 44) + (10 + 20 + 30 + 36 + 41)] = 159$ m

Hence the distance travelled is 159 m.

Revision

14 Determine the distance travelled during the interval $t = 1.0$ s to $t = 6.0$ s for the movement giving the following distance–time data:

Velocity m/s	0	11	22	28	32	34	35
Time s	0	1.0	2.0	3.0	4.0	5.0	6.0

2.4.5 Force–distance graphs

The area under a force–distance graph between two distance ordinates is the work done in moving the point of application of the force through that distance interval in the direction of the force.

Example

Determine the work done in moving the point of application of the force from 0 to 6.0 m in the direction of the force given the following force–distance values:

Force N	50	48	45	40	34	26	20
Distance m	0	1.0	2.0	3.0	4.0	5.0	6.0

Using the trapezoidal rule with strips of width 1. 0 m:

$$\text{area} = 1.0 \times [½(50 + 20) + (48 + 45 + 40 + 34 + 26)] = 228 \text{ N m}$$

Hence the work done is 228 J.

Revision

15 Determine the work done in moving the point of application of the force from 0 to 6.0 m in the direction of the force given the following force–distance values:

Force N	5.1	4.9	4.5	3.8	2.8	1.6	1.1
Distance m	0	1.0	2.0	3.0	4.0	5.0	6.0

2.5 Volumes of irregular solids

Rules that can be used for determining the volumes of irregular solids are similar to those used for estimating the areas of irregular figures; but here we have to use areas of cross-sections instead of lengths of lines.

When applying the *trapezoidal rule* we divide the solid into portions, each of constant width, as in Figure 2.31. The approximate volume of the first portion is taken as the width multiplied by the average area, i.e. $w \times ½(A_0 + A_1)$. The total volume is the sum of all the portions:

$$\text{volume} = w \times ½(A_0 + A_1) + w \times ½(A_1 + A_2) + w \times ½(A_2 + A_3)$$
$$+ w \times ½(A_3 + A_4)$$

$$= w \times [½(A_0 + A_4) + (A_1 + A_2 + A_3)]$$

Equal widths w

Figure 2.31 *Irregular solid*

In general:

$$\text{volume} = \text{width of portion} \times [½(\text{sum of first and last areas})$$
$$+ (\text{sum of the remaining areas})]$$

When applying the *mid-ordinate rule* we divide the solid into portions, each of constant width, as in Figure 2.31. The volume is then obtained by multiplying the constant width of a section by the sum of the mid cross-sectional areas.

Example

Determine the volume of a barrel which has a height of 2.0 m and its cross-sectional areas at intervals of 0.50 m are:

Area m²	1.80	2.55	3.15	2.55	1.80
Position m	0	0.50	1.00	1.50	2.00

Using the trapezoidal rule:

$$\text{volume} = 0.50 \times [\tfrac{1}{2}(1.80 + 1.80) + (2.55 + 3.15 + 2.15)]$$

$$= 5.025 \text{ m}^3$$

Revision

16 Determine the volume of a solid which has a length of 7.0 cm and its cross-sectional areas at intervals of 1.0 cm are 5.2, 3.5, 2.2, 2.2, 2.4, 2.5, 1.9, 1.0 cm².

Problems   1 Determine the shaded areas of the sections shown in Figure 2.32. All dimensions are in millimetres.

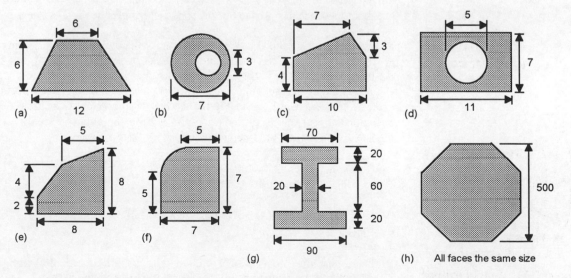

Figure 2.32   *Problem 1*

2 A cylindrical boiler has an internal diameter of 60 cm and a vertical height of 100 cm. Calculate the volume of water it contains when it is half full.

3 What length of copper wire of diameter 1.0 mm can be drawn from 1000 mm³ of copper?

4 A horizontal drainage channel has a regular cross-section of breadth at the base of 4.0 m, breadth at the top of 5.0 m and depth 4.0 m. If it has a length of 100 m, what will be the maximum volume of water it can contain?

5 Determine the volume of a prism when (a) the base is 45 mm square and its height is 50 mm, (b) the base is a a rectangle 25 mm by 60 mm and its height is 80 mm, (c) the base is circular of diameter 40 mm and its height is 50 mm.

6 Figure 2.33 shows a right-angled elbow-joint in a cylindrical duct. Determine the surface area of the metal required to form the joint.

7 Determine the volume and total surface area of (a) a cone of base diameter 600 mm and height 100 mm, (b) a pyramid with a square base of side 2.0 m and height 4.0 m, (c) a pyramid with an octagonal base of side 50 mm and perpendicular height 200 mm.

8 Determine the volume of a frustum of a cone if the diameters of the ends are 50 mm and 30 mm and the perpendicular height is 35 mm.

9 Figure 2.34 shows the form of a storage tank which is to be made from sheet metal. It has a base but no top. Determine the area of sheet metal required.

Figure 2.33   *Problem 6*

Figure 2.34 *Problem 9*

10 Determine the capacity of a bucket with base diameter 200 mm, top diameter 300 mm and vertical height 350 mm.

11 Determine the volume of a frustum of a triangular-based pyramid if the base has sides of 9.0, 12.0 and 15.0 cm, the top has sides of 3.0, 4.0 and 5.0 cm and the perpendicular height is 12 cm.

12 Estimate the areas under the graph in Figure 2.35 between the ordinates $x = 0$ and $x = 5$ by using (i) counting squares, (ii) the mid-ordinate rule, (ii) the trapezium rule.

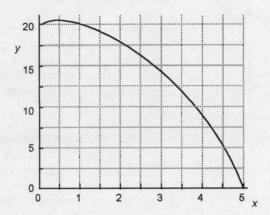

Figure 2.35 *Problem 12*

13 An irregular shape is divided into eight equal width strips, the width being 2.0 cm wide, and the mid-ordinate values measured for each strip. The values obtained where 7.8 cm, 12.2 cm, 14.8 cm, 15.7 cm, 16.4 cm, 16.3 cm, 14.8 cm and 10.0 cm. Estimate the area of the shape.

14 A river has a width of 10.0 m. The depth of the water is measured at the mid-points of ten equal intervals across the width of the river and the following values obtained at the mid-distance values of each interval: 0.50, 1.30, 2.23, 2.60, 2.77, 2.99, 2.68, 2.24, 1.30, 0.76 m. Estimate the cross-sectional area of the river.

15 The area under a graph between $x = 0$ and $x = 6$ is dived into six equal width strips. Determine the area between these ordinates if the values of ordinates are:

y	36	50	57	54	43	27	0
x	0	1.0	2.0	3.0	4.0	5.0	6.0

16 Determine the distance travelled during the interval $t = 0$ to $t = 6.0$ s for the movement giving the following distance–time data:

Velocity m/s	0	15	29	35	37	38	38
Time s	0	1.0	2.0	3.0	4.0	5.0	6.0

17 Determine the distance travelled during the interval $t = 1.0$ s to $t = 6.0$ s for the movement giving the following distance–time data:

Velocity m/s	24	19	16	12	8.0	4.0	0
Time s	0	1.0	2.0	3.0	4.0	5.0	6.0

18 Determine the work done in moving the point of application of the force from 0 to 6.0 m in the direction of the force given the following force–distance values:

Force N	8.1	6.9	4.7	3.6	2.7	1.5	0.9
Distance m	0	1.0	2.0	3.0	4.0	5.0	6.0

19 Estimate the volume of a tree trunk of length 7.0 m if the cross-sectional areas at intervals of 1.0 m are 2.2, 1.9, 1.8, 1.8, 1.7, 1.6, 1.5, 1.4 m^2.

20 Estimate the volume of gravel in a mound which has a height of 6.0 m and cross-sectional areas at intervals of 1.0 m of 40.2, 36.1, 38.2, 22.6, 12.1, 5.3, 0 m^2.

Section B
Algebra

3   Algebraic equations
4   Linear equations
5   Quadratic equations
6   Equations in action

The aims of this section are to enable the reader to:

* Manipulate algebraic equations in order to solve engineering problems.
* Substitute values into linear equations and obtain solutions.
* Solve simultaneous equations.
* Solve quadratic equations.
* Manipulate and solve equations involving in linear motion and electrical circuit problems.

This section is about developing the skills to use algebra in the solution of engineering problems. It is assumed that the reader is not completely unfamiliar with algebra and can handle fractions and powers. Chapter 3 might well be used prior to Section A.

Chapter 3 is the basic chapter for the section, revising basic concepts of algebra and applying them to the solution of equations. Chapters 4, 5 and 6 all assume the concepts given in Chapter 3. Chapter 6 shows the applications of Chapters 3, 4 and 5 in a study of straight-line motion and basic electrical circuit analysis.

3 Algebraic equations

In the study of motion in a straight line with a constant acceleration we have the equation:

$$v = u + at$$

Given v, u and t can you find a value for a? In the analysis of an electrical circuit we have the equation for resistors in parallel

$$\frac{1}{R} = \frac{1}{R_1} + \frac{1}{R_2}$$

Given values for R_1 and R_2 can you find a value for R? A study of the oscillation of a simple pendulum gives the equation

$$T = 2\pi \sqrt{\frac{L}{g}}$$

Given values of T, π and L can you find a value for g?

This chapter is about the basic techniques that can be used in manipulating algebraic expressions and equations. Later chapters in this section are concerned with more complex equations.

3.1.1 The use of letters

In algebra *letters* are used to represent quantities. Thus in $V = IR$, the V represents the value of the voltage across a resistor with a resistance value of R and I the value of the current through it. The quantities represented by the letters might vary or be constant. Thus in the equation $V = IR$ we might have a constant value for R but V and I can vary.

Letters are also used to represent units. Thus we might have a voltage of two volts and write this as 2 V. The V here does not represent a quantity but is just a shorthand way of writing the unit of voltage. In textbooks, units are written as V but when the letter is used to represent a quantity it is written in italics as V.

3.2 Manipulating equations

The term *equation* is used when there is an exact balance between what is on one side of the equals sign (=) and what is on the other side. Thus with the equation $V = IR$, the numerical value on the left-hand side of the equals sign must be the same as the numerical value on the right-hand side. Thus, if V has the value 2 then the value of IR must have the value 2.

Thus if we have the equation $x + 12 = 20$, then the numerical value of the left-hand side of the equation, i.e. $x + 12$, must be the same as the numerical value on the right-hand side of the equation, i.e. 20. It does not matter what

x represents, the expressions on the left and right of the equals sign have the same numerical value. We can then consider what value x must have if this equality is to be true. This is termed solving an equation. *Solving an equation* means finding the particular numerical value of the unknown which makes the equation balance. Thus, for the above equation, x must have the value 8 for the value of the expression on the left-hand side of the equals sign to equal the value on the right.

3.2.1 Basic rules for manipulating equations

1   *Adding the same quantity to, or subtracting the same quantity from, both sides of an equation does not change the equality.*
    Thus if we subtract 12 from both sides of the equation $x + 12 = 20$, then $x + 12 - 12 = 20 - 12$ and so $x = 20 - 12 = 8$. We might consider this rule as being: *when you move a quantity from one side of the equals sign to the other side you change its sign*. This is termed *transposition*.

2   *Multiplying, or dividing, both sides of an equation by the same non-zero quantity does not change the equality.*
    If we have the equation:

$$\frac{2x}{3} = \frac{1}{4}$$

then we can multiply both sides of the equation by 3 to give:

$$3 \times \frac{2x}{3} = 3 \times \frac{1}{4}$$

and so

$$2x = \frac{3}{4}$$

We can then divide both sides of the equation by 2 to give:

$$x = \frac{1}{2}\frac{3}{4} = \frac{3}{8}$$

You might like to think of this rule as being: *when you move a numerator from one side of an equation to the other, it becomes a denominator; when you move a denominator from one side of an equation to the other, it becomes a numerator*. This is termed *transposition*.

In general, whatever mathematical operation we do to one side of an equation, provided we do the same to the other side of the equation then the balance is not affected. Thus we can manipulate equations without affecting their balance by, for example:

1   Adding the same quantity to both sides of the equation.

2   Subtracting the same quantity from both sides of the equation.

3   Multiplying both sides of the equation by the same quantity.

4   Dividing both sides of the equation by the same quantity.

5   Squaring both sides of the equation.

6   Taking the square roots of both sides of the equation.

7   Taking logarithms of both sides of the equation.

The term *transposition* is used when a quantity is moved from one side of an equation to the other side. The following are basic rules for use with transposition:

1   A quantity which is added on the left hand side of an equation side becomes subtracted on the right hand side.

2   A quantity which is subtracted on the left hand side of an equation becomes added on the right hand side.

3   A quantity which is multiplying on the right-hand side of an equation becomes a dividing quantity on the left-hand side.

4   A quantity which is dividing on the left-hand side of an equation becomes a multiplying quantity on the right-hand side.

The following examples illustrate some of these manipulations.

Example

Solve the equation $2x - 4 = x + 1$.

If we add 4 to each side of the equation then:

$$2x - 4 + 4 = x + 1 + 4$$

and so $2x = x + 5$. If we now subtract x from both sides of the equation:

$$2x - x = x + 5 - x$$

and so we have $x = 5$ as the solution.

Alternatively we might think of the equation manipulation as: we move -4 from the left side to the right to give:

$$2x = x + 1 + 4 = x + 5$$

We then move the x from the left side to the right to give:

$$2x - x = 5$$

and so we have $x = 5$ as the solution.

We can check this result by replacing x in the original equation by this value and confirming that the equation balances: $2 \times 5 - 4 = 5 + 1$.

Example

Solve the equation $\frac{2}{3}x = 8$.

Multiplying both sides of the equation by 3 gives:

$$3 \times \frac{2}{3}x = 3 \times 8$$

and so $2x = 24$. Dividing both sides of the equation by 2 gives:

$$\frac{2x}{2} = \frac{24}{2}$$

and so the solution is $x = 12$.

Alternatively we might think of the equation manipulation as: move the denominator 3 from the left side to the right to give a numerator:

$$2x = 3 \times 8 = 24$$

Move the numerator 2 from the left side to the right side to give a denominator:

$$x = \frac{24}{2} = 12$$

We can check the result by replacing x in the original equation by this value. Then we have $\frac{2}{3} \times 12 = 8$.

Example

The voltage V across a resistance R is related to the current through it by the equation $V = IR$. When $V = 2$ V then $I = 0.1$ A. What is the value of the resistance?

Writing the equation with the numbers substituted gives $2 = 0.1R$. Multiplying both sides of the equation by 10 gives $2 \times 10 = 10 \times 0.1R$. Hence $20 = 1R$ and so the resistance R is 20 Ω. Alternatively we might consider the equation manipulation as: moving the numerator 0.1 from the right side to the left gives $2/0.1 = R$ and so R is 20 Ω. We can check the result by putting this value back in the original equation and confirm that it still balances: $2 = 0.1 \times 20$.

Example

Solve the equation $2 = \sqrt{\frac{L}{10}}$.

Squaring both sides of the equation gives is the same as multiplying both sides of the equation by the same quantity since 2 is the same as $\sqrt{(L/10)}$:

$$2 \times 2 = \sqrt{\frac{L}{10}} \times \sqrt{\frac{L}{10}}$$

$$4 = \frac{L}{10}$$

Hence, we have $L = 40$. We can check this result by putting the value in the original equation to give $2 = \sqrt{(40/10)}$.

Example

Solve the equation $100 = R^2 + 75$.

Subtracting 75 from each side of the equation gives $25 = R^2$. Taking the square root of both sides of the equation then gives $R = 5$. We can check this result by putting this value in the original equation to give $100 = 5^2 + 75$.

Revision

1   Solve the following equations:

(a) $x + 7 = 12$, (b) $2x + 3 = x + 5$, (c) $x + 3 = 4x$, (d) $\frac{1}{2}x = 12$,

(e) $\frac{2}{3}x = 4$, (f) $\frac{1}{2}x + 2 = x + 1$, (g) $\frac{2x}{5} = \frac{12}{15}$, (h) $\frac{2x}{5} = \frac{12}{15}$,

(i) $x + 4 = \frac{1}{2}$, (j) $2x = \frac{5}{3}$, (k) $20 = 5\sqrt{x}$, (l) $4 + x^2 = 20$.

2   (a) The voltage V across a resistance R is related to the current through it by the equation $V = IR$. When $V = 10$ V then $I = 2$ A. What is the value of the resistance? The resistance unit will be Ω.
(b) The extension x of a spring is related to the stretching force F by the equation $F = kx$, where k is a constant. When $F = 10$ N then $x = 20$ mm. What is the value of k? The unit of k will be N/mm.
(c) The pressure p exerted by a force F applied over an area A is given by the equation $p = F/A$. When $p = 1000$ N/m^2 and $A = 0.01$ m^2, what is the value of F? The unit of F will be N.
(d) The impedance Z of an electrical circuit is given by $Z^2 = R^2 + X^2$. With $Z = 10$ Ω and $R = 5$ Ω, what is the value of X?

3.2.2 Brackets

Brackets are used to show terms are grouped together, e.g. $2(x + 3)$ indicates that we must regard the $x + 3$ as a single term which is multiplied by 2. Thus *when removing brackets, each term within the bracket is multiplied by the quantity outside the bracket*, e.g.

$$2(x + 3) = 2x + 2 \times 3.$$

1 *Adding and subtracting bracketed quantities*
When a bracket has a + sign in front of the bracket then effectively we are multiplying all the terms in the bracket by +1, e.g.

$$a + (b + c) = a + b + c$$

When a bracket has a − sign in front of the bracket then we are multiplying all the terms in the bracket by −1, e.g.

$$a - (b + c) = a - b - c$$

2 *Multiplication of two bracketed quantities*
Consider the removal of brackets when we have $(a + b)(c + d)$. Following the above rule for the removal of brackets, each term within the first bracket must be multiplied by quantity inside the second bracket to give:

$$a(c + d) + b(c + d) = ac + ad + bc + bd$$

Note that the resulting expression is given by multiplying each term in one bracket by each term in the other bracket.

3 *Square of bracket quantities*
Another example of multiplication is the square of a bracketed term:

$$(a + b)^2 = (a + b)(a + b) = a(a + b) + b(a + b)$$

$$= a^2 + ab + ab + b^2 = a^2 + 2ab + b^2$$

The square is the sum of the squares of the two terms and twice their product.

4 *Product of the sum and difference of two terms*
Another example of multiplication is the product of the sum and difference of two terms:

$$(a + b)(a - b) = a(a - b) + b(a - b) = a^2 - ab + ab - b^2$$

$$= a^2 - b^2$$

The result is the difference of the squares of the two terms.

Example

Remove the brackets in the following:

(a) $5(x + 2)$, (b) $2 - 3(x + 5)$, (c) $2(x + 2) + 3(x - 4)$, (d) $(x + 1)(x + 3)$, (e) $(x + 5)^2$, (f) $(x + 5)(x - 5)$, (g) $(2x + 1)(5x - 2)$.

(a) $5(x + 2) = 5x + 10$

(b) $2 - 3(x + 5) = 2 - 3x - 15 = -3x - 13$

(c) $2(x + 2) + 3(x - 4) = 2x + 4 + 3x - 12 = 5x - 8$

(d) $(x + 1)(x + 3) = x(x + 3) + 1(x + 3) = x^2 + 3x + x + 3 = x^2 + 4x + 3$

(e) $(x + 5)^2 = (x + 5)(x + 5) = x(x + 5) + 5(x + 5) = x^2 + 5x + 5x + 25$
$$= x^2 + 10x + 25$$

(f) $(x + 5)(x - 5) = x(x - 5) + 5(x - 5) = x^2 - 5x + 5x - 25 = x^2 - 25$

(g) $(2x + 1)(5x - 2) = 2x(5x - 2) + 1(5x - 2) = 10x^2 - 4x + 5x - 2$
$$= 10x^2 + x - 2$$

Example

Solve the equation $\dfrac{2x + 3}{3} = \dfrac{x + 1}{4}$.

Multiplying both sides of the equation by 3 gives:

$$3\left(\frac{2x + 3}{3}\right) = 3\left(\frac{x + 1}{4}\right)$$

$$2x + 3 = 3\left(\frac{x + 1}{4}\right)$$

Multiplying both sides of the equation by 4 gives

$$4(2x + 3) = 4 \times 3\left(\frac{x + 1}{4}\right) = 3(x + 1)$$

and so $8x + 12 = 3x + 3$. Subtracting 12 from both sides of the equation gives $8x + 12 - 12 = 3x + 3 - 12$ and so $8x = 3x - 9$. Subtracting $3x$ from both sides of the equation gives

$$8x - 3x = 3x - 3x - 9$$

Hence $5x = -9$ and so $x = -9/5$.

Alternatively we could have tackled the equation by transposing numbers from side of the equals sign to the other. Thus, moving the numerator 3 from the left side to right gives:

$$2x + 3 = 3\left(\frac{x + 1}{4}\right)$$

Moving the numerator 4 from the right side to left gives:

$$4(2x + 3) = 3(x + 1)$$

Removing the brackets gives $8x + 12 = 3x + 3$ and hence $5x = -9$ and so $x = -9/5$.

Revision

3   Remove the brackets in the following:

(a) $5 + (2x + 1)$, (b) $4 - (x + 3)$, (c) $2(x + 1) + 4(x + 5)$,
(d) $5(2x - 1) + 3(x + 7)$, (e) $(x + 3)(2x + 7)$, (f) $(x - 3)^2$,
(g) $(2x + 1)^2$, (h) $(x + 1)(x - 1)$, (i) $x(2x + 1)$, (j) $x(x + 2) + 3(x + 5)$.

3.2.3 Manipulation of algebraic fractions

Consider the addition of two fractional terms a/b and c/d. We proceed by multiplying the numerator and denominator of each fraction by the same quantity; this does not change the value of a fraction. The quantity is chosen so that both fractions end up with the same denominator. Thus:

$$\frac{a}{b} \times \frac{d}{d} + \frac{c}{d} \times \frac{b}{b} = \frac{ad + cb}{bd}$$

Example

Express the following as single fractions:

(a) $\frac{2}{x} - \frac{3}{y}$, (b) $\frac{1}{x+1} + \frac{3}{x+2}$, (c) $\frac{x}{x-3} - \frac{1}{x+1}$.

(a) $\frac{2}{x} \times \frac{y}{y} - \frac{3}{y} \times \frac{x}{x} = \frac{2y - 3x}{xy}$

(b) $\frac{x}{x+1} \times \frac{x+2}{x+2} + \frac{3}{x+2} \times \frac{x+1}{x+1} = \frac{x^2 + 5x + 3}{(x+1)(x+2)}$

(c) $\frac{x}{x-3} \times \frac{x+1}{x+1} - \frac{1}{x+1} \times \frac{x-3}{x-3} = \frac{x^2 + 3}{(x-3)(x+1)}$

Revision

4   Express the following as single fractions:

(a) $\frac{2}{x} + \frac{5}{y}$, (b) $\frac{2}{x^2} - \frac{1}{y}$, (c) $\frac{2}{x+5} + \frac{1}{x-1}$, (d) $\frac{x}{x+1} - \frac{1}{x-3}$.

3.3 Manipulation of formulae

Suppose we have the equation $F = kx$ and we want to solve the equation for x in terms of the other quantities. Writing the equation as $kx = F$ and then transposing the k from the left-hand side to the right-hand side (or dividing both sides by k) gives

$$x = \frac{F}{k}$$

As another example, consider the following equation for the variation of resistance R of a conductor with temperature t:

$$R_t = R_0(1 + at)$$

where R_t is the resistance at temperature t, R_0 the resistance at 0°C and a a constant called the temperature coefficient of resistance. We might need to rearrange the equation so that we express a in terms of the other variables. As a first step we can multiply out the brackets to give:

$$R_t = R_0 + R_0 at$$

Transposing the R_0 from the left-hand to the right-hand side of the equation gives:

$$R_t - R_0 = R_0 \alpha t$$

Reversing the sides of the equation so that we have the term involving a on the left-hand side, we then have

$$R_0 at = R_t - R_0$$

Transposing the $R_0 t$ from the left-hand to right-hand sides gives

$$a = \frac{R_t - R_0}{R_0 t}$$

Example

Rearrange the following equation for resistances in parallel to obtain R_1 in terms of the other variables:

$$\frac{1}{R_1} + \frac{1}{R_2} = \frac{1}{R}$$

Transposing $1/R_2$ from the left-hand side to the right-hand side of the equation gives

$$\frac{1}{R_1} = \frac{1}{R} - \frac{1}{R_2}$$

Arranging the fractions on the right-hand side of the equation with a common denominator, then

$$\frac{1}{R_1} = \frac{R_2 - R}{RR_2}$$

We can invert the fractions provided we do the same to both sides of the equation. Effectively what we are doing here is transposing the numerators and denominators on both sides of the equation. Thus

$$R_1 = \frac{RR_2}{R_2 - R}$$

Example

The surface area A of a closed cylinder is given by the equation

$$A = 2\pi r^2 + 2\pi rh$$

where r is the radius of the cylinder and h its height. Rearrange the equation to give h in terms of the other variables.

Transposing the $2\pi r^2$ term gives:

$$A - 2\pi r^2 = 2\pi rh$$

Reversing the sides the equation gives:

$$2\pi rh = A - 2\pi r^2$$

Transposing the $2\pi r$ from the left-hand to right-hand sides gives:

$$h = \frac{A - 2\pi r^2}{2\pi r}$$

Example

The power P dissipated by passing a current I through a resistance R is given by the equation $P = I^2R$. Rearrange the equation to give I in terms of the other variables.

Writing the equation as $I^2R = P$ and then transposing the R gives:

$$I^2 = \frac{P}{R}$$

Taking the square roots of both sides of the equation then gives

$$I = \sqrt{\frac{P}{R}}$$

Revision

5   Rearrange the following equations to give the indicated variable in terms of the remaining variables:

(a) $v = u + at$, for a, (b) $F_1r_1 = F_2r_2$, for r_1, (c) $\frac{F}{A} = E \times \frac{x}{L}$, for x,

(d) $T^2 = \frac{4\pi^2 L}{g}$, for L, (e) $Z^2 = R^2 + X^2$, for X, (f) $pV = mRT$, for m,

(g) $E = \frac{1}{2}mv^2$, for v, (h) $s = ut + \frac{1}{2}at^2$, for a, (i) $Z = \frac{bd^2}{6}$, for d.

6   The length L of a bar of metal at $t°C$ is related to the length L_0 at $0°C$ by the equation $L = L_0(1 + at)$. If the length at $100°C$ is 1.004 m for a bar of length 1 m at $0°C$, what is the value of a?

7   The stress σ acting on a material is related to the force F acting over a cross-sectional area A of the material by the equation

$$\sigma = \frac{F}{A}$$

If the stress is 1000 N/mm² and the area 4 mm², what is the force?

8   The relationship between the pressure p, the volume V and the temperature T on the kelvin scale can be written as:

$$\frac{pV}{T} = mR$$

where m is the mass of the gas and R the characteristic gas constant. What will be the volume of a mass of 2.0 kg of air at a temperature of 293 K and a pressure of 1 100 000 N/m² (Pa) if the characteristic gas constant is 287 J/kg K?

3.4 Manipulation of units

It is not only numerical values that must balance when there is an equation. The units of the quantities must also balance. For example, the area A of a rectangle is the product of the lengths b and w of the sides, i.e. $A = bw$. The units of the area must therefore be the product of the units of b and w if the units on both sides of the equation are to balance. Thus if b and w are both in metres then the unit of area is metre × metre, or square metres (m²). The equation for the stress σ acting on a body when it is subject to a force F acting on a length of the material with a cross-sectional area A is

$$\sigma = \frac{F}{A}$$

Thus if the units of F are newtons (N) and the area square metres (m²) then, for the units to be the same on both sides of the equation:

$$\text{unit of stress} = \frac{\text{newtons}}{\text{square metres}} = \text{newtons/square metre}$$

or, using the unit symbols:

$$\text{unit of stress} = \frac{\text{N}}{\text{m}^2} = \text{N/m}^2$$

This unit of N/m² is given a special name of the pascal (Pa).

Velocity is distance/time and so the unit of velocity can be written as the unit of distance divided by the unit of time. For the distance in metres and the time in seconds, then the unit of velocity is metres per second, i.e. m/s or m s⁻¹. For the equation describing straight line motion of $v = at$, if the unit of v is metres/second then the unit of the at must be metres per second. Thus, if the unit of t is seconds, we must have

$$\frac{\text{metre}}{\text{second}} = \frac{\text{metre}}{\text{second}^2} \times \text{second}$$

or, in symbols,

$$\frac{\text{m}}{\text{s}} = \frac{\text{m}}{\text{s}^2} \times \text{s}$$

Hence the unit of the acceleration a must be m/s² for the units on both sides of the equation to balance.

Example

When a force F acts on a body of mass m it accelerates with an acceleration a given by the equation $F = ma$. If m has the unit of kg and a the unit m/s², what is the unit of F?

For equality of units to occur on both sides of the equation,

$$\text{unit of } F = \text{kg} \times \frac{\text{m}}{\text{s}^2} = \text{kg m/s}^2$$

This unit is given a special name of the newton (N).

Example

The pressure p due to a column of liquid of height h and density ρ is given by $p = h\rho g$, where g is the acceleration due to gravity. If h has the unit of m, ρ the unit kg/m³ and g the unit m/s², what is the unit of the pressure?

For equality of units to occur on both sides of the equation,

$$\text{unit of pressure} = \text{m} \times \frac{\text{kg}}{\text{m}^3} \times \frac{\text{m}}{\text{s}^2} = \frac{\text{kg}}{\text{m} \times \text{s}^2}$$

Note that in the previous example we obtained the unit of newton (N) as being the special name for kg m/s². Thus we can write the unit of pressure as

$$\text{unit of pressure} = \frac{\text{kg} \times \text{m}}{\text{s}^2} \times \frac{1}{\text{m}^2} = \text{N/m}^2$$

This unit of N/m² is given a special name, the pascal (Pa).

Example

For uniformly accelerated motion in a straight line, the velocity v after a time t is given by $v = u + at$, where u is the initial velocity and a the acceleration. If v has the unit m/s, u the unit m/s and t the unit s, what must be the unit of a?

For the units to balance we must have the unit of each term on the right-hand side of the equation the same as the unit on the left-hand side. Thus the unit of at must be m/s. Hence, since t has the unit of s:

$$\text{unit of } a = \frac{\text{unit of } at}{\text{unit of } t} = \frac{\text{m/s}}{\text{s}} = \frac{\text{m}}{\text{s}^2}$$

Revision

9 The pressure p acting on a surface of area A when it is subject to a force F is given by $p = F/A$. If A has the unit of m² and F the unit of N, what will be the unit of the pressure?

10 The torque T resulting from a force F being applied at the end of a lever arm of radius r is given by $T = Fr$. What is the unit of the torque when F has the unit of N and r the unit of m?

11 The specific heat capacity c of a material is defined by the equation $c = Q/mt$, where Q is the heat transfer needed for a mass m of a material to change its temperature by t. If Q has the unit of joule (J), m the unit of kg and t the unit of kelvin (K), what is the unit of c?

12 The electrical resistivity ρ of a conductor is given by the equation $\rho = RA/L$, where R is the resistance of a strip of length L and cross-sectional area A. What is the unit of resistivity if R has the unit Ω, L the unit m and A the unit m²?

13 Kinetic energy E is given by the equation $E = \frac{1}{2}mv^2$, where m is the mass in kg, v the velocity in m/s. What will be the unit of E?

Problems

1 Solve the following equations:

(a) $2x + 4 = 5 - x$, (b) $x - 4 = 12$, (c) $2x = \frac{3}{4}$, (d) $\frac{3}{2}x = \frac{2}{5}$,

(e) $\frac{x+1}{4} = 3$, (f) $x - 3 = \frac{x+1}{2}$, (g) $\frac{x-1}{2} = \frac{x+2}{3}$, (h) $\frac{1}{2}(x+2) = 5$,

(i) $4(x + 2) = 10$, (j) $\frac{x+1}{6} = \frac{2x-3}{2}$, (k) $\frac{x}{2} + \frac{x}{4} = 12$,

(l) $\frac{2x-1}{3} + 1 = 2$, (m) $x^2 + 3 = 19$, (n) $2x^2 + 5 = x^2 + 30$.

2 Simplify the following by removing the brackets:

(a) $(x + 5) + 3x$, (b) $4 - 2(x + 3)$, (c) $2(x + 1) + 5(x - 2)$,

(d) $3(x-2) + 5(x+1) - 2(2x-1)$, (e) $x(2x+3) + 5(x+1)$,

(f) $(x+3)(x-2)$, (g) $(x^2+1)(x+2)$, (h) $(x-3)^2$, (i) $(2x-1)^2$,

(j) $x^2 + (2x-3)^2$, (k) $(x+3)(x-3)$, (l) $(2x-1)(2x+1)$.

3 Express each of the following as single fractions:

(a) $\frac{5}{x} - \frac{2}{y}$, (b) $\frac{4}{x+1} - \frac{1}{2x+3}$, (c) $\frac{x}{x+5} + \frac{2x}{x-1}$, (d) $\frac{3}{x+1} - \frac{x}{x-1}$.

4 The area A of a circle is given by the equation $A = \frac{1}{4}\pi d^2$. Solve the equation for d given that the area is 5000 mm².

5 The velocity v of an object is given by the equation $v = 10 + 2t$. Solve the equation for the time t when $v = 20$ m/s.

6 The length L of a bar of metal at $t°$C is related to the length L_0 at 0°C by the equation $L = L_0(1 + 0.0002t)$. What is the length at 100°C for a bar of length 1 m at 0°C?

7 The specific latent heat L of a material is defined by the equation $L = Q/m$, where Q is the heat transfer needed to change the state of a mass m of the material. If Q has the unit of joule (J) and m the unit of kg, what is the unit of L?

8 The density ρ of a material is defined by the equation $\rho = m/V$. What is the unit of density if m has the unit kg and V the unit m³?

9 For the bending of a beam, the bending moment M is given by the equation $M = EI/R$. If E, the tensile modulus, has the unit N/m², I, the second moment of area, the unit m⁴, and R the radius of a bent beam the unit of m, what must be the unit of M?

10 Rearrange the following equations to give the indicated variable in terms of the remaining variables:

(a) $p_1 - p_2 = h\rho g$, for h, (b) $\rho = \frac{RA}{L}$, for R, (c) $E = V - Ir$, for r,

(d) $I\alpha = T - mr(a+g)$, for T, (e) $s = ut + \frac{Ft^2}{2m}$, for F,

(f) $E = \frac{mgL}{Ax}$, for x, (g) $\frac{M}{I} = \frac{E}{R}$, for R, (e) $E = \frac{1}{2}mv^2$, for m,

(i) $V = V = \frac{1}{3}\pi r^2 h$, for r, (j) $P = I^2 R$, for R,

(k) $m_1 a_1 = -m_2 a_2$, for a_2, (l) $F = \frac{mv^2}{R}$, for v,

(m) $C = \varepsilon_0 \varepsilon_r \dfrac{A}{d}$, for d.

11 For a gas that obeys Charles' law:

$$\frac{V_1}{T_1} = \frac{V_2}{T_2}$$

where V_1 is the initial volume and T_1 the initial temperature, V_2 the final volume and T_2 the final temperature. Rearrange the equation to express T_2 in terms of the other variables. Hence determine the final temperature when the volume changes from 1 m³ at 300 K to 2 m³.

12 For a beam simply supported over a span of L and carrying a central load of F plus a uniformly distributed load of w per metre of beam, then the reaction R at the beam supports is given by the equation:

$$R = \tfrac{1}{2}(W + wL)$$

Rearrange the equation so that w is expressed in terms of the other variables. Hence determine the value of w when $L = 3$ m, $W = 10$ kN and $R = 11$ kN.

4 Linear equations

4.1 Introduction

The term *linear equation* is used for an equation in which the unknown quantity is only raised to the power 1. For example, $2x + 5 = 0$ is a linear equation since the unknown, the x, is only to the power 1. An equation which is *not* linear is $2x^2 + 5 = 0$. In this equation the x is raised to the power 2. Such an equation is termed a *quadratic equation*. Quadratic equations are discussed in chapter 5.

Many equations in engineering are linear equations. For example, for motion in a straight line with a constant acceleration we have $v = u + at$, with perhaps the velocity v after a time t being the unknown, u a given initial velocity and a a given acceleration. Putting given values of, say, $u = 10$ m/s, $a = 4$ m/s^2 and $t = 2$ s in the equation we obtain as the solution $v = 10 + 4 \times 2 = 18$ m/s. There is just one value of v which fits the equation. This is always the case with a linear equation (but *not* with a quadratic equation, see chapter 5).

The above is a linear equation with just one unknown. In analysing situations in engineering we often end up with linear equations which have two unknowns. For example, in analysing an electrical circuit which has two meshes with a current I_1 in one and I_2 in the other we might obtain the equations for each mesh as $2I_1 - I_2 = 3$ and $I_1 + 3I_2 = 4$. Both these are linear equations since the unknowns are only to the power 1. We have thus two unknowns and two equations. Because both equations must give the same values for both the unknowns, they are referred to as *simultaneous equations*. Because they are linear equations we will still only obtain just one value of each unknown which fits the equations.

This chapter is about solving linear equations when they contain just one unknown and pairs of simultaneous linear equations when they contain two unknowns.

4.2 Solving linear equations

The basic rules for solving linear equations in one unknown can be summarised as:

1 Bring all the terms in the unknown to one side and all the constant terms to the other side.

2 Transpose any numbers which are multiplying or dividing the unknown.

Thus, for example, if we have:

$$3x + 4 = x + 8$$

then moving all the x terms to the left-hand side of the equation and all the constants to the right-hand side gives:

$$3x - x = 8 - 4$$

and so $2x = 4$. Transposing the 2 then gives $x = 4/2 = 2$.

We can check this value by substituting it into the original equation, i.e. the equation $3x + 4 = x + 8$. We then obtain $3 \times 2 + 4 = 2 + 8$.

Example

Solve the linear equation $3x + 4 = 8 - x$.

Moving all the x terms to the left-hand side of the equation and the constants to the right-hand side, gives

$$3x + x = 8 - 4$$

and so $4x = 4$. Transposing the 4 in front of the x from the left-hand side to the right-hand side of the equation then gives $x = 4/4 = 1$.

We can check this value by substituting it into the original equation, i.e. $3x + 4 = 8 - x$, when we have $3 \times 1 + 4 = 8 - 1$.

Example

Solve the linear equation $\dfrac{2}{x+1} = \dfrac{3}{2x-2}$.

This is a linear equation because when we transpose it to remove the denominators on both sides of the equation we end up with an equation with the unknown to just the power 1, i.e.

$$2(2x - 2) = 3(x + 1)$$

$$4x - 4 = 3x + 3$$

Hence with all the terms involving the unknown on the left-hand side and all the constants on the right, $4x - 3x = 4 + 3$ and we obtain the solution $x = 7$.

We can check this value by substituting it in the original equation to give:

$$\frac{2}{7+1} = \frac{3}{14-2}$$

Example

In a calculation of the stress σ acting on the reinforcing rod in a reinforced concrete column the following equation was obtained:

$$400 \times 10^{-6}\sigma + 2.73 \times 10^5 = 3.00 \times 10^5$$

Determine the value of the stress. Its unit will be Pa.

Rearranging the equation so that the unknown is on the left-hand side and the constants on the right-hand side gives

$$400 \times 10^{-6}\sigma = 3.00 \times 10^5 - 2.73 \times 10^5 = 0.27 \times 10^5$$

Transposing the 400×10^{-6} gives

$$\sigma = \frac{0.27 \times 10^5}{400 \times 10^{-6}} = 6.75 \times 10^7 \text{ Pa}$$

We can check this value by substituting it in the original equation. Thus:

$$400 \times 10^{-6} \times 6.75 \times 10^7 + 2.73 \times 10^5 = 3.00 \times 10^5$$

$$0.27 \times 10^5 + 2.73 \times 10^5 = 3.00 \times 10^5,$$

$$3.00 \times 10^5 = 3.00 \times 10^5.$$

Revision

1 Solve the following linear equations:

(a) $2x + 7 = x + 8$, (b) $2(x + 4) = 12$, (c) $3(x - 2) = 2(x + 1) - 5$,

(d) $2.5x + 1.2 = 4.3 - 1.1x$, (e) $2x + \frac{1}{2} = 4x - \frac{5}{2} + x$,

(f) $\frac{2}{3}x - \frac{1}{2} = \frac{7}{6}x + \frac{3}{4}$, (g) $1.6x + 0.5x + 12 = 10 - 0.3x$,

(h) $\dfrac{2}{x-1} = \dfrac{5}{x+2}$, (i) $\dfrac{x}{x+2} - 2 = \dfrac{2}{x+2}$,

(j) $\frac{1}{3}(x - 2) = \frac{4}{5}$, (k) $\frac{1}{4}x = \frac{16}{3}(2 - x)$,

(l) $0.5(x - 2) + 1.2(x + 1) = 12$, (m) $5.0 \times 10^9 x - 6.1 \times 10^{10} = 0$.

2 (a) Kirchhoff's voltage law when applied to an electrical circuit gave the equation $2I + 4I + 10I = 12$. What is the current I? The current is in amps.
(b) The resistance R_t of a conductor at a temperature t is given by the equation $R_t = R_0(1 + at)$. Determine a if $R_t = 20$ Ω, $t = 20°C$ and $R_0 = 18.4$ Ω.
(c) The resistivity ρ of a conductor is given by $\rho = RA/L$. Determine the resistance R if $\rho = 1.72 \times 10^{-8}$ Ω m, $L = 100$ m and $A = 0.8 \times 10^{-6}$ m².

4.3 Simultaneous equations

Consider the situation where we have an equation with two unknowns, e.g.

$$x + 2y = 5$$

If $y = 1$ then $x = 3$, if $y = 2$ then $x = 1$, if $y = 3$ then $x = -1$, and so on. Thus we have an infinite set of solutions. However, if we have a second equation involving the same two variables, e.g.

$$2x + y = 4$$

then there is only a single solution that will simultaneously satisfy both equations. Thus for this second equation, if $y = 1$ then $x = 3/2$, if $y = 2$ then $x = 1$, if $y = 3$ then $x = 1/2$, and so. There is only solution that satisfies both equations and that is that $y = 2$ and $x = 1$. The two equations are said to be *simultaneous equations* because they both simultaneously apply for the two variables.

There are three methods we can use to solve simultaneous equations:

1 By substitution

2 By elimination

3 By using graphs

In this chapter we will only consider the first and second methods, the third method being discussed in Section D.

4.3.1 Solution by substitution

Consider the two equations

$$x + 2y = 5 \text{ and } 2x + y = 4$$

We can rearrange the first equation to obtain the unknown x in terms of the other variable and the constants, i.e.

$$x = 5 - 2y$$

If we now substitute this expression for x into the second equation we obtain

$$2(5 - 2y) + y = 4$$

$$10 - 4y + y = 4$$

Thus, by transposition, we can obtain $4y - y = 10 - 4$. Hence $3y = 6$ and so $y = 2$. We can substitute this value of y in one of the original equations, say $x + 2y = 5$, to give $x + 2 \times 2 = 5$. Hence $x = 1$. Thus the solutions are $y = 2$ and $x = 1$. We can check that these values are correct by substituting them both in the other equation, $2x + y = 4$. Thus $2 \times 1 + 2 = 4$.

The procedure can thus be summarised as:

1 From one of the equations, obtain one of the variables, say x, in terms of constants and the other variable.

2 Substitute this variable x into the second equation.

3 Solve the second equation for x.

4 The other variable y can then be obtained by substituting the determined value of y in one of the original equations.

Example

Solve the following simultaneous equations by the substitution method:

$$x + 2y = 8 \text{ and } 2x - y = 1$$

Using the first equation to obtain x in terms of the other variable, then

$$x = 8 - 2y$$

Now, substituting this value of x into the second equation gives

$$2(8 - 2y) - y = 1$$

and so

$$16 - 4y - y = 1$$

Hence we have $5y = 15$ and so $y = 3$. Substituting this value into the first equation, $x + 2y = 8$, gives $x + 2 \times 3 = 8$ and so $x = 2$. The solutions are thus $y = 3$ and $x = 2$.

We can check this by substituting these values into the second equation, $2x - y = 1$, to give $2 \times 2 - 3 = 1$.

Example

Solve the following simultaneous equations by the substitution method:

$$\frac{x}{2} + \frac{y}{3} = 4 \text{ and } \frac{x}{2} - \frac{y}{3} = 0$$

If we multiply both sides of the first equation by 6 then we can eliminate the fractions and so give:

$$3x + 2y = 24$$

These terms can be rearranged, by transposing the $2y$ from the left-hand to right-hand side of the equation, to give:

$$3x = 24 - 2y$$

and so:

$$x = \frac{24 - 2y}{3}$$

To simplify matters, before substituting this value of x, we eliminate the fraction in the second equation by multiplying both sides of the equation by 6 to give:

$$3x - 2y = 0$$

Hence

$$3\left(\frac{24 - 2y}{3}\right) - 2y = 0$$

$$24 - 2y - 2y = 0$$

Thus $4y = 24$ and so $y = 6$. Substituting this value in the first equation, for simplicity in the form for which the fractions have been eliminated, gives $3x + 2 \times 6 = 24$ and so $3x = 24 - 12$. Hence $3x = 12$ and so $x = 4$. The solutions are thus $y = 6$ and $x = 4$.

We can check these values by substituting them into the second equation to give $\frac{4}{2} - \frac{6}{3} = 0$.

Example

When Kirchhoff's laws are applied to an electrical circuit the following two equations are produced, the currents being in amps:

$$2I_1 + 3(I_1 - I_2) = 7 \text{ and } 4I_2 + 3(I_2 - I_1) = 1$$

Determine the values of the two currents I_1 and I_2.

Rearranging the first equation to collect together all the terms involving each of the variables gives:

$$2I_1 + 3I_1 - 3I_2 = 7$$

$$5I_1 - 3I_2 = 7$$

Likewise for the second equation,

$$4I_2 + 3I_2 - 3I_1 = 1$$

$$-3I_1 + 7I_2 = 1$$

Using the first equation we can obtain:

$$I_1 = \frac{7 + 3I_2}{5}$$

Substituting this into the second equation gives

$$-3\left(\frac{7+3I_2}{5}\right) + 7I_2 = 1$$

Multiplying throughout by 5 gives:

$$-3(7 + 3I_2) + 35I_2 = 5$$

$$-21 - 9I_2 + 35I_2 = 5$$

Thus $26I_2 = 26$ and hence $I_2 = 1$ A.

Substituting this value of I_2 into the first equation, $5I_1 - 3I_2 = 7$, gives $5I_1 - 3 = 7$ and so $I_1 = 2$ A. Thus the solution is $I_1 = 2$ A and $I_2 = 1$ A.

We can check these values by substituting then into the second equation, $-3I_1 + 7I_2 = 1$, to give $-3 \times 2 + 7 \times 1 = 1$.

Revision

3   Solve, by substitution, the following sets of simultaneous equations:

(a) $2x + y = 3$ and $4x - y = 3$, (b) $3x + y = 7$, $2x + 3y = 7$,

(c) $x + 4y = 13$, $2x + y = 5$, (d) $x + 2y = 2$, $3x + y = 11$,

(e) $3x + 4y = 5$, $x + y = 1$, (f) $x - y = 7$, $2x + y = 8$,

(g) $4x - y = -1$, $2x + y = 7$, (h) $2x + 3y = 4$, $6x - y = 2$,

(i) $\frac{1}{x+y} = \frac{1}{3}$, $\frac{2}{3x-y} = \frac{4}{10}$, (j) $x + y = 6$, $\frac{1}{2}x + \frac{3}{4}y = 2$.

4   For a balanced beam we have when considering the equilibrium of the forces:

$$R_1 + R_2 = 5.5$$

and as a result of taking moments:

$$0.4R_1 + 3.4R_2 - 0.85 = 6.3$$

R_1 and R_2 are the reaction forces at the supports and are in units of kN. Determine the reaction forces.

5   When applying Kirchhoff's laws to the analysis of an electrical circuit, the following equations were obtained:

$$25I_1 + 20I_2 = 5 \text{ and } 5I_1 - 10I_2 = -15$$

Determine the currents I_1 and I_2.

6   The forces acting on a beam are such that

$$5F_1 + 2F_2 = 7 \text{ and } 4F_1 - F_2 = 3$$

Determine the values of F_1 and F_2.

7   The velocity v of a moving object is given by the equation $v = u + at$. If $v = 5$ m/s at $t = 1$ s and $v = 11$ m/s at $t = 3$ s, what are the values of u and a?

4.3.2 Solution by elimination

Solving simultaneous equations by elimination relies on the rule that: *one equation can be added to another or subtracted from one another without changing the simultaneous solutions.* This is because if one equation balances and a second equation balances then the sum of the two equations also balances.
   Consider, as in the previous section, the two equations

$$x + 2y = 5 \text{ and } 2x + y = 4$$

We want to add, or subtract, them in such a way that we eliminate one of the variables. Suppose we double the second equation and then subtract it from the first equation.

$$
\begin{array}{ll}
& x + 2y\ \ = 5 \\
\text{minus} & \underline{4x + 2y\ \ = 8} \\
& -3x\ \ \ \ \ \ \ = -3
\end{array}
$$

Thus $x = 1$. We can substitute this value in the first equation to give $1 + 2y = 5$ and hence $2y = 5 - 1$ and $y = 2$. The solution is thus $x = 1$ and $y = 2$.
   We can check these values by substituting them into the second equation (the second because we already have substituted into the first equation), $2x + y = 4$ to give $2 \times 1 + 2 = 4$.
   The procedure for solving simultaneous equations by elimination can thus be summarised as:

1   Multiply or divide by a constant both sides of one of the equations so that one of the terms has the same form in both equations.

2   Add or subtract the two equations to eliminate one of the variables and give the value of the other.

3   Substitute the value of the variable into one of the simultaneous equations to obtain the other variable.

Example

Solve the following simultaneous equations by the elimination method:

$$x + 2y = 8 \text{ and } 2x - y = 1$$

This is the example considered in the previous section and solved by the substitution method.

We can eliminate the y variable term by doubling the second equation to give $-2y$ in that equation and adding it to the first equation which has $+2y$.

$$
\begin{array}{llr}
& x + 2y & = 8 \\
\text{plus} & \underline{4x - 2y} & \underline{= 2} \\
& 5x & = 10
\end{array}
$$

Thus $x = 2$. We can substitute this in the first equation, $x + 2y = 8$, to give $2 + 2y = 8$ and hence $y = 3$. The solution is thus $x = 2$ and $y = 3$.

We can check these values by substituting them in the second equation, $2x - y = 1$, to give $2 \times 2 - 3 = 1$.

Example

Solve the following simultaneous equations by the elimination method:

$$\frac{x}{2} + \frac{y}{3} = 4 \text{ and } \frac{x}{2} - \frac{y}{3} = 0$$

This is the example considered in the previous section and solved by the substitution method.

It generally simplifies matters if we multiply both sides of the first equation by 6 to eliminate the fractions. Thus:

$$3x + 2y = 24$$

Likewise, we can eliminate the fractions in the second equation by multiplying by 6 to give:

$$3x - 2y = 0$$

We can eliminate the y term by adding these two equations. Thus:

$$
\begin{array}{llr}
& 3x + 2y & = 24 \\
\text{plus} & \underline{3x - 2y} & \underline{= 0} \\
& 6x & = 24
\end{array}
$$

Thus $x = 4$. We can substitute this into the first equation, $3x + 2y = 24$, to give $3 \times 4 + 2y = 24$. Hence $2y = 12$ and so $y = 6$. The solution is $x = 4$ and $y = 6$.

We can check these values by substituting them into the second equation, $3x - 2y = 0$, to give $3 \times 4 - 2 \times 6 = 0$.

Example

When Kirchhoff's laws are applied to an electrical circuit the following two equations are produced, the current being in amps:

$$2I_1 + 3(I_1 - I_2) = 7 \text{ and } 4I_2 + 3(I_2 - I_1) = 1$$

Determine, by elimination, the values of the two currents I_1 and I_2.

This is the example considered in the previous section and solved by substitution.

Rearranging the first equation to collect together all the terms involving each of the variables gives:

$$2I_1 + 3I_1 - 3I_2 = 7$$

$$5I_1 - 3I_2 = 7$$

Likewise for the second equation,

$$4I_2 + 3I_2 - 3I_1 = 1$$
$$-3I_1 + 7I_2 = 1$$

If we multiply the first equation by 7 and the second equation by 3 then we obtain:

$$35I_1 - 21I_2 = 49$$
$$\text{plus } \underline{-9I_1 + 21I_2 = \ 3}$$
$$26I_1 \qquad = 52$$

Thus $26I_1 = 52$ and so we have $I_1 = 2$ A. Substituting this value into the first equation, $5I_1 - 3I_2 = 7$, gives $10 - 3I_2 = 7$. Hence $I_2 = 1$ A. Thus the solution is $I_1 = 2$ A and $I_2 = 1$ A.

We can check these values by substituting them into the second equation, $-3I_1 + 7I_2 = 1$. Thus $-3 \times 2 + 7 \times 1 = 1$.

Revision

8 Solve, by elimination, revision problems 3, 4 ,5, 6 and 7.

Problems

1 Solve the following linear equations:

(a) $x + 5 - 2x = x + 3$, (b) $0.4x + 1.2x = 6 - 0.1x$,

(c) $3(x + 5) = 7(x + 1)$, (d) $23.2x + 10.9 = 1.6x - 4.3$,

(e) $2.5 \times 10^6 x - 2.1 \times 10^9 = 0$, (f) $\frac{1}{2}(x + 6) = \frac{1}{3}(x - 12)$,

(g) $\frac{1}{5}x - \frac{5}{8} = \frac{9}{10}x - \frac{9}{20}$, (h) $\frac{x+5}{2x-2} = 12$, (i) $\frac{1}{x} + \frac{1}{4} = \frac{1}{12}$.

2 For resistances in parallel the following equation was obtained:

$$\frac{1}{R} + \frac{1}{10} = \frac{1}{5}$$

What is the value of R? The resistance is in ohms.

3 When applying the principle of moments to a beam the following equation was obtained:

$$4F + 2 \times 10 - 12 \times 2 = 0$$

What is the value of F?

4 Solve the following simultaneous equation by (i) substitution, and (ii) elimination:

(a) $4x + 3y = 5$ and $x - 3y = 5$, (b) $2x - y = 5$ and $x + 4y = 7$,

(c) $2x + 3y = 10$ and $x + y = 3$, (d) $4x + y = 10$ and $x + 5y = 12$,

(e) $3x + 2y = 7$ and $x - 4y = 7$, (f) $x - 2y = 6$ and $2x + 3y = 5$,

(g) $x + 2y = 9$ and $5x - y = 1$, (h) $x + y = 8$ and $4x - 2y = 2$,

(i) $\frac{1}{2}x + \frac{2}{3}y = 5$ and $\frac{1}{4}x + y = \frac{13}{2}$,

(j) $\frac{1}{x+2y} = \frac{2}{14}$ and $\frac{5}{3x-2y} = -\frac{10}{6}$.

5 With a machine the effort E that is required to overcome a load L is given by the equation $E = a + bL$, where a and b are constants. If an effort of 6 N is required to overcome a load of 8 N and an effort of 8 N to overcome a load of 12 N, determine the values of a and b.

6 Kirchhoff's laws applied to a circuit gave the following equations for the currents in two parts of the circuit:

$$2I_1 + 5I_2 = 4.5 \text{ and } 5I_1 + 3I_2 = 6.5$$

Determine the values of the currents I_1 and I_2, both currents being in amps.

7 The forces acting on a beam are such that

$$2F_1 + 3F_2 = 13 \text{ and } 4F_1 - 2F_2 = 2$$

Determine the values of F_1 and F_2.

8 The velocity v of a moving object is given by the equation $v = u + at$. If $v = 3$ m/s at $t = 2$ s and $v = 4$ m/s at $t = 4$ s, what are the values of u and a?

9 The velocity v of a moving object is given by the equation $v = u + at$. If $v = 5$ m/s at $t = 2$ s and $v = 7$ m/s at $t = 3$ s, what are the values of u and a?

10 For a beam in equilibrium we have

$$R_1 + R_2 = 80 \text{ and } 1.5R_1 + 120 = 4.5R_2$$

Determine the reactive forces R_1 and R_2.

5 Quadratic equations

5.1 Introduction
A *linear equation* is one where the highest power of a variable is 1 (see chapter 4). Such equations have the general form

$$ax + b = 0$$

where x is the variable and a and b are constants. There is just one solution, $x = -b/a$.

The term *quadratic equation* is used for an equation where the highest power of a variable is 2. Such equations have the general form

$$ax^2 + bx + c = 0$$

where x is the variable and a, b and c are constants. Such equations have two solutions and the obtaining of such solutions is discussed in this chapter.

Quadratic equations occur often in engineering. An example of such an equation in engineering occurs with uniformly accelerated motion in a straight line, namely

$$s = ut + \tfrac{1}{2}at^2$$

where s is the distance travelled in time t, u the initial velocity and a the acceleration. This equation can be rewritten in the general form indicated above, of $ax^2 + bx + c = 0$, as:

$$\tfrac{1}{2}at^2 + ut - s = 0$$

The linear equation and the quadratic equation are just two examples of what are termed *polynomials*. A polynomial is the term used for any equation involving powers of the variable which are positive integers. Such powers can be 1, 2, 3, 4, 5, etc. For example, $x^4 + 4x^3 + 2x^2 + 5x + 2 = 0$ is a polynomial with the highest power being 4. In this chapter the discussion is restricted to just the quadratic equation. The constants a, b, c, etc. that are used to multiply the x terms are termed *coefficients*. Thus we might have with a polynomial having the highest power 3:

$$ax^3 + bx^2 + cx^1 + dx^0 = 0$$

Example

Which of the following are quadratic equations?

(a) $2x + 6 = 0$, (b) $x^2 + 3x + 2 = 0$, (c) $3x = 5$, (d) $x^2 = 4 + 2x$,

(e) $2x^3 + 5x^2 + 3x + 1 = 0$, (f) $x^2 = 20$, (g) $\frac{1}{x} + 2x = 4$.

Equations (a) and (c) are linear equations since they can be written in the form $ax + b = 0$. Thus (c) can be written as $3x - 5 = 0$. Equations (b), (d), (f) and (g) are quadratic equations since they can be written in the form $ax^2 + bx + c = 0$, even if b or c are zero (a must not be zero if it is to be a quadratic). Equation (d) can be written as $x^2 - 2x - 4 = 0$ and equation (f) as $x^2 + 0x - 20 = 0$. When both sides of equation (g) are multiplied by x we obtain $1 + 2x^2 = 4x$ which rearranges to give the equation $2x^2 - 4x + 1 = 0$. Equation (e) has the highest power of the variable as 3 and so is not a linear or quadratic equation. All the equations are polynomials.

Revision

1   Which of the following equations are (i) linear equations, (ii) quadratic equations?

(a) $x = 3$, (b) $2x + 5 = 0$, (c) $x^2 = 4$, (d) $x^2 + 5x + 2 = 0$, (e) $x^3 + x^2 = 4$,

(f) $4x^2 + 3x = 5$, (g) $\frac{1}{x+1} = 3$, (h) $\frac{1}{x} - 3x = 5$.

5.2 Factors

To factorise a number means to write it as the product of smaller numbers. Thus, for example, we can factor 12 to give $12 = 3 \times 4$. The 3 and 4 are factors of 12. If the 3 and the 4 are multiplied together then 12 is obtained. To *factorise a polynomial* means to write it as the product of simpler polynomials. Thus for the quadratic expression $x^2 + 5x + 6$ we can write:

$$x^2 + 5x + 6 = (x + 2)(x + 3)$$

$(x + 2)$ and $(x + 3)$ are factors. If the two factors are multiplied together then the $x^2 + 5x + 6$ is obtained. Note that, in general:

$$(a + b)(c + d) = a(c + d) + b(c + d) = ac + ad + bc + bd$$

For example:

$$(x + 2)(x + 3) = x(x + 3) + 2(x + 3) = x^2 + 3x + 2x + 6 = x^2 + 5x + 6$$

Note that when we multiply the first terms in each bracket we obtain the first term of the quadratic expression. When the last term in each bracket are multiplied we obtain the last term of the quadratic expression. Thus, in general, we look for the factors of a quadratic expression by finding all the possible factors of the first and last terms and finding which combination will give the correct middle term.

The following are some commonly encountered forms of equations and their factors:

$x^2 + (a + b)x + ab \quad (x + a)(x + b)$ $\qquad\qquad$ [1]

$x^2 + 2ax + a^2 \qquad (x + a)(x + a) \text{ or } (x + a)^2$ $\qquad$ [2]

$x^2 - 2ax + a^2 \qquad (x - a)(x - a) \text{ or } (x - a)^2$ $\qquad$ [3]

$x^2 - a^2 \qquad\qquad (x + a)(x - a)$ $\qquad\qquad$ [4]

$x^2 + ax \qquad\qquad x(x + a)$ $\qquad\qquad\qquad$ [5]

Example

Factorise: (a) $x^2 + 4x + 3$, (b) $2x^2 + 13x + 15$, (c) $x^2 + 4x$, (d) $x^2 - 25$.

(a) The first term has the only two factors of $1x$ and $1x$. The last term has the only two factors of 3 and 1. Thus the factors are $(x + 3)$ and $(x + 1)$, the product of these two giving the correct middle term.

(b) The first term has the only two factors of $2x$ and $1x$. The last term has the only two factors of 5 and 3. Combinations of these factors will give $(2x + 5)(x + 3)$, $(2x + 3)(x + 5)$. The product of the two bracketed terms are $2x^2 + 11x + 15$ and $2x^2 + 13x + 15$. Thus the correct factors are $(2x + 3)(x + 5)$.

(c) We can extract an x from both terms in the expression to give $x(x + 4)$ and so these are the factors.

(d) Since 25 is 5^2 we have an expression which is the difference of two squares and so of the form given in equation [4] above. Thus the factors are $(x + 5)(x - 5)$.

Revision

1 Factorise:

(a) $x^2 - 3x + 2$, (b) $x^2 + 7x + 10$, (c) $2x^2 - 5x - 3$, (d) $15x^2 + 19x + 6$, (e) $9x^2 + 12x + 4$, (f) $4x^2 - 9$, (g) $5x^2 + 9x$, (h) $6x^2 + 11x - 10$.

5.2.1 Roots

If we have $u \times v = 0$ then we must have either $u = 0$ or $v = 0$ or both u and v are 0. This is because 0 times any number is 0. Thus if we have the quadratic equation $x^2 + 5x + 6 = 0$ and rewrite it as $(x + 2)(x + 3) = 0$, then we must have either $x + 2 = 0$, or $x + 3 = 0$ or both equal to 0. This means that the solutions to the quadratic equation are the solutions of these two linear equations, i.e. $x = -2$ and $x = -3$. These values are called the *roots* of the equation. We can check these values by substituting them into the quadratic equation. Thus for $x = -2$ we have $4 - 10 + 6 = 0$ and thus $0 = 0$. For $x = -3$ we have $9 - 15 + 6 = 0$ and thus $0 = 0$.

Thus we can *solve a quadratic equation* by the following:

1 Factorise the quadratic

2 Set each factor equal to 0

3   Solve the resulting linear equations

Example

Factorise and hence solve the quadratic equation $x^2 - 3x + 2 = 0$.

This equation is of the form given by equation [1] above. To factorise this equation we need to find the two numbers which when multiplied together will give 2 and which when added together will give -3. If we multiply -1 and -2 we obtain 2 and the addition of -1 and -2 gives -3. Thus we can write:

$$(x - 1)(x - 2) = 0$$

The solutions are thus given by $x - 1 = 0$, i.e. $x = 1$, and $x - 2 = 0$, i.e. $x = 2$.
   We can check these values by substituting them into the original equation, $x^2 - 3x + 2$. Thus, for $x = 1$ we have $1 - 3 + 2 = 0$ and so $0 = 0$. For $x = 2$ we have $4 - 6 + 2 = 0$ and so $0 = 0$.

Example

Factorise and hence solve the quadratic equation $x^2 - 4 = 0$.

This equation is of the form given by equation [4] above. We can write this equation in the form $x^2 + 0x - 4 = 0$. We thus need to find a product of two numbers which gives -4 and a sum which gives 0. Hence the factors are:

$$(x - 2)(x + 2) = 0$$

Thus the solutions are $x - 2 = 0$, i.e. $x = 2$, and $x + 2 = 0$, i.e. $x = -2$. These values can be checked by substituting them in the original equation, $x^2 - 4 = 0$. With $x = 2$ we have $4 - 4 = 0$ and with $x = -2$ we have $4 - 4 = 0$.

Example

Factorise and hence solve the quadratic equation $x^2 + 6x + 9 = 0$.

This is of the form given by equation [2] above. The product of two numbers is 9 and the sum is 6. Hence

$$(x + 3)(x + 3) = 0$$

The solutions are thus given by $x + 3 = 0$, i.e. $x = -3$, and $x + 3 = 0$, i.e. $x = -3$. There are two roots with the same value. We can check these values by substituting them in the original equation, $x^2 + 6x + 9 = 0$. Thus $9 - 18 + 9 = 0$.

Example

Factorise and hence solve the quadratic equation $x^2 + 5x = 0$.

This is of the form of equation [5] given above. We can rewrite this equation in the form $x(x + 5) = 0$. Hence the solutions are given by $x = 0$ or $x + 5 = 0$, i.e. $x = -5$. we can check these values by substituting them in the original equation, $x^2 + 5x = 0$. Thus with $x = 0$ we have $0 + 0 = 0$ and with $x = -5$ we have $25 - 25 = 0$.

Example

If the roots of a quadratic equation are $+2$ and $+1$, what is the equation?

We have $x = +2$ and so $x - 2 = 0$, and $x = +1$ and so $x - 1 = 0$. Thus we can write the equation as

$$(x - 2)(x - 1) = 0$$

Multiplying this out gives $x^2 - 3x + 2 = 0$.

Revision

3 Factorise and hence solve the following quadratic equations:

(a) $x^2 + x - 2 = 0$, (b) $x^2 + 2x - 8 = 0$, (c) $x^2 - 5x + 4 = 0$,
(d) $x^2 - 4 = 0$, (e) $x^2 + 4x + 4 = 0$, (f) $x^2 - 2x + 1 = 0$, (g) $x^2 + 7x = 0$,
(h) $x^2 - 1 = 0$, (i) $x^2 + 2x - 3 = 0$, (j) $x^2 - 5x + 6 = 0$.

4 Determine the quadratic equations giving the following roots:

(a) $x = 1$, $x = 4$, (b) $x = -1$, $x = 2$, (c) $x = 0$, $x = 6$, (d) $x = -2$, $x = -3$,
(e) $x = 2$, $x = 0$, (f) $x = -1$, $x = -2$, (g) $x = 3$, $x = 1$.

5.3 Completing the square

Consider the equation $x^2 + 6x + 9 = 0$. This equation can be factorised to give $(x + 3)(x + 3) = 0$, i.e. $(x + 3)^2 = 0$. It is a perfect square, both the roots being the same.

Now consider the equation $x^2 + 6x + 2 = 0$. What are the factors? We can rewrite the equation as:

$$x^2 + 6x = -2$$

If we add 9 to both sides of the equation then we obtain

$$x^2 + 6x + 9 = -2 + 9 = 7$$

The left-hand side of the equation has been made into a perfect square by the adding of the 9. Thus we can write:

$$(x + 3)^2 = 7$$

This means that $x + 3$ must be one of the square roots of 7, i.e.

$$x + 3 = \pm\sqrt{7}$$

The plus or minus is because every positive number has two square roots, one positive and one negative. Thus we have $(+\sqrt{7}) \times (+\sqrt{7}) = 7$ and $(-\sqrt{7}) \times (-\sqrt{7}) = 7$. Hence:

$$x = -3 \pm \sqrt{7}$$

The two solutions are thus $x = -3 + \sqrt{7}$ and $x = -3 - \sqrt{7}$.

This method of determining the roots of a quadratic equation is known as *completing the square*. In the above discussion the left-hand side of the equation was made into a perfect square by the adding of 9. How do we determine what number to add in order to make a perfect square? Any expression of the form $x^2 + ax$ becomes a perfect square when we add $(a/2)^2$, since:

$$x^2 + ax + \left(\frac{a}{2}\right)^2 = \left(x + \frac{a}{2}\right)^2$$

Thus for $x^2 + 6x$ we have $a = 6$ and so $(a/2) = 3$. Hence we add $3^2 = 9$.

The procedure for determining the roots of a quadratic equation by completing the square can thus be summarised as:

1   Put the equation into the form $x^2 + ax = b$

2   Determine the value of $(a/2)$.

3   Then add $(a/2)^2$ to both sides of the equation to give:

$$x^2 + ax + (a/2)^2 = b + (a/2)^2$$

4   Hence obtain the equation:

$$(x + a/2)^2 = b + (a/2)^2$$

5   Determine the two roots by taking the square root of both sides of the equation, i.e.

$$x + \frac{a}{2} = \pm \sqrt{b + \left(\frac{a}{2}\right)^2}$$

The above rule for completing the square only works if the coefficient of x^2, i.e. the number in front of x^2, is 1. However, if this is not the case we can simply divide throughout by that coefficient in order to make it 1.

Example

Use the method of completing the square to solve the quadratic equation $x^2 + 10x - 4 = 0$.

The quadratic equation can be written as:

$$x^2 + 10x = 4$$

Adding $(10/2)^2$, i.e. 25, to both sides of the equation gives:

$$x^2 + 10x + 25 = 4 + 25$$

Thus:

$$(x + 5)^2 = 29$$

Hence, $x + 5 = \pm\sqrt{29} = \pm 5.39$ and the solutions of the quadratic equation are $x = +5.39 - 5 = 0.39$ and $x = -5.39 - 5 = -10.39$.

We can check these values by substituting them in the equation $x^2 + 10x - 4 = 0$. Thus, for $x = 0.39$ we have $0.39^2 + 3.9 - 4 = 0.05$, which, because of the rounding used to limit the number of decimal places in determining the root, is effectively zero. For the other solution of $x = -10.39$ we have $(-10.39)^2 + 103.9 - 4 = 0.05$, which, because of the rounding used to limit the number of decimal places in determining the root, is effectively zero.

Example

Use the method of completing the square to solve the quadratic equation $2x^2 + 5x - 4 = 0$.

The coefficient of the x^2 is 2. Thus to make the coefficient 1 we divide throughout by 2.

$$x^2 + 2.5x - 2 = 0$$

Thus we can write:

$$x^2 + 2.5x = 2$$

Completing the square gives:

$$x^2 + 2.5x + \left(\frac{2.5}{2}\right)^2 = 2 + \left(\frac{2.5}{2}\right)^2$$

and so:

$$(x + 1.25)^2 = 2 + 1.25^2 = 3.56$$

This then gives $x + 1.25 = \pm\sqrt{3.56} = \pm 1.89$. The solutions of the equation are $x = +1.89 - 1.25 = 0.64$ and $x = -1.89 - 1.25 = -3.14$.

We can check these values by substituting them in the equation $2x^2 + 5x - 4 = 0$. Thus, with $x = 0.64$ we have $2(0.64)^2 + 3.2 - 4 = 0.02$ and so effectively zero. With $x = -3.14$ we have $19.7 - 15.7 - 4 = 0$.

Revision

5 Use the method of completing the square to solve the following quadratic equations:

(a) $x^2 + 8x + 12 = 0$, (b) $x^2 + 2x + 8 = 0$, (c) $3x^2 - x - 5 = 0$,
(d) $3x^2 - 6x - 11 = 0$, (e) $3x^2 + 4x - 1 = 0$, (f) $2x^2 + 5x - 4 = 0$,
(g) $x^2 + 4x - 8 = 0$, (h) $x^2 - 4x + 2 = 0$.

5.4 Solving by formula

Consider the quadratic equation $ax^2 + bx + c = 0$. To obtain the roots by the completing the square method, we divide throughout by a to give:

$$x^2 + \frac{b}{a}x + \frac{c}{a} = 0$$

This can be written as:

$$x^2 + \frac{b}{a}x = -\frac{c}{a}$$

We complete the square by adding $(b/2a)^2$ to both sides of the equation. Hence:

$$x^2 + \frac{b}{a}x + \left(\frac{b}{2a}\right)^2 = -\frac{c}{a} + \left(\frac{b}{2a}\right)^2$$

and so:

$$\left(x + \frac{b}{2a}\right)^2 = -\frac{c}{a} + \left(\frac{b}{2a}\right)^2 = \frac{-4ac + b^2}{4a^2}$$

Taking the square root of both sides of the equation gives

$$x + \frac{b}{2a} = \pm\frac{\sqrt{b^2 - 4ac}}{2a}$$

Thus:

$$x = -\frac{b}{2a} \pm \frac{\sqrt{b^2 - 4ac}}{2a}$$

and so we have the general formula for the solution of a quadratic equation:

$$x = \frac{-b \pm \sqrt{b^2 - 4ac}}{2a}$$

Example

Use the formula to solve the quadratic equation $x^2 - 6x + 2 = 0$.

Comparing the equation with $ax^2 + bx + c = 0$, then $a = 1$, $b = -6$ and $c = 2$. Thus the formula gives

$$x = \frac{-b \pm \sqrt{b^2 - 4ac}}{2a} = \frac{-(-6) \pm \sqrt{(-6)^2 - 4 \times 1 \times 2}}{2 \times 1}$$

$$= \frac{6 \pm \sqrt{36 - 8}}{2} = \frac{6 \pm \sqrt{28}}{2} = \frac{6 \pm 5.29}{2} = 3 \pm 2.65$$

Hence the solutions are $x = 5.65$ and $x = 0.35$. We can check these values by substituting them in the equation $x^2 - 6x + 2 = 0$. Thus for the root $x = 5.65$ we have $31.9 - 33.9 + 2 = 0$ and for $x = 0.35$ we have $0.1 - 2.1 + 2 = 0$.

Example

Use the formula to solve the quadratic equation $3x^2 - 5x - 7 = 0$.

Comparing the equation with $ax^2 + bx + c = 0$, then $a = 3$, $b = -5$ and $c = -7$. Thus the formula gives:

$$x = \frac{-b \pm \sqrt{b^2 - 4ac}}{2a} = \frac{-(-5) \pm \sqrt{(-5)^2 - 4 \times 3 \times (-7)}}{2 \times 3}$$

$$= \frac{5 \pm \sqrt{25 + 84}}{6} = \frac{5 \pm \sqrt{109}}{6} = \frac{5 \pm 10.44}{6} = 0.83 \pm 1.74$$

Hence the solutions are $x = 2.57$ or -0.91.

We can check these values by substitution in the original equation, $3x^2 - 5x - 7 = 0$. Thus, for $x = 2.57$ we have $19.81 - 12.85 - 7 = -0.04$. This is effectively zero, bearing in mind the rounding to two decimal places of the solution. For $x = -0.91$ we have $2.48 + 4.55 - 7 = 0.03$. This is effectively zero, bearing in mind the rounding to two decimal places of the solution.

Example

When an object is thrown vertically upwards, its height h varies with the time t according to the equation $h = 20t - 5t^2$. At what times will the height be 10 m?

This requires a solution of the equation $10 = 20t - 5t^2$. This, when rearranged, is $5t^2 - 20t + 10 = 0$. Comparing this with the quadratic equation $ax^2 + bx + c = 0$, then $a = 5$, $b = -20$ and $c = 10$. Using the formula we have

$$t = \frac{20 \pm \sqrt{20^2 - 4 \times 5 \times 10}}{2 \times 5} = \frac{20 \pm \sqrt{400 - 200}}{10} = \frac{20 \pm 14.1}{10}$$

$$= 2 \pm 1.41$$

Hence the solutions are $t = 3.41$ s and 0.59 s. we can check these values by substitution in the equation $10 = 20t - 5t^2$. Thus for $t = 3.41$ we have $68.2 - 58.1 = 10.1$, i.e. effectively 10 bearing in mind the rounding of the values of the roots in working to just one decimal place. For $t = 0.59$ we have $11.8 - 1.74 = 10.06$, i.e. effectively 10 bearing in mind the rounding of the values of the roots in working to just one decimal place.

Example

The total surface area A of a cylinder of radius r and height h is given by the equation $A = 2\pi r^2 + 2\pi rh$. If $h = 6$ cm, what will be the radius required to give a surface area of 88/7 cm^2? Take π as 22/7.

Putting the numbers in the equation gives

$$\frac{88}{7} = 2 \times \frac{22}{7} r^2 + 2 \times \frac{22}{7} \times 6r$$

Multiplying throughout by 7 and dividing by 44 gives

$$2 = r^2 + 6r$$

Hence we can write

$$r^2 + 6r - 2 = 0$$

Comparing this with the quadratic equation $ax^2 + bx + c = 0$, then $a = 1$, $b = 6$ and $c = -2$. Using the formula we have

$$r = \frac{-6 \pm \sqrt{36 - 4 \times 1 \times (-2)}}{2 \times 1} = \frac{-6 \pm \sqrt{44}}{2} = \frac{-6 \pm 6.63}{2}$$

Hence the solutions are $r = -6.32$ cm and $r = 0.32$ cm. The negative solution has no physical significance. Hence the solution is a radius of 0.32 cm. Note that often in physical problems involving quadratics, the negative solution may have no physical significance.

 We can check this value of 0.32 cm by substitution in the equation $2 = r^2 + 6r$. Hence $0.10 + 1.92 = 2.02$, which is effectively 2 bearing in mind the rounding of the root value to two decimal places that has occurred.

Revision

6 Use the formula to solve the quadratic equations given in revision problem 5.

7 Use the formula to solve the following quadratic equations:

(a) $x^2 - 3x - 2 = 0$, (b) $x^2 + 8x + 12 = 0$, (c) $3x^2 + 12x - 1 = 0$,
(d) $3x^2 - 6x - 11 = 0$, (e) $x^2 + 3x - 7 = 0$.

8 The height h risen by an object, after a time t, when thrown vertically upwards with an initial velocity u is given by the equation $h = ut - \frac{1}{2}gt^2$, where g is the acceleration due to gravity. Solve the quadratic equation for t if $u = 100$ m/s, $h = 150$ m and $g = 9.81$ m/s^2.

9 A rectangle has one side 3 cm longer than the other. What will be the dimensions of the rectangle if the diagonals have to have lengths of 10 cm? Hint: let one of the sides have a length x, then the other side has a length of $3 + x$. The Pythagoras theorem can then be used.

10 In an electrical circuit the reactance X is given by the equation:

$$X = \omega L - \frac{1}{\omega C}$$

Solve the equation for ω, the unit being rad/s with the following given units, when $L = 0.5$ H, $C = 20$ μF (1μF $= 10^{-6}$ F) and $X = 100$ Ω.

11 The bending moment M a distance x along a beam is given by the equation:

$$M = \frac{wx(L - x)}{2}$$

Determine the values of x when $M = 50$ N m, $w = 3$ N/m and $L = 20$ m.

12 The motion of an object is described by the equation $s = ut + \frac{1}{2}at^2$. Determine t when $s = 96$ m, if $u = 10$ m/s and $a = 2$ m/s^2.

13 A rectangular sheet of metal 8 cm by 10 cm has a rectangle cut from its centre to leave a border all round the hole of width x. Determine the value of x if the area of the border is to be 20 cm^2.

14 A square sheet of metal of side 80 cm has a rectangle cut from one corner to leave an L-shaped piece with arms of width x. Determine the value of x if the area of the L-shaped piece is to be 1776 cm^2.

5.4.1 The nature of the roots

The formula for the roots of the quadratic equation is

$$x = \frac{-b \pm \sqrt{b^2 - 4ac}}{2a}$$

Consider the following three situations:

1 If we have $(b^2 - 4ac) > 0$, then the square root is of a positive number. There are then *two distinct roots*.

2 If we have $(b^2 - 4ac) = 0$, then the square root is zero and the formula gives just one value for x. Since a quadratic equation must have two roots, we say that the equation has *two coincident roots*.

3 If we have $(b^2 - 4ac) < 0$, then the square root is of a negative number. A new type of number has to be invented to enable such expressions to be solved. The number is referred to as a *complex number* and the roots are said to be *imaginary* (the roots in 1 and 2 above are said to be *real*). Such numbers are not discussed further in this book.

Example

Have the following quadratic equations real roots?

(a) $x^2 - 5x - 2 = 0$, (b) $x^2 + 3x + 5 = 0$, (c) $x^2 - 4x + 4 = 0$.

(a) Comparing the equation with $ax^2 + bx + c = 0$, then $a = 1$, $b = -5$ and $c = -2$. Thus $(b^2 - 4ac)$ is $25 + 8 = +33$. It is thus greater than 0 and so there are two distinct real roots.
(b) Comparing the equation with $ax^2 + bx + c = 0$, then $a = 1$, $b = 3$ and $c = 5$. Thus $(b^2 - 4ac)$ is $9 - 20 = -11$. It is thus less than 0 and so the roots are imaginary.
(c) Comparing the equation with $ax^2 + bx + c = 0$, then $a = 1$, $b = -4$ and $c = 4$. Thus $(b^2 - 4ac)$ is $16 - 16 = 0$. Because it is zero there are two coincident real roots.

Revision

15 Have the following quadratic equations real roots?

(a) $2x^2 + 3x + 4 = 0$, (b) $x^2 - 2x - 5 = 0$, (c) $x^2 - 3x + 8 = 0$,
(d) $x^2 + 2x + 1 = 0$, (e) $2x^2 - 5x + 1 = 0$.

Problems

1 Factorise and hence solve the following quadratic equations:

(a) $x^2 + x - 2 = 0$, (b) $x^2 + 4x + 3 = 0$, (c) $x^2 - 4x + 3 = 0$,
(d) $x^2 + 2x + 1 = 0$, (e) $x^2 - 9 = 0$, (f) $x^2 - 25 = 0$, (g) $x^2 + 3x = 0$,

(h) $2x^2 + 7x + 5 = 0$, (i) $6x^2 + 5x + 1 = 0$, (j) $6x^2 + 19x + 10 = 0$, (k) $2x^2 - 7x - 15 = 0$, (l) $9x^2 - 33x + 28 = 0$.

2   Determine the quadratic equations giving the following roots:

(a) $x = 1$, $x = -1$, (b) $x = 2$, $x = 2$, (c) $x = -3$, $x = -1$, (d) $x = 0$, $x = 1$.

3 Solve the following quadratic equations using (a) the method of completing the square, (b) the formula:

(a) $x^2 - 2x - 8 = 0$, (b) $x^2 + x - 3 = 0$, (c) $2x^2 + x - 3 = 0$, (d) $3x^2 - 4x + 1 = 0$, (e) $x^2 + 2x - 5 = 0$, (f) $5x^2 + x - 3 = 0$, (g) $4x^2 + 4x - 15 = 0$, (h) $5x^2 - 8x + 2 = 0$, (i) $4x^2 + 5x - 3 = 0$.

4 The area of a rectangle is 100 cm². If it has a width which is 5 cm less than its length, what are the dimensions of the rectangle? Hint: let the length be x and the width is then $(x - 5)$.

5 A rectangle has a perimeter of 26 cm and an area of 30 cm². What are its dimensions? Hint: let x equal the length of one side, then the length of the other is $13 - x$.

6 An object is thrown vertically upwards with a velocity of 50 m/s. The height h of the object after a time t is given by $h = 50t - 4.9t^2$. Hence determine the times when the object is at a height of 100 m.

7 A thermocouple gives an e.m.f. E in μV which depends on the temperature θ in degrees C according to the equation $E = 6.88\theta - 0.0192\theta^2$. Solve the equation for a thermocouple for which the e.m.f. is 200 μV.

8 The bending moment M at a point a distance x along a beam is given by:

$$M = \frac{wx(L - x)}{2}$$

Determine the values of x when $M = 50$ N m, $w = 5$ N/m and $L = 20$ m.

9 The surface area A of a cylinder is given by $A = 2\pi r^2 h + 2\pi r^2$, where r is the radius and h the height. What height cylinder is required if, with $r = 5$ cm, the area is to be 125 cm²?

10 The current i, in amperes, through a circuit element is related to the voltage v, in volts, across it by the equation $i = 0.001v^2 + 0.010v$. Determine the value of v when i is 0.004 A.

11 A square sheet of metal has a square hole cut from it so that the border all round the hole has a width of 4 mm. If the hole has an area of one quarter of the area of the original square, determine the length of the side of the square.

12 A right-angled triangle has a hypotenuse of length 18 mm and a total perimeter of length 40 mm. Determine the lengths of the other two sides.

13 A right-angled triangle is to have a hypotenuse of length 10 cm. If, for the other two sides one of them is to be longer than the other by 3 cm, what are the lengths of the sides?

14 Determine the two numbers which have a difference of 5 and a product of 266.

15 A rectangle has a length which is 3 cm greater than its width. If the area of the rectangle is 28 cm², what are its length and width?

16 A piece of wire of length 40 cm is to be bent into a rectangular shape to enclose an area of 100 cm². What will be the lengths of the sides of the rectangle?

17 Determine the number which added to its square gives 90.

18 A rectangular sheet of metal is 14 cm by 17 cm. Strips of the same width are cut off one side and one end. If the area remaining is 108 cm², what is the width of the pieces removed?

19 What are the two consecutive numbers which have a product of 156?

20 The length of a factory floor is to be 50 m more than its width and the total area 266 m². What will be the width and length?

21 Have the following quadratic equations real roots?

(a) $x^2 + 3x + 1 = 0$, (b) $x^2 - 5x - 2 = 0$, (c) $x^2 + 2x + 4 = 0$,

(d) $x^2 + 3x + 4 = 0$, (e) $2x^2 + 4x + 2 = 0$.

6 Equations in action

In the earlier chapters in this section, the algebraic techniques developed have been illustrated by showing their use in engineering problems. In this chapter the emphasis is on the engineering with the algebraic techniques used to support the development of an engineering topic. The aim is to put the algebraic techniques in an engineering context. The engineering topics in this chapter are the development and use of the equations for straight-line motion and basic electrical circuit analysis.

6.2 Straight line motion

Consider an object moving along a straight line with a uniformly accelerated motion. If u is the initial velocity, i.e. at time $t = 0$, and v the velocity after some time t, then the change in velocity in the time interval t is $(v - u)$. Hence the acceleration a is:

$$a = \frac{v - u}{t}$$

Multiplying both sides of the equation by t gives $at = v - u$. Hence the equation can be manipulated to give

$$v = u + at \qquad\qquad \text{[Equation 1]}$$

If the object, in its straight line motion, covers a distance s in a time t, then the average velocity in that time interval is s/t. With an initial velocity of u and a final velocity of v at the end of the time interval, the average velocity is:

$$\text{average velocity} = \frac{u + v}{2}$$

Hence

$$\frac{s}{t} = \frac{u + v}{2}$$

Multiplying both sides of the equation by t gives:

$$s = \left(\frac{u + v}{2}\right)t$$

Substituting for v by using the equation $v = u + at$ gives:

$$s = \left(\frac{u + (u + at)}{2}\right)t = \left(\frac{2u + at}{2}\right)t$$

Hence:

$$s = ut + \tfrac{1}{2}at^2 \qquad\qquad \text{[Equation 2]}$$

Consider the equation $v = u + at$. Squaring both sides of this equation gives:

$$v^2 = (u + at)^2 = u^2 + 2uat + a^2t^2$$

We can rewrite this equation as

$$v^2 = u^2 + 2a(ut + \tfrac{1}{2}at^2)$$

Hence:

$$v^2 = u^2 + 2as \qquad\qquad \text{[Equation 3]}$$

The equations [1], [2] and [3] are referred to as the equations for straight-line motion. The following examples illustrate their use in solving engineering problems.

Example

An object moves in a straight line with a uniform acceleration. If it starts from rest and takes 12 s to cover 100 m, what is the acceleration?

We have $u = 0$, $s = 100$ m, $t = 12$ s and are required to obtain a. Using $s = ut + \tfrac{1}{2}at^2$:

$$100 = 0 + \tfrac{1}{2}a \times 12^2 = 0 + 72a$$

Rearranging this equation to give $72a = 100$, then dividing both sides by 72, gives $a = 100/72 = 1.4$ m/s^2.

Example

A package slides down a slope with a constant acceleration. Its velocity increases from 100 mm/s to 320 mm/s in a distance of 600 mm. What is the acceleration?

The question gives $u = 100$ mm/s, $v = 320$ mm/s, $s = 600$ mm and we are required to obtain a. Using $v^2 = u^2 + 2as$:

$$320^2 = 100^2 + 2a \times 600$$

Transposing terms, we can rewrite this equation as:

$$1200a = 320^2 - 100^2 = 102\,400 - 10\,000 = 92\,400$$

Hence, dividing both sides of the equation by 1200 gives the answer as $a = 92\,400/1200 = 777$ mm/s^2.

Example

A car travelling at 25 m/s brakes and slows down with a uniform retardation of 1.2 m/s^2. How long will it take to come to rest?

A retardation is a negative acceleration in that the final velocity is less than the starting velocity. Thus we have u = 25 m/s, a final velocity of v = 0, retardation a = −1.2 m/s^2 and are required to obtain t. Using $v = u + at$, then:

$$0 = 25 + (-1.2)t = 25 - 1.2t$$

Transposing the 1.2t to the other side of the equation gives 1.2t = 25. Dividing both sides of the equation by 1.2 gives t = 25/1.2 = 20.8 s.

Example

A car is moving with a velocity of 10 m/s. It then accelerates at 0.2 m/s^2 for 100 m. What will be the time taken for the 100 m to be covered?

We have u = 10 m/s, a = 0.2 m/s^2 and s = 100 m and are required to obtain t. Using $s = ut + \frac{1}{2}at^2$:

$$100 = 10t + \frac{1}{2} \times 0.2t^2$$

We can write this equation as:

$$0.1t^2 + 10t - 100 = 0$$

This is a quadratic equation. Using the formula

$$t = \frac{-b \pm \sqrt{b^2 - 4ac}}{2a}$$

then:

$$t = \frac{-10 \pm \sqrt{100 - 4 \times 0.1(-100)}}{2 \times 0.1} = \frac{-10 \pm \sqrt{140}}{0.2} = \frac{-10 \pm 11.83}{0.2}$$

Hence t = −50 − 59.2 = −109.2 s or t = −50 + 59.2 = 9.2 s. Since the negative time has no significance, the answer is 9.2 s. The answer can be checked by substituting it back in the original equation.

Revision

1 The following all refer to an object moving in a straight line with uniform acceleration.

(a) Initially at rest, acceleration 2 m/s^2 for 8 s. Determine the distance travelled in that time.
(b) Initial velocity 3 m/s, acceleration 2 m/s^2 for 6 s. Determine the velocity at the end of that time.
(c) Initial velocity 3 m/s, final velocity 5 m/s after 2 s. Determine the distance travelled in that time.
(d) Initial velocity 3 m/s, distance travelled 20 m, final velocity 7 m/s. Determine the time taken.
(e) Initial velocity 10 m/s, acceleration −4 m/s^2, final velocity 2 m/s. Determine the distance travelled.
(f) Initial velocity 10 m/s, acceleration −4 m/s^2, final velocity 2 m/s. Determine the time taken.
(g) Acceleration 0.5 m/s^2 for 10 s, final velocity 10 m/s. Determine the initial velocity.

2   A train is moving with a velocity of 10 m/s. It then accelerates at a uniform rate of 2.5 m/s^2 for 8 s. What is the velocity after the 8 s?

3   An object starts from rest and moves with a uniform acceleration of 2 m/s^2 for 20 s. What is the distance moved by the object?

4   A cyclist is moving with a velocity of 1 m/s. He/she then accelerates at 0.4 m/s^2 for 100 m. What will be the time taken for the 100 m?

5   Rearrange the equation $v^2 = u^2 + 2as$ to give (a) s in terms of the other quantities, (b) a in terms of the other quantities, (c) u in terms of the other quantities.

6   A car accelerates from 7.5 m/s to 22.5 m/s at 2 m/s^2. What is (a) the time taken, (b) the distance travelled during this acceleration?

6.2.1 Simultaneous equations

The following example and revision problems involve straight-line motion when simultaneous equations can occur.

Example

A stationary car A is passed by car B moving with a uniform velocity of 15 m/s. Two seconds later, car A starts moving with a constant acceleration of 1 m/s^2 in the same direction. How long will it take for car A to draw level with car B?

Both the cars will have travelled the same distance s when they have drawn level. We will measure time t from when car B initially passes car A. For car B, since $u = 15$ m/s and $a = 0$, using $s = ut + \frac{1}{2}at^2$:

$$s = 15t + 0$$

For car A, since $u = 0$ and $a = 1$ m/s^2, we can write:

$$s = 0 + \tfrac{1}{2} \times 1 \times (t-2)^2$$

We write $(t - 2)$ for the time since car A is in motion for a time 2 s less than t. These two simultaneous equations can be solved by substituting for s in the second equation. Thus

$$15t = 0.5(t-2)^2 = 0.5(t^2 - 4t + 4) = 0.5t^2 - 2t + 2$$

Transposing the $15t$ to the other side of the equation gives the quadratic equation

$$0.5t^2 - 17t + 2 = 0$$

Using the formula for the quadratic equation:

$$t = \frac{-b \pm \sqrt{b^2 - 4ac}}{2a} = \frac{17 \pm \sqrt{17^2 - 4 \times 0.5 \times 2}}{2 \times 0.5} = 17 \pm \sqrt{285}$$

Hence $t = 17 + 16.9 = 33.9$ s or $t = 17 - 16.9 = 0.1$ s. With the data given, this second answer is not feasible and so the time is 33.9 s after car B has passed car A.

Revision problems

7   Car A, travelling with a uniform velocity of 25 m/s, overtakes a stationary car B. Two seconds later, car B starts and accelerates at 6 m/s^2. How far will B have to travel before it catches up A?

8   Car A and car B are parked along the same straight road with car B 27 m in front of A. Car A moves off with a uniform acceleration of 1 m/s^2 and then 6 s later car B starts off with a uniform acceleration of 1.5 m/s^2. How long will it take, from car A starting off, for the cars to draw level?

6.2.2 Motion under the action of gravity

All freely falling objects in a vacuum fall with the same uniform acceleration directed towards the surface of the earth as a result of a gravitational force acting between the object and the earth. This acceleration is termed the *acceleration due to gravity g*. For most practical purposes, the acceleration due to gravity at the surface of the earth is taken as being 9.81 m/s^2. The equations for motion of a falling object are those for motion in a straight line with the acceleration as g.

When an object is thrown vertically upwards it suffers an acceleration directed towards the surface of the earth. An acceleration directed in the opposite direction to which an object is moving is a retardation, i.e. a negative acceleration since it results in a final velocity less than the initial velocity. The result is that the object slows down. The object slows down until its velocity upwards eventually becomes zero, it then having attained its greatest height above the ground. Then the object reverses the direction of its motion and falls back towards the earth, accelerating as it does. Figure 6.1 illustrates these points.

Figure 6.1 *Vertical motion under gravity*

The equations for motion due to gravity are those for straight-line motion with uniform acceleration, the acceleration being g. Care has to be taken with the signs used with velocities, displacements and the acceleration to ensure that all the directions involved for the elements in an equation are the same. Thus when an object is falling then the velocities, displacements and accelerations are all measured in a downward direction. This means that g is a positive quantity since the acceleration is in that direction. When an object is moving upwards then the velocities, displacements and accelerations must all be in an upward direction. Thus for the acceleration we must use $-g$.

Example

A stone is dropped down a vertical shaft and reaches the bottom in 5 s. How deep is the shaft? Take g as 9.8 m/s^2.

We have $u = 0$, $t = 5$ s and $a = g = 9.8$ m/s^2. We have to determine the distance fallen, h. Thus, using $s = ut + \frac{1}{2}at^2$:

$$h = 0 + \frac{1}{2} \times 9.8 \times 5^2 = 122.5 \text{ m}$$

This equation is consistent with regard to directions since the acceleration and h are both in the same direction, i.e. downwards.

Example

An object is thrown vertically upwards with a velocity of 8.0 m/s. What will be the greatest height reached and the time taken to reach that height? Take g as 9.8 m/s^2.

We have $u = 8.0$ m/s, $a = -g = -9.8$ m/s^2 and, at the greatest height H, $v = 0$. We have initially to determine H. The velocities and displacement are measured in an upward direction and so the acceleration due to gravity is negative. Thus, using $v^2 = u^2 + 2as$:

$$0 = 8.0^2 + 2(-9.8)H$$

Thus $0 = 64 - 19.6H$. Transposing the $19.6H$ term to the other side of the equation gives $19.6H = 64$, and so $H = 64/19.6 = 3.27$ m.

To determine t we can use $v = u + at$:

$$0 = 8.0 + (-9.8)t$$

Thus $8.0 - 9.8t = 0$ and, transposing the $-9.8t$ to the other side of the equation, gives $9.8t = 8.0$, and so $t = 8.0/9.8 = 0.82$ s.

Revision

9   A stone is thrown vertically upwards with an initial velocity of 15 m/s. What will be the distance from the point of projection and the velocity after 1 s? Take g as 9.8 m/s^2.

10 A stone is thrown vertically upwards with an initial velocity of 5 m/s. What will be the time taken for it to reach the greatest height? Take g as 9.8 m/s^2.

11 A stone falls off the edge of a cliff. With what velocity will it hit the beach below if the height of the cliff above the beach is 50 m? Take g as 9.8 m/s^2.

6.3 Electrical circuits

For many resistors, if the temperature does not change, the potential difference across a resistor is proportional to the current through it. Thus:

$$V = RI$$

where R is the constant of proportionality and is termed the resistance. This is called *Ohm's law*. Circuit elements which obey Ohm's law have a constant value of resistance which does not change when the current changes.

The power P developed when a current passes through a resistance is given by $P = IV$, hence we can write:

$$P = I^2R = \frac{V^2}{R}$$

Example

If the current through a resistor of resistance 10 Ω is 0.2 A, determine (a) the potential difference across it and (b) the power dissipated in it.

(a) Using Ohm's law,

$$V = RI = 10 \times 0.2 = 20 \text{ V}$$

(b) The power P is given by

$$P = I^2R = 0.2^2 \times 10 = 0.4 \text{ W}$$

Example

Rearrange the equation $P = I^2R$ to give the current I in terms of the other quantities.

We can write the equation as $I^2R = P$. Dividing both sides of the equation by R gives:

$$I^2 = \frac{P}{R}$$

Then taking the square root of both sides gives:

$$I = \sqrt{\frac{P}{R}}$$

Revision

12 Rearrange the following equation to give V in terms of the other quantities.

$$P = \frac{V^2}{R}$$

13 What is the resistance of a resistor if the potential difference across it is 3 V and the current through it is 0.2 A?

14 If the current through a 10 Ω resistor is 0.5 A, what is the potential difference across it?

15 If the current through a resistor of resistance 100 Ω is 20 mA, what will be (a) the potential difference across it and (b) the power dissipated in it?

6.3.1 Resistors in series and parallel

Figure 6.2 *Resistors in series*

When resistors are in *series*, as in Figure 6.2, then we have the same current I through each element. Thus the potential difference V_1 across resistor R_1 is IR_1. The potential difference V_2 across resistor R_2 is IR_2. The potential difference V across the series arrangement is the sum of the potential differences, i.e.

$$V = V_1 + V_2 = IR_1 + IR_2 = I(R_1 + R_2)$$

Dividing both sides of the equation by I gives

$$\frac{V}{I} = R_1 + R_2$$

We could replace the two resistors by a single, equivalent, resistor R giving the same current I and overall potential difference V if we have $R = V/I$. Hence:

$$R = R_1 + R_2$$

For resistors in series the total resistance R is the sum of their resistances.

When circuit elements are in *parallel*, as the resistors are in Figure 6.3, then the same potential difference occurs across each element. Since charge does not accumulate at a junction, then the current entering a junction must equal the sum of the current leaving it, i.e. $I = I_1 + I_2$. But for resistor R_1 we have $V = I_1R_1$ and for resistor R_2 we have $V = I_2R_2$. Thus, substituting these values in the above equation gives:

$$I = \frac{V}{R_1} + \frac{V}{R_2} = V\left(\frac{1}{R_1} + \frac{1}{R_2}\right)$$

Dividing both sides of the equation by V gives:

$$\frac{I}{V} = \frac{1}{R_1} + \frac{1}{R_2}$$

We could replace the two resistors by a single, equivalent, resistor R if we have $R = I/V$. Hence:

$$\frac{1}{R} = \frac{1}{R_1} + \frac{1}{R_2}$$

The reciprocal of the equivalent resistance is equal to the sum of the reciprocals of the parallel resistances.

Example

A circuit consists of four resistors in series, they having resistances of $2\ \Omega$, $10\ \Omega$, $5\ \Omega$ and $8\ \Omega$. A voltage of $12\ V$ is applied across them. Determine (a) the total resistance, (b) the circuit current, (c) the power dissipated in each resistor, (d) the total power dissipation.

(a) The total resistance is the sum of the four resistances, i.e.

$$R = R_1 + R_2 + R_3 + R_4$$

and so we have $R = 2 + 10 + 5 + 8 = 25\ \Omega$.

Figure 6.3 *Resistors in parallel*

(b) The current I through the circuit is given by $V = IR$. Dividing both sides of this equation by R gives:

$$\frac{V}{R} = I$$

Hence:

$$I = \frac{V}{R} = \frac{12}{25} = 0.48 \text{ A or } 480 \text{ mA}$$

(c) The power developed in each resistance is given by $P = I^2R$, the current I being the same through each resistor. Thus, for each resistor:

$$P = 0.48^2 \times 2 = 0.46 \text{ W}$$

$$P = 0.48^2 \times 10 = 2.30 \text{ W}$$

$$P = 0.48^2 \times 5 = 1.15 \text{ W}$$

$$P = 0.48^2 \times 8 = 1.84 \text{ W}$$

(d) The total power developed can be obtained, either by adding together the powers dissipated by each resistor or determining the power dissipated by the equivalent resistor. Thus:

$$P = 0.46 + 2.30 + 1.15 + 1.84 = 5.75 \text{ W}$$

or $P = I^2R = 0.48^2 \times 25 = 5.76$ W. The slight difference in the answers is due to rounding errors.

Example

A circuit consists of two resistors of 5 Ω and 10 Ω in parallel with a voltage of 2 V applied across them. Determine (a) the total circuit resistance, (b) the current through each resistor, (c) the total current through the circuit.

(a) The total resistance R is given by

$$\frac{1}{R} = \frac{1}{R_1} + \frac{1}{R_2} = \frac{1}{5} + \frac{1}{10} = \frac{3}{10}$$

We thus have $R = 10/3 = 3.3 \ \Omega$.

(b) The same potential difference V will be across each resistor. Since $V = IR$, then if we divide both sides of this equation by R we obtain:

$$I = \frac{V}{R}$$

Thus the currents are:

$$I = \frac{V}{R_1} = \frac{2}{5} = 0.4 \text{ A or } 400 \text{ mA}$$

$$I = \frac{V}{R_2} = \frac{2}{10} = 0.2 \text{ A or } 200 \text{ mA}$$

(c) The total current entering the arrangement can be obtained by either adding the currents through each of the branches or determining the current taken by the equivalent resistance. Thus, adding the currents through the branches gives:

$$I = 0.4 + 0.2 = 0.6 \text{ A or } 600 \text{ mA}$$

or, considering the current taken by the equivalent resistance,

$$I = \frac{V}{R} = \frac{2}{3.3} = 0.61 \text{ A or } 610 \text{ mA}$$

Rounding errors account for the slight differences in these answers.

Revision

16 Calculate the equivalent resistance of three resistors of 5 Ω, 10 Ω and 15 Ω which are connected in (a) series, (b) parallel.

17 Three resistors of 15 kΩ, 20 kΩ and 24 kΩ are connected in parallel. What is (a) the total resistance and (b) the power dissipated if a voltage of 24 V is applied across the circuit?

18 A circuit consists of two resistors in parallel. The total resistance of the parallel arrangement is 4 Ω. If one of the resistors has a resistance of 12 Ω, what will be the resistance of the other one?

19 A circuit consists of two resistors in series. If they have resistances of 4 Ω and 12 Ω, what will be (a) the circuit current and (b) the potential difference across each resistor when a potential difference of 4 V is applied across the circuit?

20 A circuit consists of two resistors in parallel. If they have resistances of 4 Ω and 12 Ω, what will be (a) the current through each resistor and (b) the total power dissipated when a potential difference of 4 V is applied across the circuit?

6.3.2 Series–parallel circuits

Many circuits are more complicated than just the simple series or parallel connections of resistors and have varied combinations of both series and parallel connected resistors. In general, such circuits are analysed by

(a)

(b)

(c)

Figure 6.4   *Example*

beginning with the innermost parts of the circuit and block-by-block reducing the elements to single equivalent resistances, before finally simplifying the circuit to just one equivalent resistance. The following example illustrates the procedure.

Example

Determine the total resistance of the circuits shown in Figure 6.4.

(a) The circuit consists of a parallel arrangement of 4 Ω and 2 Ω which is then in series with resistance of 6 Ω. For the two resistors in parallel, the equivalent resistance is given by:

$$\frac{1}{R} = \frac{1}{4} + \frac{1}{2} = \frac{3}{4}$$

Hence $R = 4/3 = 1.33$ Ω. This is in series with the 6 Ω resistor and so the total resistance is given by $R = 1.33 + 6 = 7.33$ Ω.

(b) The circuit consists of 2 Ω and 4 Ω in series. Thus the equivalent resistance R is $R = R_1 + R_2 = 2 + 4 = 6$ Ω. This is then in parallel with 3 Ω. Thus the total resistance is given by:

$$\frac{1}{R} = \frac{1}{3} + \frac{1}{6} = \frac{3}{6}$$

and so $R = 6/3 = 2$ Ω.

(c) The 4 Ω and the 8 Ω resistors are in parallel and can be replaced by an equivalent resistance R, where:

$$\frac{1}{R} = \frac{1}{4} + \frac{1}{8} = \frac{3}{8}$$

and so $R = 8/3 = 2.67$ Ω. This is in series with the 5 Ω resistor and so their combined resistance is $5 + 2.67 = 7.67$ Ω. This is in parallel with 3 Ω and so the combined resistance is given by

$$\frac{1}{R} = \frac{1}{7.67} + \frac{1}{3} = \frac{3+7.67}{7.67 \times 3} = \frac{10.67}{23.01}$$

Thus $R = 23.01/10.67 = 2.16$ Ω. This is in series with the 2 Ω resistance and so the total resistance of the circuit $2.16 + 2 = 4.16$ Ω.

Revision

21 Determine the total resistances of the circuits and the circuit currents I shown in Figure 6.5.

22 A resistor of 16 Ω is connected in series with an arrangement of a 12 Ω resistor in parallel with a 6 Ω resistor. If the supply voltage to the circuit is 12 V, what is (a) the total circuit resistance, (b) the current through each of the resistors, (c) the total power dissipated?

(a)

(b)

(c)

(d)

Figure 6.5   *Revision problem 21*

23 A resistor of 10 Ω is connected in series with an arrangement of a 20 Ω resistor in parallel with a 30 Ω resistor. If the supply voltage is 12 V, what is (a) the total circuit resistance, (b) the potential difference across each resistor, (c) the total power dissipated?

6.3.3 Kirchhoff's laws

Kirchhoff's current law states that at any node in an electrical circuit, the current entering it equals the current leaving it. The term *node* is used for a point in an electrical circuit at which conductors join to form a junction. Thus for the node shown in Figure 6.6(a), we must have

$$I_1 = I_2 + I_3 + I_4$$

This law is often stated as: the algebraic sum of the currents flowing towards a node is zero. The algebraic sum is the sum obtained when account is taken of the direction of the currents, with perhaps currents into a node being taken as positive and currents leaving it as negative. Then, for the node in Figure 6.6(a) we have, with transposition of terms in the above equation:

$$I_1 - I_2 - I_3 - I_4 = 0$$

Kirchhoff's voltage law states that around any closed path in a circuit, the algebraic sum of the voltages across all the components and the algebraic sum of the applied voltages is zero. A closed path, often called a *loop*, is exactly what is implied. If we start at one node and follow the circuit through perhaps a number of nodes we end up at the starting node without encountering a node more than once. Figure 6.6(b) shows such a closed path. The term 'algebraic sum' means that we have to take account of the directions of the potential differences and applied voltages when we add them. The potential difference across a resistor R when a current I flows through it is IR and is in the opposite direction to the current. Thus moving round the loop in Figure 6.6(b) in a clockwise direction gives

$$E - IR_1 - IR_2 - IR_3 = 0$$

We can analyse a network by using both Kirchhoff's current law and Kirchhoff's voltage law and obtaining a set of simultaneous equations. The first of the examples that follows illustrates this. However, we can simplify the analysis and effectively only use the voltage law if we use what are termed *mesh currents*. The term *mesh* is used for a loop which does not contain any other loop. Thus for the circuit shown in Figure 6.7, there are three loops ABCF, CDEF and ABCDEF but only the first two are meshes. For each mesh a mesh current is defined, such a current being considered to circulate round the mesh. The same direction of circulation must be taken for all the mesh currents in a circuit, the usual convention being for all mesh currents to circulate in a clockwise direction.

(a)

(b)

Figure 6.6 *Using Kirchhoff's laws*

Figure 6.7 *Mesh currents*

Consider mesh 1. The current through R_1 is the mesh current I_1. The current through resistor R_2, which is common to the two meshes, is the algebraic sum of the two mesh currents, i.e. $I_1 - I_2$ in the direction of I_1. Thus, applying Kirchhoff's voltage law to the mesh gives:

$$E_1 - I_1R_1 - (I_1 - I_2)R_2 = 0$$

For mesh 2 we have a current of I_2 through R_3 and a current of $(I_2 - I_1)$, in the direction of I_2, through R_2. Thus applying Kirchhoff's voltage law to this mesh gives:

$$-E_2 - I_2R_2 - (I_2 - I_1)R_2 = 0$$

Thus, with a two-mesh circuit, we end up with two simultaneous equations. With a three-mesh circuit we would have obtained three simultaneous equations, with a four-mesh circuit four simultaneous equations.

Example

Determine the current through the 20 Ω resistor in the circuit shown in Figure 6.8.

Figure 6.8 *Example*

Consider node a in the circuit. With the currents as labelled, Kirchhoff's current law gives:

$$I_1 + I_2 + I_3 = 0 \qquad [1]$$

Now apply Kirchhoff's voltage law to loop 1. The current through the 5 Ω resistor is I_1 and so the potential difference across the resistor, in the opposite direction to the current, is $5I_1$. The current through the 20 Ω resistor is I_3 and so the potential difference across the resistor, in the opposite direction to the current, is $20I_3$. Thus for loop 1 we can write:

$$5 - 5I_1 + 20I_3 = 0 \qquad [2]$$

Now consider loop 2. The current through the 10 Ω resistor is I_2 and so the potential difference across the resistor, in the opposite direction to the current, is $10I_2$. The current through the 20 Ω resistor is I_3 and so the potential difference across the resistor, in the opposite direction to the current, is $20I_3$. Thus for loop 2 we can write:

$$10I_2 - 20 - 20I_3 = 0 \qquad [3]$$

We thus have three simultaneous equations with three unknowns.
Adding 5 times equation [1] to equation [2] gives:

$$5I_1 + 5I_2 + 5I_3 = 0$$
$$\text{plus } \underline{5 - 5I_1 \qquad + 20I_3 = 0}$$
$$5 \qquad + 5I_2 + 25I_3 = 0$$

If we multiply this equation by 2 and subtract equation [3] from it then we obtain:

$$10 + 10I_2 + 50I_3 = 0$$
$$\text{minus } \underline{-20 + 10I_2 - 20I_3 = 0}$$
$$+30 \qquad + 70I_3 = 0$$

Hence $70I_3 = -30$ and so $I_3 = -30/70 = -0.43$ A. The minus sign is because the direction of the current is in the opposite direction to that indicated by the arrow for the current direction in the figure.

Example

Determine, using mesh analysis, the current through the 20 Ω resistor in the circuit shown in Figure 6.9.

This question is a repeat of the previous example but tackled by mesh analysis.

Figure 6.9 *Example*

With a current of I_1 through the 5 Ω resistor then the potential difference across that resistor is $5I_1$. For the 20 Ω resistor, the current through it is $(I_1 - I_2)$ and so the potential difference across that resistor is $20(I_1 - I_2)$. For mesh 1, applying Kirchhoff's voltage law, we thus obtain:

$$5 - 5I_1 - 20(I_1 - I_2) = 0$$

This can be rewritten as

$$5 = 25I_1 - 20I_2 \qquad [1]$$

For mesh 2, the current through the 10 Ω resistor is I_2 and so the potential difference across it is $10I_2$. The current through the 20 Ω resistor is $(I_2 - I_1)$ and so the potential difference across that resistor is $20(I_2 - I_1)$. Thus, applying Kirchhoff's voltage law to mesh 2:

$$-10I_2 - 20 - 20(I_2 - I_1) = 0$$

This can be rewritten as:

$$20 = 20I_1 - 30I_2 \qquad [2]$$

We now have in equations [1] and [2] a pair of simultaneous equations. Multiplying equation [2] by 4 and subtracting from it five times equation [1] gives

$$20 = 100I_1 - 80I_2$$
$$\text{minus } 100 = 100I_1 - 150I_2$$
$$\overline{-80 = \quad 0 \quad + \quad 70I_2}$$

Thus $I_2 = 80/70 = -1.14$ A. Substituting this value in equation [1] gives $5 = 25I_1 - 20(-1.14)$ and so $-17.8 = 25I_1$ and so $I_1 = -0.71$ A. The minus signs indicate that the currents are in the opposite directions to those indicated in the figure. The current through the 20 Ω resistor, which is $(I_1 - I_2)$ in the direction of I_1, is thus $-0.71 - (-1.14) = 0.43$ A.

Figure 6.10 *Revision problem 24*

Revision

24 Determine the currents labelled as I in the circuits given in Figure 6.10.

Problems

1 A conveyor belt is used to move an object along a production line with an acceleration of 1.5 m/s². What is the velocity after it has moved 3 m?

2 A train starts from rest and moves with a uniform acceleration so that it takes 300 s to cover 9000 m. What is the acceleration?

3 Sound travels in air at 0°C at 330 m/s. How far will the sound of a shot travel in 5 s?

4 A car is initially at rest. It then accelerates at 2 m/s² for 6 s. What will be the velocity after that time?

5 A car accelerates at a uniform rate from 15 m/s to 35 m/s in 20 s. How far does the car travel in this time?

6 An object has an initial velocity of 10 m/s and is accelerated at 3 m/s² for a distance of 800 m. What is the time taken to cover this distance?

7 Rearrange the equation $v = u + at$ to give (a) a in terms of the other quantities, (b) t in terms of the other quantities, (c) u in terms of the other quantities.

8 Rearrange the equation $s = ut + \frac{1}{2}at^2$ to give (a) a in terms of the other quantities, (b) u in terms of the other quantities.

9 What is the velocity attained by an object falling from rest through a distance of 4.9 m? Take g as 9.8 m/s².

10 An object is thrown vertically upwards with an initial velocity of 8 m/s. How long will it take for the object to return to the same point from which it was thrown? Take g as 9.8 m/s^2.

11 A fountain projects water vertically to a height of 5 m. What is the velocity with which the water must be leaving the fountain nozzle? Take g as 9.8 m/s^2.

12 A stone is thrown vertically upwards with an initial velocity of 9 m/s. What is the greatest height reached by the stone and the time taken? Take g as 9.8 m/s^2.

13 An object is thrown vertically upwards with a velocity of 34.2 m/s. How long will it take to reach a height, on its way up and its way back down, of 49 m above the point of projection? Take g as 9.8 m/s^2.

14 An object is thrown vertically upwards with a velocity of 20 m/s from an initial height h above the ground. It takes 5 s from the time of being projected upwards before the object hits the ground. Determine h. Take g as 9.8 m/s^2.

15 An object is thrown vertically upwards with a velocity of 30 m/s and, from the same point at the same time, another object is thrown vertically downwards with a velocity of 30 m/s. How far apart will the objects be after 3s? Take g as 9.8 m/s^2.

16 A ball is held 2.0 m above the ground and then released. The ball hits the ground and then rebounds with half the speed it had prior to the impact. Find the greatest height the ball reaches after bouncing. Take g as 9.8 m/s^2.

17 An electrical lamp filament dissipates 20 W when the current through it is 0.10 A. What is (a) the potential difference across the lamp and (b) its resistance at that current?

18 A 100 Ω resistor is specified as being able to be used up to a maximum power dissipation of 250 mW. What should be the maximum current used with the resistor?

19 If the current through a resistor of resistance 100 Ω is 50 mA, what will be (a) the potential difference across it and (b) the power dissipated in it?

20 If the current through a resistor of resistance 1 kΩ is 0.02 A, what will be (a) the potential difference across it and (b) the power dissipated in it?

21 Resistors of resistances 20 Ω, 30 Ω and 50 Ω are connected in series across a d.c. supply of 20 V. What is the current in the circuit and the potential difference across each resistor?

22 Determine the power dissipated in each resistor and the total power dissipated when a voltage of 10 V is applied across (a) a series-connected (b) a parallel-connected pair of resistors with resistances of 20 Ω and 50 Ω.

23 A circuit consists of two parallel-connected resistors of 2.2 kΩ and 3.9 kΩ in series with a 1.5 kΩ resistor. If the supply voltage connected to the circuit is 20 V, what is the current drawn from the supply?

24 A circuit consists of three parallel-connected resistors of 330 Ω, 560 Ω and 750 Ω in series with a 800 Ω resistor. If the supply voltage connected to the circuit is 12 V, what are the voltages across the parallel resistors and the series resistor?

25 Three resistors of 6 Ω, 12 Ω and 24 Ω are connected in parallel across a 12 V voltage supply. What is the total resistance and the current through each resistor?

26 A circuit consists of a resistor of 8 Ω in series with a an arrangement of another 8 Ω resistor in parallel with a 12 Ω resistor. If the voltage input to the circuit is 24 V, what will be the current taken from the voltage source?

27 A circuit consists of a resistor of 2 Ω in series with an arrangement of three parallel resistors, these having resistances of 4 Ω, 10 Ω and 20 Ω. If the voltage input to the circuit is 10 V, what will be the current through the 2 Ω resistor?

28 Determine the equivalent resistances of the circuits shown in Figure 6.11.

(a)

(b)

Figure 6.11 *Problem 28*

29 Resistors with resistances of 10 Ω, 20 Ω and 30 Ω are connected (a) in series, (b) in parallel across a 12 V supply. Determine the current taken from the supply in each case.

30 Determine, using Kirchhoff's laws, the current I in each of the circuits shown in Figure 6.12.

Figure 6.12 *Problem 30*

Section C
Further algebra

7   Sequences and series
8   Exponentials
9   Logarithms

The aims of this section are to enable the reader to:

- Use sequences and the arithmetic, geometric and binomial series in the solution of engineering problems.
- Use the binomial theorem to simplify algebraic expressions.
- Manipulate and solve equations involving logarithmic and exponential expressions.
- Solve engineering problems involving exponential growth and decay.

This section extends the coverage of algebra given in Section B. It is assumed that Section B has been covered and the parts of Chapter 1 concerned with powers fully comprehended. There is a case for studying parts of Section E on graphs prior to Section C and others parts immediately following it.

7 Sequences and series

7.1 Introduction

This chapter is an introduction to sequences and series. A *sequence* is a set of numbers arranged in a definite order. Consider the following sequence of numbers: 1, 3, 5, 7, 9. Such a set of numbers is termed a *sequence* because they are stated in a definite order, i.e. 1 followed by 3 followed by 5, etc. In the above sequence of 1, 3, 5, 7, 9, each successive member is formed by adding 2 to the previous member. Another example of a sequence is money in an interest-earning bank account. Initially there might be, say, £100 in the account. If it earns interest at the rate of, say, 10% per year then after 1 year there will be £110 in the account. At the end of the next year there will be £121 in the account. At the end of the next year £132.1. The sequence of numbers describing the amount in the account at the ends of each year are:

£100 £110 £121 £132.1

We have a sequence of numbers where there is a definite relationship between successive values.

The term *series* is used for the sum of the terms of a sequence. There are many instances in mathematics were we have values which we can express as series and so aid calculations. A particular useful series is termed the *binomial series* (or binomial theorem) and is frequently used to simplify engineering expressions. In this chapter we consider the arithmetic, geometric and binomial series.

7.2 Sequences

The term *sequence* is used for a set of numbers in a definite order. Thus 1, 2, 3, 4, 5 is a sequence where it is easy to see there is a definite order in forming each member of the set. We can describe the terms in the sequence by stating the law by which the sequence is formed as: add 1 to the previous term. Thus the sequence 1, 2, 3, 4, 5 is obtained by starting with a 1 and then adding 1 to it to obtain the next term, then adding 1 to it for the next term and so on for each successive term in the sequence. The sequence 3, 6, 12, 24, 48 is obtained by starting with a 3 and then multiplying it by 2 to give the next term, then multiplying by 2 for the next term and so on for each successive term. The sequence 2, 4, 16, 256 is obtained by starting with a 2, then squaring it to obtain the next term and so squaring each term in sequence. The following sections discuss some of the common forms of sequences.

When a series finishes after a certain number of terms it is said to be *finite*; when there is an unending number of terms it is said to be *infinite*.

Revision

1 Determine the laws governing the numbers in the following sequences:

(a) 2, 4, 6, 8, (b) 5, 10, 20, 40, (c) 1, –1, 1, –1.

7.2.1 Arithmetic sequence

An *arithmetic sequence* (sometimes called an *arithmetic progression*) is one where each term is formed from the previous one by simply adding a constant value. Thus if a is the first term and d the common difference between successive terms, i.e. the constant value that has to be added, then the terms are:

$$a, (a + d), (a + 2d), (a + 3d), (a + 4d), ..., \text{etc.}$$

An example of an arithmetic sequence is 4, 7, 10, 13, 16, ..., etc. where $a = 4$ and $d = 3$.

The first terms of an arithmetic series is a, the second term is $(a + d)$, the third term is $(a + 2d)$ and thus the nth term of an arithmetic series is:

$$n\text{th term} = a + (n - 1)d$$

Example

State the first five terms of the arithmetic sequence when the first term is 3 and the common difference 4.

The terms are 3, 3 + 4 = 7, 7 + 4 = 11, 11 + 4 = 15, 15 + 4 = 19.

Example

State the 10th term of the arithmetic sequence for which the first three terms are 2.0, 4.5, 7.0.

For this sequence we have $a = 2.0$ and $d = 2.5$. Hence the 10th term is $2.0 + (10 - 1)2.5 = 24.5$.

Revision

2 State the first five terms of the arithmetic sequence when (a) the first term is 6 and the common difference 2, (b) the first term is 3 and the common difference 4.

3 State the 7th term of the arithmetic sequence for which the first three terms are 1, 5, 9.

7.2.2 Geometric sequence

A *geometric sequence* (sometimes called a *geometric progression*) is one where each term is formed from the previous one by multiplying it by a constant factor. Thus if a is the first term and r the common ratio between successive terms, i.e. the multiplying factor, then the terms are:

$a, ar, ar^2, ar^3, ar^4, \dots$, etc.

An example of a geometric sequence is 4, 8, 16, 32, ..., etc. where $a = 4$ and $r = 2$. The first term of a geometric sequence is a, the second term ar, the third term ar^2, and so the nth term is:

$$n\text{th term} = ar^{n-1}$$

As an illustration of an engineering situation which approximates to a geometrical sequence, consider the gear ratios for a machine tool such as a lathe. Suppose the machine is required to turn work pieces with diameters ranging from, say, 10 mm to 150 mm with the surface speed of the work piece to be as near as possible 40 000 mm/min. The speed of rotation required to give this for the 150 mm work piece is $40\ 000/(\pi \times 150) =$ 85 rev/min. The speed of rotation required for the 10 mm work piece is $40\ 000/(\pi \times 10) = 1273$ rev/min. If we intend to have, say, 8 gear ratios to span the range of work pieces and give, as near as possible, this cutting speed, then we use a geometric sequence to determine the required speeds of rotation. The initial term in the sequence is thus $a = 85$ and the eighth term $1273 = ar^{8-1} = 85r^7$. Thus $15 = r^7$ and so r is approximately 1.47 (Note, since $15^{1/7} = r$, you can determine r using a calculator by keying in 15, selecting $x^{1/y}$, then keying in 7). The eight speeds are thus 85, 85×1.47 $= 125$, $125 \times 1.47 = 184$, $184 \times 1.47 = 270$, $270 \times 1.47 = 397$, 397×1.47 $= 584$, $584 \times 1.47 = 858$, $858 \times 1.47 = 1261$ rev/min (this discrepancy from 1273 results from rounding).

Example

State the first five terms of the geometric sequence when the first term is 3 and the common ratio 2.

The terms are 3, $3 \times 2 = 6$, $6 \times 2 = 12$, $12 \times 2 = 24$, $24 \times 2 = 48$.

Example

State the 10th term of the geometric sequence for which the first three terms are 4, 2, 1.

The series has $a = 4$ and $r = \frac{1}{2}$. Thus the 10th term is $4 \times (\frac{1}{2})^9 = 4/512$ $= 1/128$.

Example

The rate of growth of a tree is found to be such that at the end of each successive year it has a height which is 1.2 times that of its height in the previous year. If initially it has a height of 4.0 m, list the heights in the succeeding three years.

The heights form a geometric sequence with $a = 4.0$ m and $r = 1.2$. Thus the heights are $4.0 \times 1.2 = 4.8$ m, $4.8 \times 1.2 = 5.8$ m, $5.8 \times 1.2 = 7.0$ m.

Revision

4   State the first five terms of the geometric sequence when (a) the first term is 2 and the common ratio is 4, (b) the first term is 4 and the common ratio is $-\frac{1}{2}$.

5   State the 6th term of the geometric sequence for which the first three terms are 1, –3, 9.

7.3 Series

Associated with sequences are series. The term *series* is used for the sum of the terms of a sequence. Thus 1, 4, 7, 10 is a sequence in which the law determining a term is that each number exceeds the previous one by 3 and $1 + 4 + 7 + 10$ is the series associated with the sequence. The term *finite* is applied to a series when it has a finite number of terms, the term *infinite* when it does not.

7.3.1 Arithmetic series

An *arithmetic series* has each term formed from the previous term by simply adding on a constant value. Such a series can be written as:

$$a + \{a + d\} + \{a + 2d\} + \{a + 3d\} + \ldots + \{a + (n-1)d\}$$

There are n terms in the above series and the sum S_n of the first n terms is:

$$S_n = a + \{a + d\} + \{a + 2d\} + \{a + 3d\} + \ldots + \{a + (n-1)d\}$$

If we write the terms in the reverse sequence we have:

$$S_n = \{a + (n-1)d\} + \{a + (n-2)d\} + \{a + (n-3)d\} + \ldots + a$$

If we add corresponding terms in these two equations we obtain:

$$2S_n = [a + \{a + (n-1)d\}] + [(a + d) + \{a + (n-2)d\}]$$
$$+ \{(a + 3d) + \{a + (n-3)d\}] + \ldots + [a + \{a + (n-1)d\}]$$

$$= [2a + (n-1)d] + [2a + (n-1)d] + [2a + (n-1)d] + \ldots$$
$$+ [2a + (n-1)d]$$

$$= n [2a + (n-1)d]$$

Hence:

$$S_n = \tfrac{1}{2}n [2a + (n-1)d]$$

Example

Determine the sum of the first 9 terms of the arithmetic series which has the first term as −2 and the common difference as 5.

Using $S_n = \frac{1}{2}n\,[2a + (n-1)d]$:

$$S_9 = \frac{1}{2} \times 9[2 \times (-2) + (9-1)5] = 162$$

Example

How many terms of an arithmetic series with a first term of 2 and a common difference of 0.4 will be required to give a total of 170?

Using $S_n = \frac{1}{2}n\,[2a + (n-1)d]$:

$$170 = \frac{1}{2}n[4 + (n-1)0.4] = 2n + 0.2n^2 - 0.2n$$

Rearranging gives:

$$0.2n^2 + 1.8n - 170 = 0$$

Multiplying by 5 gives:

$$n^2 + 9n - 850 = 0$$

$$(n - 25)(n + 34) = 0$$

Thus $n = 25$ or -34. Since n must be a positive number the answer is 25 terms.

Revision

6   Determine the sums to ten terms of the following series: (a) $1.6 + 2.5 + 3.4 + ...$, (b) $1 + 5 + 9 + ...$, (c) $4 + 8 + 12 + ...$

7   How many terms of the series $2 + 2.4 + 2.8 + 3.2 + ...$ will be needed to give a total of 170?

7.3.2 Geometric series

A *geometric series* has each term formed from the previous terms by multiplying it by a constant factor. Such a series can be written as:

$$a + ar + ar^2 + ar^3 + ... + ar^{n-1}$$

The sum to the nth term is:

$$S_n = a + ar + ar^2 + ar^3 + ... + ar^{n-1}$$

Multiplying by r gives:

$$rS_n = ar + ar^2 + ar^3 + ar^4 + \ldots + ar^n$$

Hence:

$$S_n - rS_n = a - ar^n$$

$$S_n = \frac{a(1 - r^n)}{1 - r}$$

The geometric series $1 + 2 + 4 + \ldots$ has terms which continually increase in size and so the sum increases as we sum to more terms. The sum of an infinite number of terms is thus infinite. However, if we consider the series $1 + \frac{1}{2} + \frac{1}{4} + \ldots$ then each term is decreasing in size. The sum of n terms of this series is:

$$S_n = \frac{a(1 - r^n)}{1 - r} = \frac{1(1 - \frac{1}{2^n})}{1 - \frac{1}{2}} = 2 - \frac{1}{2^{n-1}}$$

The bigger we make n the smaller becomes the term $1/2^{n-1}$ and so as we increase the number of terms the sum will come closer and closer to the value 2. We can say that the sum to infinity of the series is 2.

In general we can write:

$$S_n = \frac{a(1 - r^n)}{1 - r} = \frac{a}{1 - r} - \frac{ar^n}{1 - r}$$

If n is a fraction then a^n will get smaller and smaller as n increases. Thus, for a series that has terms that decrease in value, the sum to infinity is:

$$S_\infty = \frac{a}{1 - r}$$

Example

Determine the sum to the 8th term of the geometric series with a first term of -2 and a constant ratio of $-\frac{1}{2}$.

Using the above equation:

$$S_n = \frac{a(1 - r^n)}{1 - r} = \frac{(-2)\{1 - (\frac{1}{2})^8\}}{1 - (-\frac{1}{2})} = -\frac{85}{64}$$

Example

Determine the sum to infinity of the series $8 + 4 + 2 + 1 + \ldots$

$$S_\infty = \frac{a}{1 - r} = \frac{8}{1 - \frac{1}{2}} = 16$$

Revision

8 Determine the sum to the 10th term of the following geometric series: (a) $4 + 6 + 9 + ...$, (b) $12 + 6 + 3 + ...$, (c) $1 + 2 + 4 + ...$

9 Determine the sum to infinity of the series $12 + 6 + 3 + ...$

7.4 Binomial series

The term *binomial* is used to indicate that an expression contains two terms. Thus $(1 + x)$ and $(a + b)$ are binomial terms. Consider the expansion of powers of a binomial term:

$$(1 + x)^0 = 1$$

$$(1 + x)^1 = 1 + x$$

$$(1 + x)^2 = (1 + x)(1 + x) = 1 + 2x + x^2$$

$$(1 + x)^3 = (1 + x)(1 + 2x + x^2) = 1 + 3x + 3x^2 + x^3$$

$$(1 + x)^4 = (1 + x)(1 + 3x + 3x^2 + x^3) = 1 + 4x + 6x^2 + 4x^3 + x^4$$

Multiplying out the binomial terms is very laborious. However, if you look at the above results you can see that:

1 The first term is always 1.

2 The expansion is a series of terms with increasing powers of x, terminating with x^n, where n is the power to which the $(1 + x)$ is raised.

3 Each term is multiplied by some number (termed a coefficient). The second term is always nx.

4 The series terminates after $(n + 1)$ terms.

What we need is the rule to generate the series. The general formula we can use to write the series is:

$$(1 + x)^n = 1 + nx + \frac{n(n - 1)}{2!}x^2 + \frac{n(n - 1)(n - 2)}{3!}x^3 + ... + x^n$$

Such an expression is valid with any positive whole number value for n and any value of x. The product 1×2 is known as factorial 2 and written as 2!; the product $1 \times 2 \times 3$ is factorial 3 and written as 3!. Hence factorial 5, i.e. 5!, is $1 \times 2 \times 3 \times 4 \times 5$. Many calculators have a function key which allows factorials to be easily obtained; the sequence to find factorial 5 is: key in 5, then select the function $x!$.

If we have $(a + b)^n$, we can put such a binomial expression in the form used above:

$$(a+b)^n = \left[a\left(1 + \tfrac{b}{a}\right) \right]^n = a^n \left(1 + \tfrac{b}{a}\right)^n$$

Expanding the bracketed term then gives:

$$(a+b)^n = a^n \left[1 + n\tfrac{b}{a} + \frac{n(n-1)}{2!}\left(\tfrac{a}{b}\right)^2 + \ldots + \left(\tfrac{b}{a}\right)^n \right]$$

$$= a^n + na^{n-1}b + \frac{n(n-1)}{2!}a^{n-2}b^2 + \ldots + b^n$$

The above is a more general form of the binomial theorem; however, it is often simpler to just remember the simpler form and multiply the result by the a^n factor. This is illustrated in the following examples.

Example

Determine the series expansion of $(1 + x)^4$.

$$(1+x)^4 = 1 + 4x + \frac{4\times3}{2\times1}x^2 + \frac{4\times3\times2}{3!}x^3 + \frac{4\times3\times2\times1}{1\times2\times3\times4}x^4$$

$$= 1 + 4x + 6x^2 + 4x^3 + x^4$$

Example

Determine the series expansion of $(2y + x)^4$.

$$(2y+x)^4 = (2y)^4\left(1 + \frac{x}{2y}\right)^4$$

$$= (2y^4)\left[1 + 4\left(\frac{x}{2y}\right) + \frac{4\times3}{2\times1}\left(\frac{x}{2y}\right)^2 + \frac{4\times3\times2}{3!}\left(\frac{x}{2y}\right)^3 \right.$$

$$\left. + \frac{4\times3\times2\times1}{1\times2\times3\times4}\left(\frac{x}{2y}\right)^4 \right]$$

$$= 16y^4 + 32y^3x + 24y^2x^2 + 8y^3x^3 + x^4$$

Example

Determine the series expansion of $(4 - 3x)^3$.

This can be written as $4^3[1 + (-\tfrac{3}{4}x)]^3$. Hence:

$$4^3\left[1 + \left(-\frac{3x}{4}\right) \right]^3 = 4^3\left[1 + 3\left(-\frac{3x}{4}\right) + \frac{3\times2}{2\times1}\left(-\frac{3}{4}x\right)^2 \right.$$

$$\left. + \frac{3\times2\times1}{1\times2\times3}\left(-\frac{3x}{4}\right)^4 \right]$$

$$= 64 - 144x^3 + 108x^2 - 27x^2$$

Example

Evaluate 2.5^4.

We can write 2.5^4 as $(1 + 1.5)^4 = (1 + x)^4$ with $x = 1.5$. Hence, since:

$$(1+x)^4 = 1 + 4x + \frac{4\times3}{2\times1}x^2 + \frac{4\times3\times2}{3!}x^3 + \frac{4\times3\times2\times1}{1\times2\times3\times4}x^4$$

$$= 1 + 4x + 6x^2 + 4x^3 + x^4$$

then $2.5^4 = 1 + 4 \times 1.5 + 6 \times 1.5^2 + 4 \times 1.5^3 + 1.5^4 = 39.0625$.

Revision

10 Determine the series expansions of:

(a) $(1 + x)^7$, (b) $(2 + x)^4$, (c) $(2x + 3y)^4$, (d) $(1 - \frac{1}{2}x)^6$.

7.4.1 Expansion for a fractional positive index

We can use the binomial theorem with fractional positive indices. For example:

$$(1+x)^{1/2} = 1 + \left(\frac{1}{2}\right)x + \frac{\left(\frac{1}{2}\right)\left(-\frac{1}{2}\right)}{2!}x^2 + \frac{\left(\frac{1}{2}\right)\left(-\frac{1}{2}\right)\left(-\frac{3}{2}\right)}{3!}x^3 + ...$$

$$= 1 + \frac{1}{2}x - \frac{1}{8}x^2 + \frac{1}{16}x^3 - ...$$

The series does not terminate. Suppose we try to use this expansion to determine the square root of 4. Since $4^{1/2} = (1 + 3)^{1/2}$ then we have $1 + 1.5 - 1.125 + 1.6875 - ...$; the expansion is clearly not giving the answer 2.

The binomial theorem can *only* be used with fractional positive indices when x is less than 1.

Revision

11 Determine the series expansion of (a) $(1 + x)^{3/2}$, (b) $(1 + 3x)^{1/3}$, stating the limits of x for which the expansion is valid.

12 Use the binomial theorem to determine the cube root of 1.03 as a series. Hint: the cube root of 1.03 can be written as $(1 + 0.03)^{1/3}$.

7.4.2 Expansion for a negative index

In the above section we only considered the binomial theorem when n was a positive whole number. We can, however, use the theorem when n is a negative number, subject to some restrictions.

Consider $(1 + x)^{-1}$. Using the binomial theorem we obtain:

$$(1+x)^{-1} = 1 + (-1)x + \frac{(-1)(-2)}{2!}x^2 + \frac{(-1)(-2)(-3)}{3!}x^3 + \ldots$$

$$= 1 - x + x^2 - x^3 + \ldots$$

The series does not terminate. The above result is the same as we could have obtained by writing $(1 + x)^{-1} = 1/(1 + x)$ and dividing 1 by $(1 + x)$. However, suppose we try to use the expansion to evaluate 1/3. Since $1/3 = 3^{-1} = (1 + 2)^{-1} = (1 + x)^{-1}$ where $x = 2$, then the series is, as above, $1 - 2 + 2^2 - 2^3 + \ldots = 1 - 2 + 4 - 8 + \ldots$ But this is plainly not 1/3 which is 0.333. Thus the binomial theorem cannot be used in this case.

We can use the binomial theorem with negative indices provided x is greater than –1 and less than +1. We can write this as $-1 < x < +1$ or $|x| < +1$, the $|x|$ indicating that x can be either positive or negative. Enclosing a term between vertical lines indicates we are concerned with the *modulus* of the term, the modulus of a positive number being the same as the number while the modulus of a negative number is a positive number of the same magnitude, e.g. $|3| = 3$ and $|-3| = 3$. In using the binomial theorem with a negative index there is no restriction as to whether the index is a whole number, fractional numbers can be used.

Example

Determine the series expansion of $(1 - 2x)^{-1/2}$ and state the limits of x for which the expansion is valid.

$$[1 + (-2x)]^{-1/2} = 1 + \left(-\frac{1}{2}\right)(-2x) + \frac{\left(-\frac{1}{2}\right)\left(-\frac{3}{2}\right)}{2!}(-2x)^2$$

$$+ \frac{\left(-\frac{1}{2}\right)\left(-\frac{3}{2}\right)\left(-\frac{5}{2}\right)}{3!}(-2x)^3 + \ldots$$

$$= 1 + x + \frac{3}{2}x^2 + \frac{5}{2}x^3 + \ldots$$

The expansion is valid if $|-2x| < +1$, i.e. $|x| < 1/2$. Note that in simplifying this, we can transpose quantities from either side of the $<$ sign in the same way as with an $=$ sign. If x has a positive value we can write $|x| < \frac{1}{2}$ as $x < \frac{1}{2}$ and if x has a negative value as $-x < \frac{1}{2}$ or $-\frac{1}{2} < x$. Thus $|x| < \frac{1}{2}$ can be written as $-\frac{1}{2} < x < \frac{1}{2}$.

Revision

13 Use the binomial theorem to write 5/6 as a series Hint: $5/6 = 1/1.2$.

14 Determine the series expansion of (a) $(1 + 2x)^{-3}$, (b) $1/(1 - 4x)^2$, stating the limits of x for which the expansion is valid.

7.4.3 The binomial theorem: a summary

The theorem can be stated as:

$$(1 + x)^n = 1 + nx + \frac{n(n-1)}{2!}x^2 + \frac{n(n-1)(n-2)}{3!}x^3 + \ldots + x^n$$

If n is a positive whole number:

1   The series terminates after $(n + 1)$ terms.
2   The series is valid for any value of x.

If n is other than a positive whole number:

1   The series never terminates.
2   The series is only valid if x lies between 1 and −1.

7.5 Approximations

In the binomial series of $(1 + x)^n$, if x is small when compared with 1 then often the effect of the x^2, x^3, x^4, etc. terms can be discounted and thus we can make the approximation that:

$$(1 + x)^n \simeq 1 + nx$$

Example

The area of a circle is $\frac{1}{4}\pi d^2$. What is the accuracy of the area if the measurement of the diameter d can be in error by ±1%.

If the correct diameter is d then the measured value can be $d + 1\%$ of d $= d + 1d/100 = d(1 + 0.01)$. The area of the circle given by this measured value is thus:

$$\text{measured value} = \frac{1}{4}\pi[d(1 + 0.01)]^2 = \frac{1}{4}\pi d^2(1 + 0.01)^2$$

$$\simeq \frac{1}{4}\pi d^2(1 + 0.01 \times 2) \simeq \frac{1}{4}\pi d^2 \times 1.02$$

The area is thus in error by +2%. Likewise if we consider the measured value to be $d - 1\%$ of d, the error is −2%. Thus the accuracy is ±2%.

Revision

15 If an error of +2% was made in the measurement of the diameter of a sphere, what is the percentage error in the calculated volume?

7.5.1 Approximations when multiplying or dividing quantities

Suppose we have $(1 + a)^n(1 + b)^m$ where both a and b are small compared with 1. We can then use the binomial theorem to make the approximation:

$$(1 + a)^n(1 + b)^m \simeq (1 + an)(1 + bm)$$

Since $(1 + an)(1 + bm) = 1 + an + bm + abnm$ and we can neglect $abnm$ as being small because it involves the products of two numbers which are smaller than 1, we can further approximate this to:

$$(1 + a)^n(1 + b)^m \simeq (1 + an)(1 + bm) \simeq 1 + an + bm$$

Suppose we have $(1 + a)^n/(1 + b)^m$ where both a and b are small compared with 1. We can then use the binomial theorem to make the approximation:

$$(1 + a)^n(1 + b)^{-m} \simeq (1 + an)(1 - bm) \simeq 1 + an - bm$$

Example

Determine the approximate percentage change in the volume of a cylinder when its diameter d increases by 2% and its length h decreases by 3%.

The volume V of a cylinder is $\frac{1}{4}\pi d^2 h$. The new value of the diameter will be $d + 2\%$ of $d = d + 2d/100 = d(1 + 0.02)$. The new value of h will be $h - 3\%$ of $h = h - 3h/100 = h(1 - 0.03)$. The new volume is thus:

$$\text{new volume} = \frac{1}{4}\pi[d(1 + 0.02)^2 h(1 - 0.03)]$$

$$= \frac{1}{4}\pi d^2 h(1 + 0.02)^2)(1 - 0.03)^1$$

$$\simeq \frac{1}{4}\pi d^2 h(1 + 2 \times 0.02)(1 - 0.03)$$

$$\simeq \frac{1}{4}\pi d^2 h(1 + 0.04 - 0.03)$$

$$\simeq \frac{1}{4}\pi d^2 h \times 1.01$$

The volume thus increases by a factor of about 1.01. The percentage change in volume is thus about 1%.

Example

Simplify the expression $(1 - 2x)^2/(1 + x)^4$ if powers of x greater than the first can be neglected.

We can write the expression as $(1 - 2x)^2(1 + x)^{-4}$ and use the binomial expression to expand each binomial term, neglecting powers of x greater than the first. Thus:

$$(1 - 2x)^2(1 + x)^{-4} \simeq [1 + 2(-2x)][1 + (-4)x]$$

$$\simeq (1 - 4x)(1 - 4x) \simeq 1 - 8x$$

Revision

16 Kinetic energy is $\frac{1}{2}mv^2$. Determine the percentage change in kinetic energy when the mass m increases by 2% and the velocity is reduced by 4%.

17 Simplify the expression $(1 + x)^{1/2}/(1 - 2x)^2$ if powers of x greater than the first can be neglected.

Problems

1 State the rules determining the terms in the following sequences: (a) 2, 3, 4, 5, ..., (b) –3, 6, –12, 24, ..., (c) 5, 6, 7, 8, ..., (d) 1, 4, 16, 64, ..., (e) –1, 1, –1, 1, ...

2 Find the 10th term of the following arithmetic sequences: (a) 2, 5, 8, ..., (b) 1, 6, 11, ..., (c) 5, 9, 13, ...

3 State the first five terms of the arithmetic sequence when: (a) the first term is 2 and the common difference 3, (b) the first term is –4 and the common difference is 2.

4 Find the 6th term of the following geometric sequences: (a) 6, 18, 54, ..., (b) 1, 3, 9, ..., (c) 1, –½ , ¼, ...

5 State the first five terms of the geometric sequence when: (a) the first term is 24 and the common ratio is ½, (b) the first term is 10 and the common ratio is 0.1.

6 A machine is bought for £5000. If its value depreciates at the rate of £300 per year, what is its value after 4 years?

7 A ball dropped onto level ground rebounds each time to 0.7 of the height from which it fell. If it starts from a height of 200 cm, list the heights of the first four bounces.

8 A machine tool is to have five rotational speeds with the slowest being 20 rev/min and the fastest 200 rev/min. If the speeds are to be in a geometrical sequence, determine the common ratio.

9 Determine the sum of the first 20 terms of the following arithmetic series: (a) $3 + 6 + 9 + 12 + ...$, (b) $2 + 5 + 8 + 11 + ...$, (c) $1.2 + 2.2 + 3.2 + 4.2 + ...$

10 Determine the number of terms required in the series $11 + 15 + 19 + 24 + ...$ to give a total of 341.

11 An arithmetic series has a first term of 27 and a common difference of -2. How many terms will be needed to give a sum of 192?

12 An arithmetic series has a first term of 3 and a common difference of 4. How many terms will be needed to give a sum of 2775?

13 Determine the sums of the following geometric series: (a) $128 + 64 + 32 + ...$ for the first 8 terms, (b) $2 + 8 + 32 + ...$ for the first 4 terms, (c) $8.00 + 8.15 + 8.30 + ...$ for the first 30 terms.

14 Determine the sums to infinity of the following series: (a) $9 - 6 + 4 - ...$, (b) $12 + 6 + 3 + ...$, (c) $5 + 0.5 + 0.05 + ...$

15 A ball falls vertically onto a horizontal surface and, on each bounce, rebounds to 4/5 of the height from which it fell. If the ball starts at a height of 3.5 m, determine the total distance it travels before coming to rest.

16 £100 is invested and earns compound interest of 2% per year. What will be the amount in the account after 8 years?

17 Determine, using the binomial series, the expansion of the following as series and state where there are any restrictions on the validity of the result:

(a) $(3 + x)^5$, (b) $(2 + x)^7$, (c) $(1 + 2x)^4$, (d) $(2 - x)^6$, (e) $(y - 2x)^3$,

(f) $(x + 2y)^4$, (g) $(1 + x)^{-1/2}$, (h) $(2 + x)^{1/2}$, (i) $(1 - 2x)^{-1/2}$, (j) $(1 + 3x)^{1/3}$,

(k) $(1 + x)^{-2}$, (l) $(3 - 2x)^{-1}$.

18 The periodic time T of a simple pendulum is related to its length L and the acceleration due to gravity g by $T = 2\pi\sqrt{(L/g)}$. What will be the percentage change in the periodic time if the length increases by 2%?

19 The maximum bending stress σ at the surface of a cantilever when subject to bending is related to its length L, breadth b, depth d and the load F applied at the free end by $\sigma = 6FL/(bd^2)$. Determine the

percentage change in σ if (a) F increases by 1% and the dimensions remain unchanged, (b) b is reduced by 1% and d is reduced by 1 %, the other terms remaining unchanged.

20 Evaluate, using the binomial theorem, the following to five decimal places (a) 1.002^4, (b) $1/0.98$, (c) 0.98^7.

21 Simplify the following expressions if powers of x greater than the first can be neglected:

(a) $(1 + x)^3/(1 - x)^{1/2}$, (b) $(1 + 2x)^{1/2}/(1 - x)$, (c) $\sqrt{[(1 - x)/(1 + x)]}$.

8 Exponentials

8.1 Introduction

There are many situations in engineering where we are concerned with the rate at which some quantity grows or the rate at which it decays. Examples are:

1   The variation with time of the temperature of a cooling object.

2   The variation with time of the charge on a capacitor when it is being charged and when it is being discharged.

3   The variation with time of the current in a circuit containing inductance when the current is first switched on and then when it is switched off.

4   The decay with time of the radioactivity of a radioactive isotope.

5   The growth of bacteria or indeed any uncontrolled population.

This chapter is about equations we can use to describe such growth or decay of quantities.

8.2 Exponentials

We can describe growth and decay processes by an equation of the form:

$$y = a^t$$

where a is some constant, and y the value of the quantity at a time t. Such equations are called *exponentials* with a being called the *base*. An exponential is *not* a polynomial. The powers of a polynomial are constants, e.g. the 2 in x^2, whereas the power of an exponential is the variable. An exponential could be 2^t, 3^t, 4^t, etc.

8.2.1 Growth

To illustrate the use of $y = a^t$ for a growing quantity, consider how the value of y changes with time when we have $a = 2$ and so the equation:

$$y = 2^t$$

At $t = 0$ we have $y = 2^0 = 1$, at $t = 1$ we have $y = 2^1 = 2$, at $t = 2$ we have $y = 2^2 = 4$, at $t = 3$ we have $y = 2^3 = 8$, and so on. Thus we have the data:

$y = 2^t$	1	2	4	8	16	32
t	0	1	2	3	4	5

Figure 8.1 $y = 2^t$

This data describes a quantity that increases with time, starting at $t = 0$ with the value 1 and rapidly increasing as t increases. Figure 8.1 shows this data plotted as a graph. Note that the graph describes a function which doubles for every successive interval of $t = 1$.

8.2.2 Decay

For a quantity that decreases with time, consider how the value of y changes with time when we have $a = 2$ and the equation

$$y = 2^{-t}$$

Figure 8.2 $y = 2^{-t}$

At $t = 0$ we have $y = 2^0 = 1$, at $t = 1$ we have $y = 2^{-1} = 0.5$, at $t = 2$ we have $y = 2^{-2} = 0.25$, at $y = 3$ we have $y = 2^{-3} = 0.125$, and so on. Thus we have the data:

$y = 2^{-t}$	1	0.5	0.25	0.125	0.0625	0.03125
t	0	1	2	3	4	5

This data describes a quantity that decreases with time, starting with a value 1 at $t = 0$ and rapidly decreasing in size as t increases. Figure 8.2 shows this data plotted as a graph. Note that the graph describes a function which halves for every successive interval of $t = 1$.

8.2.3 Using a calculator with exponentials

To find the value of a^x the x^y key can be used. For example, to find 2^x with $x = 4$, the key sequences would be:

press the 2 key,
select and press the x^y key,
press the 4 key,
press the = key

The result is 16. To find the value of 2^{-x} with $x = 4$, the key sequences would be:

press the 2 key,
select and press the x^y key,
press the 4 key,
press the +/− key to obtain a display of −4,
press the = key

The result is 0.0625.

8.3 The exponential

This chapter is about quantities that can be described by the general equation $y = a^x$, where the varying quantity x in engineering is often time. An exponential could be 2^x, 3^x, 4^x, etc. We could write equations for growth or decay process with different values of the base. However, we usually standardise the base to one particular value. The most widely used exponential is e^x, where e is a constant with the value 2.728 281 828 ... and is often referred to as *the exponential*. Whenever an engineer refers to an exponential change he or she is almost invariably referring to an equation written in terms of e^x.

The following shows the values of e^x and e^{-x} for various values of x and Figure 8.3 the resulting graphs The e^x graph describes a growth curve, the e^{-x} a decay curve. Note that both graphs have $y = 1$ when $x = 0$.

$y = e^x$	0.14	0.37	1	2.72	7.39	20.09	... infinite
x	−2	−1	0	1	2	3	... infinite

$y = e^{-x}$	7.39	2.72	1	0.37	0.14	0.05	... 0
x	−2	−1	0	1	2	3	... infinite

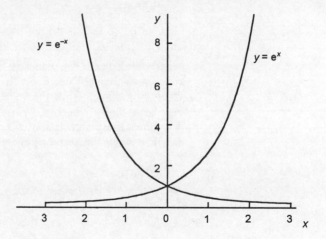

Figure 8.3 $y = e^t$ *and* $y = e^{-t}$

In a more general form we can write the exponential equation in the form $y = e^{kx}$, or $y = e^{-kx}$, where k is some constant. This constant k determines how fast y changes with x. The following data illustrates this:

x	0	1	2	3	... infinite
$y = e^{-1x}$	1	0.368	0.135	0.050	... 0
$y = e^{-2x}$	1	0.135	0.018	0.003	... 0

x	0	1	2	3	... infinite
$y = e^{1x}$	1	2.718	7.389	20.086	... infinite
$y = e^{2x}$	1	7.389	54.598	403.429	... infinite

The bigger k is the faster y decreases, or increases, with x.

When $x = 0$ then for $y = e^{kx}$, or $y = e^{-kx}$, $y = e^0$ and so $y = 1$. This is thus the value of y that occurs when x is zero. Since we may often have an initial value other than 1, we write the equation in the form:

$$y = A \, e^{kx}$$

where A is the initial value of y at $x = 0$. For example, for the discharging of a capacitor in an electrical circuit we have, for the charge q on the capacitor at a time t, the equation

$$q = Q_0 \, e^{-t/CR}$$

When $t = 0$ then $q = Q_0$. The constant k is $1/CR$. The bigger the value of CR the smaller the value of $1/CR$ and so the slower the rate at which the capacitor becomes discharged.

One form of equation involving exponentials that is quite common is of the form:

$$y = A - A \, e^{-kx}$$

When $x = 0$ then $e^0 = 1$ and so $y = A - A = 0$. The initial value is thus 0. As x increases then e^{-kx} decreases from 1 towards 0, eventually becoming zero when x is infinite. Thus the value of y increases as x increases. When x is very large then e^{-kx} becomes virtually 0 and so y becomes equal to A. Figure 8.4 shows the graph. It shows a quantity y which increases rapidly at first and then slows down to become eventually A.

Figure 8.4 $y = A - A \, e^{-kx}$

For example, for a capacitor which starts with zero charge on its plates and is then charged we have the equation:

$$q = Q_0 - Q_0\, e^{-t/CR}$$

When $t = 0$ then $e^0 = 1$ and so $q = Q_0 - Q_0 = 0$. As t increases, so the value of $e^{-t/CR}$ decreases and so q becomes more and more equal to Q_0.

Example

For an object cooling according to Newton's law, the temperature θ of the object varies with time t according to the equation $\theta = \theta_0\, e^{-kt}$, where θ_0 and k are constants. (a) Explain why this equation represents a quantity which is decreasing with time. (b) What is the value of the temperature at $t = 0$? (c) How will the rates at which the object cools change if in one instance $k = 0.01$ and in another $k = 0.02$ (the units of k are per °C)?

(a) If we assume that t and k will only have positive values, then the $-kt$ means that the power is negative and so the temperature decreases with time.
(b) When $t = 0$ then $e^{-kt} = 1$ and so $\theta = \theta_0$. Thus θ_0 is the initial value at the time $t = 0$.
(c) Doubling the value of k means that the object will cool faster, in fact it will cool twice as fast.

Example

The current i in amperes in an electrical circuit varies with time t according to the equation $i = 10(1 - e^{-t/0.4})$. What will be (a) the initial value of the current when $t = 0$, (b) the final value of the current at infinite time?

(a) When $t = 0$ then $e^{-t/0.4} = e^0 = 1$. Thus $i = 10(1 - 1) = 0$.
(b) When t becomes very large then $e^{-t/0.4}$ becomes 0. Thus we have $i = 10(1 - 0)$ and so the current becomes 10 A.

Revision

1   The number N of radioactive atoms in a sample of radioactive material decreases with time t and is described by the equation $N = N_0\, e^{-\lambda t}$, where N_0 and λ are constants. (a) Explain why this equation represents a quantity which is decreasing with time. (b) What will be the number of radioactive atoms at time $t = 0$? (c) For a radioactive material that decreases only very slowly with time, will λ have a large or smaller value than with a radioactive material which decreases quickly with time?

2   The length L of a rod of material increases from some initial length with the temperature θ above that at which the initial length is measured and is described by the equation $L = L_0\, e^{a\theta}$, where L_0 and θ are constants. (a) Explain why the equation represents a quantity which increases with

time. (b) What will be the length of the rod when $\theta = 0$? (c) What will be the effect of a material having a higher value of α than some other material?

3　For an electrical circuit involving inductance, the current in amperes is related to the time t by the equation $i = 3(1 - e^{-10t})$. What is the value of the current when (a) $t = 0$, and (b) t is very large?

4　What are the values of y in the following equations when (i) $x = 0$, (ii) x is very large, i.e. infinite?

(a) $y = 2\ e^{3x}$, (b) $y = 10\ e^{-5x}$, (c) $y = 2(1 - e^{-2x})$, (d) $y = 2\ e^{-0.2x}$,
(e) $y = -4\ e^{-x/3}$, (f) $y = 0.5(1 - e^{-x/5})$, (g) $y = 4(1 - e^{-x/2})$, (h) $y = 10\ e^{4x}$,
(i) $y = 0.2 - 0.2\ e^{-3x}$.

8.3.2 Using a calculator with exponentials

To find the value of e^x the e^x key can be used. For example, to find e^4, the key sequences would be:

press the 4 key,
select and press the e^x key (there is usually no need to press the = key).

The result is 54.598... To find e^{-4}, the key sequences would be:

press the 4 key,
select and press the +/− key,
press the e^x key.

The result is 0.0183...

Example

Use a calculator to determine the value of $2\ e^{t/3}$, for when $t = 5$.

The key sequence might be:

press the 5 key,
press the ÷ key,
press the 3 key,
press the = key,
select and press the e^x key,
press the × key,
press the 2 key,
press the = key.

The result is 10.59.

Example

Newton's law of cooling for an object can be written as $\theta = \theta_0\,e^{-kt}$, where k is a constant, θ is the temperature at a time t and θ_0 the original temperature at $t = 0$. For $\theta_0 = 50°C$ and $k = 0.002$ (unit s^{-1}), what will be the value of the temperature after a time of 100 s?

The equation to be solved is thus $q = 50\,e^{-0.002t}$, with $t = 100$. The key sequences with a calculator might be:

    press the keys for 0.002,
    press the × key,
    press the keys for 100,
    press the = key,
    press the +/− key,
    select and press the e^x key,
    press the × key,
    press the keys for 50,
    press the = key.

The result is 40.9°C

Revision

5 Use a calculator to determine the values of:

    (a) e^6, (b) e^{-6}, (c) $e^{0.4}$, (d) $e^{-0.3}$, (e) $e^{12.2}$, (f) $e^{-5.6}$.

6 Use a calculator to complete, to two decimal places, the following table of values:

x	0	+1	+2	+3	infinite
$y = 5\,e^{1x}$					
$y = 5\,e^{-1x}$					
$y = 5(1 - e^{-1x})$					

7 Use a calculator to complete, to two decimal places, the following table of values:

x	0	+1	+2	+3	infinite
$y = 10\,e^{2x}$					
$y = 10\,e^{-2x}$					
$y = 10(1 - e^{-2x})$					

8 Use a calculator to complete, to two decimal places, the following table of values:

x	0	+1	+2	+3	infinite
$y = 2\,e^{x/2}$					
$y = 2\,e^{-x/2}$					
$y = 2(1 - e^{-x/2})$					

9 Use a calculator to complete, to two decimal places, the following table of values:

x	0	+1	+2	+3	infinite
$y = 10\,e^{0.1x}$					
$y = 10\,e^{-0.1x}$					
$y = 10(1 - e^{-0.1x})$					

10 The voltage, in volts, across a capacitor is given by $20\,e^{-0.1t}$, where t is the time in seconds. Determine the voltage when t is (a) 1 s, (b) 10 s.

11 The atmospheric pressure p is related to the height h above the ground at which it is measured by the equation $p = p_0\,e^{-h/c}$, where c is a constant and p_0 the pressure at ground level where $h = 0$. Determine the pressure at a height of 1000 m if p_0 is 1.01×10^5 Pa and $c = 70\,000$ (unit m^{-1}).

12 The current i, in amperes, in a circuit involving an inductor in series with a resistor when a voltage is E is applied to the circuit at time $t = 0$ is given by the equation

$$i = \frac{E}{R}(1 - e^{-Rt/L})$$

If R/L has the value 2 Ω/H (actually the same unit as seconds), what is the current when (a) $t = 0$, (b) $t = 1$ s?

13 The voltage v across a resistor in series with an inductor when a voltage E is applied to the circuit at time $t = 0$ is given by the equation

$$v = E(1 - e^{-t/T})$$

where T is the so-called time constant of the circuit. If $T = 0.5$ s, what is the voltage when (a) $t = 0$, (b) $t = 1$ s?

8.3.3 Why e?

Why choose this strange number 2.718... for the base? The reason is linked to the properties of expressions written in this way. For $y = e^x$, the rate of change of y with x, i.e. the slope of a graph of y against x, is equal to e^x:

slope of graph of y against $x = y = e^x$

This topic is discussed in more detail in the chapter concerned with differentiation.

8.3.4 The exponential series

The number e has another property; we can describe e^x by the series:

$$e^x = 1 + x + \frac{x^2}{2!} + \frac{x^3}{3!} + \frac{x^4}{4!} + \dots$$

Example

Use the power series to determine the value of $e^{0.1}$.

$$e^{0.1} = 1 + 0.1 + \frac{0.1^2}{2!} + \frac{0.1^3}{3!} + \frac{0.1^4}{4!} + \dots$$

$$= 1 + 0.1 + 0.005 + 0.000\,167 + 0.000\,004 + \dots$$

$$= 1.105\,171$$

Example

Use the power series to expand $e^{2x}(1 - 2x)$ as a series.

Expanding the exponential term as a series gives:

$$e^{2x} = 1 + 2x + \frac{(2x)^2}{2!} + \frac{(2x)^3}{3!} + \frac{(2x)^4}{4!} + \dots$$

Hence:

$$(1 - 2x)e^{2x} = 1 + 2x + 2x^2 + \frac{4x^3}{3} + \frac{2x^4}{3} + \dots$$

$$- 2x - 4x^2 - 4x^3 - \frac{8x^4}{3} - \frac{4x^5}{3} - \dots$$

$$= 1 - 2x^2 - \frac{8x^3}{3} - 2x^4 + \dots$$

Revision

14 Use the power series for e^x to determine $e^{-0.1}$ to five significant figures.

15 Use the power series for e^x to expand the expression $e^x(1 + 2x)^2$.

8.4 Manipulating exponentials

The techniques used for the manipulation of exponentials are the same as those described in chapter 1 for manipulating powers. Thus:

$$a^x a^y = a^{x+y} \qquad [1]$$

$$\frac{a^x}{a^y} = a^{x-y} \qquad [2]$$

$$(a^x)^y = a^{xy} \qquad [3]$$

$$(ab)^x = a^x b^x \qquad [4]$$

$$\left(\frac{a}{b}\right)^x = \frac{a^x}{b^x} \qquad [5]$$

where a and b are bases.

Example

Simplify the following:

(a) $2^x 2^{2x}$, (b) $e^{2t} e^{4t}$, (c) $(e^{2t})^{-3}$, (d) $\dfrac{e^{5x}}{e^{2x}}$, (e) $\dfrac{10\,e^{t/2}}{2\,e^{t/3}}$, (f) $\dfrac{1}{e^{2t}} + \dfrac{2}{e^{3t}}$

(a) Using equation [1] from the above list, then $2^x 2^{2x} = 2^{x+2x} = 2^{3x}$.
(b) Using equation [1] from the above list, then $e^{2t} e^{4t} = e^{2t+4t} = e^{6t}$.
(c) Using equation [3] from the above list, then $(e^{2t})^{-3} = e^{-6t}$.
(d) Using equation [2] from the above list, then:

$$\frac{e^{5x}}{e^{2x}} = e^{5x-2x} = e^{3x}$$

(e) Using equation [2] from the above list, then:

$$\frac{10\,e^{t/2}}{2\,e^{t/3}} = 5\,e^{\frac{t}{2}-\frac{t}{3}} = 5\,e^{t/6}$$

(f) Bringing the fraction to a common denominator:

$$\frac{1}{e^{2t}} + \frac{2}{e^{3t}} = \frac{e^{3t} + 2\,e^{3t}}{e^{2t}e^{3t}} = \frac{e^{3t} + 2\,e^{3t}}{e^{5t}} = e^{-2t} + 2\,e^{-3t}$$

Alternatively we could takes the reciprocals of each terms and write the equation as:

$$\frac{1}{e^{2t}} + \frac{2}{e^{3t}} = e^{-2t} + 2\,e^{-3t}$$

Revision

16 Simplify the following:

(a) $2^{3x} 2^{5x}$, (b) $3^{2x} 3^{-4x}$, (c) $e^3 e^5$, (d) $e^{3t} e^{5t}$, (e) $e^{-5t} e^{3t}$, (f) $(e^{-4t})^3$, (g) $(1 + e^{2t})^2$,

(h) $\dfrac{1}{e^{3t}}$, (i) $\dfrac{e^{3t}}{e^{5t}}$, (j) $\left(\dfrac{1}{e^{4t}}\right)^2$, (k) $\dfrac{10\,e^{4t}}{2\,e^{t}}$, (l) $\dfrac{2\,e^{3t}e^{5t}}{5\,e^{4t}}$.

Problems

1 If the voltage v, in volts, in an electrical circuit is given by the equation $v = 100\ e^{-2t}$, what will be the voltage when the time t is 0? Does the equation describe a voltage which is increasing or decreasing with time?

2 In a chemical reaction the amount C of the starting material left after a time t is given by the equation $C = C_0\ e^{-kt}$, where C_0 and k are constants. (a) Explain why this equation describes a decrease with time of C. (b) What was the amount of the material at $t = 0$? (c) If one reaction has a higher value of k than another, how will this affect the rate at which the amount of material C changes with time?

3 Determine the values of y in the following equations when (i) $x = 0$, (ii) x is infinite:

(a) $y = 2\ e^{x}$, (b) $y = 12\ e^{x/2}$, (c) $y = 3\ e^{-x}$, (d) $y = 10\ e^{-x/10}$, (e) $y = 2\ e^{-5x}$,

(f) $y = 2(1 - e^{-x/4})$, (g) $y = 12(1 - e^{-4x})$, (h) $y = 2(1 - e^{-x/1000})$.

4 Use a calculator to determine, to two decimal places, the values of the following:

(a) e^{4}, (b) e^{-4}, (c) $e^{0.4}$, (d) $e^{-0.4}$, (e) $10\ e^{2/3}$, (f) $5\ e^{-5/9}$, (g) $-12\ e^{0.5}$,

(h) $10(1 - e^{-3})$, (i) $2(1 - e^{-0.1})$, (j) $4(1 - e^{-1/2})$.

5 Use a calculator to complete, to two decimal places, the following table of values of y:

x	0	+1	+2	+3	infinite
$y = 10\ e^{0.2x}$					
$y = 10\ e^{-0.2x}$					
$y = 10(1 - e^{-0.2x})$					
$y = 10\ e^{x/10}$					
$y = 10\ e^{-x/10}$					
$y = 10(1 - e^{-x/10})$					

6 The charge q on a discharging capacitor is related to the time t by the equation $q = Q_0\ e^{-t/CR}$, where Q_0 is the charge at $t = 0$, R is the resistance in the circuit and C the capacitance. Determine the charge on a capacitor after a time of 0.2 s if initially the charge was 1 µC (1 µC = 10^{-6} C), R is 1 MΩ (1 MΩ = 10^{6} Ω) and C is 4 µF (1 µF = 10^{-6} F). Note that with the units in seconds (s), coulombs (C), ohms (Ω) and farads (F), the resulting charge will be in coulombs.

7 The current i, in amperes, in a circuit with an inductor in series with a resistor is given by the equation $i = 4(1 - e^{-10t})$, where the time t is in seconds. Determine the current when (a) $t = 0$, (b) $t = 0.05$ s, (c) $t = 0.10$ s, (d) $t = 0.15$ s, (e) $t = $ infinity.

8 The voltage v, in volts, across a capacitor after a time t, in seconds, is given by the equation $v = 10\,e^{-t/3}$. Determine the value of the voltage v after 2 s.

9 The resistance R, in ohms, of an electrical conductor at a temperature of $\theta°$C is given by the equation $R = R_0\,e^{a\theta}$. Determine the resistance at a temperature of 1000°C if R_0 is 5000 Ω and a is 1.2×10^{-4} (unit per °C).

10 The current i, in amperes, in an electrical circuit varies with time t and is given by the equation $i = 2(1 - e^{-10t})$. Determine the current after times of (a) 0.1 s, (b) 0.2 s, (c) 0.3 s.

11 The amount N of a radioactive material decays with time t and is given by the equation $N = N_0\,e^{-0.7t}$, where t is in years. If at time $t = 0$ the amount of radioactive material is 1 g, what will be the amount after five years?

12 The atmospheric pressure p, in pascals, varies with the height h, in kilometres, above sea level according to the equation $p = p_0\,e^{-0.15h}$. If the pressure at sea level is 10^5 Pa, what will be the pressure at heights of (a) 1 km, (b) 2 km?

13 The voltage v, in volts, across an inductor in an electrical circuit varies with time t, in milliseconds, according to the equation $v = 200\,e^{-t/10}$. Determine the voltage after times of (a) 0.1 ms, (b) 0.5 ms.

14 When the voltage E to a circuit consisting of an inductor in series with a resistor is switched off, the voltage across the inductor varies with time t according to the equation $v = -E\,e^{-t/T}$, where T is the time constant of the circuit. If $T = 2$ s, determine the voltage when (a) $t = 0$, (b) $t = 1$ s.

15 When a voltage E is applied to a circuit consisting of a capacitor in series with a resistor at time $t = 0$, the voltage v across the capacitor varies with time according to the equation $v = E(1 - e^{-t/T})$, where T is the time constant of the circuit. If $T = 0.1$ s, determine the voltage when (a) $t = 0$, (b) $t = 0.1$ s.

16 The temperature θ, in °C, of a cooling object varies with time t, in minutes, according to the equation $\theta = 200\,e^{-0.04t}$. Determine the temperature when (a) $t = 0$, (b) $t = 10$ minutes, (c) t is infinite.

17 The concentration C, as a percentage, of the starting material in a chemical reaction varies with time t, in seconds, according to the

equation $C = 100\ e^{-0.005t}$. Determine the concentration when (a) $t = 0$, (b) $t = 10$ s, (c) $t = 100$ s, (d) t is infinite.

18 Under one set of conditions the amplitude A of the oscillations of a system varies with time t according to the equation $A = A_0\ e^{kt}$. Under other conditions the amplitude varies according to the equation $y = A_0\ e^{-kt}$. If k is a positive number, how do the oscillations differ?

19 Simplify the following:

(a) $4^x 4^{2x}$, (b) $5^{-x} 5^{3x}$, (c) $e^{3x} e^{6x}$, (d) $e^{-x} e^{-3x}$, (e) $(e^{2t})^3$, (f) $(1 + e^{2t})^2$,

(g) $(2 + 3\,e^{4t})^2$, (h) $\dfrac{4}{e^{4t}}$, (i) $\dfrac{6\,e^{3t}}{2\,e^{5t}}$, (j) $\left(\dfrac{4}{2\,e^{3t}}\right)^2$.

20 Use the power series for e^x to determine $e^{0.05}$ to six significant figures.

21 Use the power series for e^x to expand as a series: (a) $e^x(x^2 - 1)$, (b) $e^x/(1 - x^2)^{1/2}$.

9 Logarithms

9.1 Introduction

Consider the equation $y = 2^x$. If we are given a value for x then we can calculate y. Thus, if we have $x = 3$ then $y = 2^3 = 8$. But suppose we are given a value for y and have to calculate x. How do we solve such an equation? For example, if we have $2.5 = 2^x$, what is x? How do we find the number to which we have to raise a power in order to get a value of 2.5? This chapter introduces a technique for dealing with such problems. It is called *logarithms*.

9.2 Logarithms

If $y = a^x$, then x is called the *logarithm to base a of y*. This can be written as

$$\log_a y = x$$

A logarithm of a number is thus the power to which a base has to be raised to be equal to the number. Logarithms to base 10 are called *common logarithms* and are written as $\log_{10}$ or just log or more commonly lg. In the absence of any special mention of base, lg should be taken to be to base 10. Logarithms to base e are called *natural logarithms* or *Napierian logarithms* and are written as $\log_e$ or more commonly ln.

Consider some logarithms to base 10. When $x = 0$ then $10^0 = 1$ and so $\lg_{10} 1 = 0$. When $x = 1$ then $10^1 = 10$ and so $\lg_{10} 10 = 1$. When $x = 100$ then $10^2 = 100$ and so $\lg_{10} 100 = 2$. Now consider negative values for x. When $x = -1$ then $10^{-1} = 0.1$ and so $\lg_{10} 0.1 = -1$. When $x = -2$ then $10^{-2} = 0.01$ and so $\lg_{10} 0.01 = -2$. Thus we have:

y	0.01	0.1	1	10	100
$\lg_{10} y$	-2	-1	0	1	2

Now consider natural logarithms. When $x = 0$ then $e^0 = 1$ and so we have $\ln 1 = 0$. When $x = 1$ then $e^1 = 2.73$ and so $\ln 2.73 = 1$. When $x = 2$ then $e^2 = 7.39$ and so $\ln 7.39 = 2$. Now consider negative values for x. When we have $x = -1$ then $e^{-1} = 0.37$ and so $\ln 0.37 = -1$. When $x = -2$ then $e^{-2} = 0.14$ and so $\ln 0.14 = -2$. Hence we have:

y	0.14	0.37	1	2.73	7.39
$\ln y$	-2	-1	0	1	2

In general, since $a^1 = a$ then $\lg_a a = 1$ and since $a^0 = 1$ then $\lg_a 1 = 0$. For values of y greater than 1 we have positive values for the logarithm of y; for values of y less than 1 we have negative values for the logarithm of y. Note that we cannot have a value of $\lg_{10} y$, or $\ln y$, or indeed logarithms to any base, which is 0 or negative.

Example

Write the following equations in logarithmic form:

(a) $3.0 = 10^x$, (b) $0.28 = 10^x$, (c) $4.5 = e^x$, (d) $0.54 = e^x$, (e) $\frac{1}{2} = e^x$.

(a) We can write this as a logarithm to base 10 and so as $x = \lg_{10} 3.0$.
(b) We can write this as a logarithm to base 10 and so as $x = \lg_{10} 0.28$.
(c) We can write this as a logarithm to base e and so as $x = \ln 4.5$.
(d) We can write this as a logarithm to base e and so as $x = \ln 0.54$.
(e) We can write this as a logarithm to base e and so as $x = \ln \frac{1}{2}$.

Example

Write the following logarithms in exponential form:

(a) $x = \lg_{10} 5.0$, (b) $x = \lg_{10} 0.68$, (c) $x = \ln 3.0$, (d) $x = \ln 0.23$.

(a) We can write this as $5.0 = 10^x$.
(b) We can write this as $0.68 = 10^x$.
(c) We can write this as $3.0 = e^x$.
(d) We can write this as $0.23 = e^x$.

Example

Determine x in the equations (a) $\lg_{10} x = 2$, (b) $\lg_3 x = 3$.

(a) We can write this as $x = 10^2 = 100$.
(b) We can write this as $x = 3^3 = 27$.

Revision

1 Write the following equations in logarithmic form:

 (a) $7.8 = 10^x$, (b) $0.97 = 10^x$, (c) $1.54 = e^x$, (d) $0.99 = e^x$, (e) $4.5 = 2^x$,

 (f) $1.3 = 5^x$, (g) $4.5 = 10^x$, (h) $1.8 = e^x$.

2 Write the following equations in exponential form:

 (a) $x = \lg_{10} 7.5$, (b) $x = \lg_{10} 0.45$, (c) $x = \ln 9.9$, (d) $x = \ln 0.88$,

 (e) $x = \lg_2 20$, (f) $x = \lg_5 12$, (g) $x = \ln 3$.

3 Determine x in the following equations:

 (a) $x = \lg_2 16$, (b) $x = \lg_{10} 100$, (c) $x = \lg_{10} 0.01$, (d) $\lg_{10} x = 4$,

 (e) $\lg_3 x = 2$, (f) $\lg_2 x = 4$.

9.2.1 Using a calculator for logarithms

Logarithms can be obtained using calculators. Thus to determine $\lg_{10} 5$ the procedure would be: press key 5, press key log (there is no need then to press =), to give the result 0.699. This means that:

$$\lg_{10} 5 = 0.699 \text{ and so } 5 = 10^{0.699}$$

To determine $\ln 5$ the procedure would be: press key 5, press ln (there is no need then to press =), to give the result 1.609. This means that:

$$\ln 5 = 1.609 \text{ and so } 5 = e^{1.609}$$

Example

Determine, using logarithms, the value of x in the following equations:

(a) $3.0 = 10^x$, (b) $0.28 = 10^x$, (c) $4.5 = e^x$, (d) $0.54 = e^x$

(a) We can write this as a logarithm to base 10 and so as $x = \lg_{10} 3.0$. Hence, using a calculator, we key in 3.0, then press key log. The result is 0.477. ✓
(b) We can write this as a logarithm to base 10 and so as $x = \lg_{10} 0.28$. Hence, using a calculator, we key in 0.28, then press key log. The result is -0.553.
(c) We can write this as a logarithm to base e and so as $x = \ln 4.5$. Hence, using a calculator, we key in 4.5, then press key ln. The result is 1.504.
(d) We can write this as a logarithm to base e and so as $x = \ln 0.54$. Hence, using a calculator, we key in 0.54, then press key ln. The result is -0.616.

Example

Determine x in the equation $1.54 = \ln (3x)$.

From the definition of the logarithm, then e to the power 1.54 must be $3x$.

$$3x = e^{1.54} = 4.665$$

Hence $x = 1.555$.

Example

Determine x in the equation $1.563 = \ln\left(\frac{2.645}{x}\right)$.

From the definition of the logarithm, then e to the power 1.563 must be $2.645/x$.

$$e^{1.563} = \frac{2.645}{x}$$

Hence:

$$x = \frac{2.645}{e^{1.563}} = 2.645\,e^{-1.563} = 2.645 \times 0.210 = 0.555$$

Revision

4   Determine, using logarithms, the values of x to three decimal places in the following equations:

   (a) $7.8 = 10^x$, (b) $0.97 = 10^x$, (c) $1.54 = e^x$, (d) $0.99 = e^x$, (e) $x = \lg_{10} 7.5$,

   (f) $x = \lg_{10} 0.45$, (g) $x = \ln 9.9$, (h) $x = \ln 0.88$, (i) $\ln x = 5$,

   (j) $\ln x = 0.1$, (k) $1.5 = \ln\left(\frac{x}{1.8}\right)$, (l) $2.76 = \ln\left(\frac{53.4}{x}\right)$,

   (m) $1.89 = \lg_{10}\left(\frac{45}{x}\right)$, (n) $0.310 = \ln(5x)$, (o) $10 = \ln(2x)$.

9.3 Properties of logarithms

The properties of logarithms follow directly from the properties of powers discussed in chapter 1. Consider the multiplication of two numbers. If we have $A = a^x$ and $B = a^y$, then using logarithms we have $\lg_a A = x$ and $\lg_a B = y$. Since $A \times B = a^{x+y}$ then the logarithm of the product is:

$$\lg_a(A \times B) = x + y$$

Thus:

$$\lg_a(A \times B) = \lg_a A + \lg_a B \qquad [1]$$

For example, $\lg(5 \times 2) = \lg 5 + \lg 2$.
   Consider the division of two numbers. If we have $A = a^x$ and $B = a^y$, then $A \div B = a^{x-y}$. Thus

$$\lg_a\left(\frac{A}{B}\right) = x - y$$

and so:

$$\lg_a\left(\frac{A}{B}\right) = \lg_a A - \lg_a B \qquad [2]$$

For example, $\lg(5/2) = \lg 5 - \lg 2$.
   Consider raising a number to a power. If we have $A = a^x$, then $A^n = a^{nx}$. Thus:

$$\lg_a A^n = nx$$

Since $\lg_a A = x$, then:

$$\lg_a A^n = n \lg_a A \qquad [3]$$

For example, $\lg 5^2 = 2 \lg 5$. You can think of this in terms of the multiplication of two numbers, i.e. $\lg 5^2 = \lg (5 \times 5) = \lg 5 + \lg 5 = 2 \lg 5$.

Example

Evaluate the following logarithms:

(a) $\lg_{10} (2 \times 30)$, (b) $\ln (12 \times 0.02)$, (c) $\lg_{10} (20/7)$, (d) $\ln (2/0.03)$,

(e) $\lg_{10} 4^3$, (f) $\ln 3^7$, (g) $\ln \sqrt{6e}$.

(a) Using the rule for multiplication given above,

$$\lg_{10} (2 \times 30) = \lg_{10} 2 + \lg_{10} 30 = 0.301 + 1.477 = 1.778$$

(b) Using the rule for multiplication given above:

$$\ln (12 \times 0.02) = \ln 12 + \ln 0.02 = 2.485 - 3.912 = -1.427$$

(c) Using the rule for division given above:

$$\lg_{10}\left(\frac{20}{7}\right) = \lg_{10} 20 - \lg_{10} 7 = 1.301 - 0.845 = 0.456$$

(d) Using the rule for division given above:

$$\ln\left(\frac{2}{0.03}\right) = \ln 2 - \ln 0.03 = 0.694 - (-3.507) = 4.200$$

(e) Using the rule for powers given above:

$$\lg_{10} 4^3 = 3 \lg_{10} 4 = 3 \times 0.602 = 1.806$$

(f) Using the rule for powers given above,

$$\ln 3^7 = 7 \ln 3 = 7 \times 1.099 = 7.693$$

(g) We can write $\ln \sqrt{(6e)}$ as $\ln (6e)^{1/2}$ and so, using the rule for powers, as ½ $\ln 6e$. Using the rule for multiplication this can be written as

$$\tfrac{1}{2} \ln 6e = \tfrac{1}{2}(\ln 6 + \ln e) = \tfrac{1}{2}(1.796 + 1) = 1.396$$

Revision

5   Evaluate the following:

(a) $\lg_{10}(123 \times 513)$, (b) $\ln(0.003 \times 125)$, (c) $\lg_{10}(0.015/0.013)$,

(d) $\ln(3.18/578)$, (e) $\lg_{10} 0.012^6$, (f) $\ln 1.56^4$, (g) $\ln e^2$, (h) $\lg 10^{-6}$.

9.3.1 Change of base

Suppose we have a logarithm to base a and require its value to a different base b. Let $\lg_b x = y$, then $b^y = x$. Taking logarithms to base a of both sides of this equation gives:

$$\lg_a b^y = \lg_a x$$

$$y \lg_a b = \lg_a x$$

Hence:

$$y = \frac{\lg_a x}{\lg_a b}$$

But $\lg_b x = y$, and so:

$$\lg_b x = \frac{\lg_a x}{\lg_a b}$$

9.4 Solving exponential equations

The following examples illustrate how logarithms, in particular natural logarithms, can be used to solve equations.

Example

Determine t in the equation $0.5 = 1 - e^{-t/3}$.

The equation can be rearranged to give $e^{-t/3} = 0.5$ and hence, since we have $e^{-t/3} = 1/e^{t/3}$, then:

$$e^{t/3} = \frac{1}{0.5} = 2$$

Taking natural logarithms of both sides of the equation gives:

$$\frac{t}{3} = \ln 2$$

Hence $t = 3 \ln 2 = 3 \times 0.693 = 2.079$.

Example

Determine x in the equation $5^x = 1.8$.

Taking natural logarithms of both sides of the equation gives:

$$\ln (5^x) = \ln 1.8$$

Hence $x \ln 5 = \ln 1.8$ and so

$$x = \frac{\ln 1.8}{\ln 5} = \frac{0.588}{1.609} = 0.365$$

Alternatively, we could have used logarithms to base 10. Then we obtain $x \lg_{10} 5 = \lg_{10} 1.8$ and so $x = \lg_{10} 1.8/\lg_{10} 5 = 0.255/0.699 = 0.365$.

Example

Determine x in the equation $4^{x+5} = 0.7$.

Taking natural logarithms of both sides of the equation gives:

$$(x + 5) \ln 4 = \ln 0.7$$

Hence:

$$x + 5 = \frac{\ln 0.7}{\ln 4} = \frac{-0.357}{1.386} = -0.258$$

Thus $x = -0.258 - 5 = -5.258$.

Revision

6 Determine t in the following equations:

(a) $12 = 5\, e^{-2t}$, (b) $4 = 9\, e^{-3t}$, (c) $12 = 100(1 - e^{-t/2})$, (d) $2 = 15(1 - e^{3t})$.

7 Determine x in the following equations:

(a) $2^x = 1.9$, (b) $4^x = 9.7$, (c) $4^{3x-1} = 12$, (d) $2^{x+1} = 0.5$, (e) $0.9^{2x+3} = 4.5$,

(f) $2^{2x-1} = 13$, (g) $e^{2x} = 4$, (h) $e^{x-1} = 4$.

8 The amount N of a radioactive substance decays with time t according to the equation $N = N_0\, e^{-\lambda t}$. What will be the time taken for the amount of the radioactive material to drop to half its initial value?

9 The resistance R of an electrical conductor increases with the temperature θ, being given by the equation $R = R_0\, e^{a\theta}$, where R_0 and a are constants. Determine the value of a if $R_0 = 500\ \Omega$ and R is $520\ \Omega$ at a temperature of $300°C$.

Problems

1 Write the following equations in logarithmic form:

(a) $2.4 = 10^x$, (b) $2.5 = 10^x$, (c) $0.68 = 10^x$, (d) $1.3 = e^x$, (e) $55.6 = e^x$,

(f) $5.8 = e^x$, (g) $0.35 = e^x$.

2 Write the following equations in exponential form:

(a) $x = \lg_{10} 34$, (b) $x = \lg_{10} 0.54$, (c) $x = \lg_{10} 345$, (d) $x = \ln 37$,

(e) $x = \ln 0.86$, (f) $x = \ln 397$.

3 Determine the values of x, to three decimal places, for the following:

(a) $2.4 = 10^x$, (b) $2.5 = 10^x$, (c) $0.68 = 10^x$, (d) $1.3 = e^x$, (e) $55.6 = e^x$,

(f) $5.8 = e^x$, (g) $0.35 = e^x$, (h) $x = \lg 34$, (i) $x = \lg_{10} 0.54$, (j) $x = \lg_{10} 345$,

(k) $x = \ln 37$, (l) $x = \ln 0.86$, (m) $x = \ln 397$.

4 Evaluate the following:

(a) $\ln (12 \times 0.019)$, (b) $\lg_{10} (13 \times 57)$, (c) $\ln (193/27)$,

(d) $\lg_{10} (0.019/0.178)$, (e) $\ln 24.6^5$, (f) $\lg_{10} 18.2^5$.

5 Determine x in the following equations:

(a) $1.67 = \ln\left(\frac{4.56}{x}\right)$, (b) $6.78 = \ln\left(\frac{x}{234}\right)$, (c) $0.189 = \ln(5x)$,

(d) $0.013 = \ln (2x)$.

6 Determine t in the following equations:

(a) $3 = 5\, e^{-t/3}$, (b) $0.1 = 10\, e^{-4t}$, (c) $12 = 100(1 - e^{-t/4})$,

(d) $0.6 = 2(1 - e^{-t/5})$.

7 The current i, in amperes, in an electrical circuit varies with time t, in seconds, and is given by the equation $i = I(1 - e^{-t/CR})$. If $I = 10$ A, what will be the value of t when $i = 1$ A if $CR = 5$ s?

8 Determine x in the following equations:

(a) $3^x = 100$, (b) $5^x = 17$, (c) $2^{3x+1} = 5$, (d) $4^{4x-5} = 0.56$,

(e) $10^{4x+5} = 0.056$, (f) $e^{3x} = 2\, e^x$, (g) $e^{-(2x+1)} = 4$.

9 The atmospheric pressure p varies with the height h above sea level, being given by the equation $p = p_0\, e^{-h/C}$, where p_0 and C are constants. If $p_0 = 1.01 \times 10^5$ Pa and $p = 9.96 \times 10^4$ Pa at $h = 1000$ m, determine C.

10 The voltage drop v across an inductor in an electrical circuit varies with time t and is given by the equation $v = 10\ e^{-Rt/L}$. What will be the time taken for the voltage to drop to 4 V if $R = 100\ \Omega$ and $L = 0.001$ H?

11 When a hot object cools from a temperature θ_0 above its surroundings, then its temperature θ varies with time t according to the equation $\theta = \theta_0\ e^{-kt}$. Determine k if for $\theta_0 = 500°C$, the object cools to $100°C$ in a time of 3000 s.

12 Determine the value of T_1 when $T_2 = 20$, $\mu = 0.5$ and $\theta = 1.2$ rad, given:

$$\ln\left(\frac{T_1}{T_2}\right) = \mu\theta$$

Section D
Trigonometry

10 Trigonometric ratios
11 Further trigonometry
12 Vectors

The aims of this section are to enable the reader to:

- Convert angles in degrees to radians and vice versa.
- Calculate trigonometric ratios.
- Calculate unknown lengths and angles of triangles or compound shapes formed from triangles.
- Use the sine and cosine rules.
- Determine the areas of triangles.
- Use trigonometry relationships, including $\sin^2 A + \cos^2 A = 1$, the relationships such as $\sin (A + B)$ and $\cos (A + B)$ and those for double angles, e.g. $\sin 2A$, in the solution of problems.
- Solve engineering problems involving trigonometric ratios.
- Solve engineering problems involving vectors.

This section is concerned with the development of the skills to use trigonometry in the solution of engineering problems. Chapter 12 is devoted to the applications of such skills in the use of vectors in engineering.

Chapter 10 assumes some familiarity with angles, triangles and the Pythagoras theorem, though the topics are revised in the chapter. Chapter 10 is the basic chapter with Chapter 11 extending the coverage of trigonometry to situations requiring more dexterity with trigonometric ratios. Chapter 12 applies trigonometry concepts to the solution of problems involving vectors.

10 Trigonometric ratios

10.1 Introduction

Trigonometry is concerned with the relationships between the sides and angles of triangles. Such relationships are involved in a wide range of engineering problems. For example, in considering the equilibrium of a structure the angles at which forces are applied have to be taken into account and this can be done by the use of trigonometry. Quantities, such as forces, where both the direction and the size are necessary to determine their effect are termed vector quantities. Another example involving trigonometry is the description of alternating currents and voltages by equations involving sines or cosines and their representation by phasors.

In this opening section, the units used for angles are reviewed and then later sections develop the trigonometric ratios and their inter-relationships. The trigonometric ratios are then considered for any magnitude angle and graphs of the trigonometric ratios plotted against angle. The chapter then considers the use of the sine and cosine rules with triangles which are not right-angled. Finally the chapter considers the determination of the areas of triangles. Chapter 11 continues the use of the trigonometric ratios.

10.2 Degrees and radians

Figure 10.1 *Degrees*

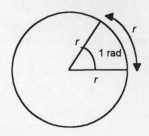

Figure 10.2 *The radian*

The basic units in which angles are measured are *degrees* and *radians*. With degrees, one complete rotation of a radial line (Figure 10.1) is defined as being a rotation through 360°. A rotation through a quarter of a revolution is a rotation through 90°, such an angle being referred to as a *right angle*. With radians, when the tip of the radial line has moved through a distance equal to the length of the line, i.e. the radius, then the angle swept out is said to be 1 radian (Figure 10.2), the radian being defined as the angle subtended at the centre of a circle by an arc whose length is equal to the radius. Since the circumference of a circle is $2\pi r$, then the number of radians swept out by the rotating radial line in moving through one complete circle is 2π radians. Thus:

2π radians is the same angle as 360°

and so:

$$1 \text{ radian} = \frac{360°}{2\pi} = 57.3°$$

A rotation through a quarter of a revolution, i.e. a right angle, is a rotation through a quarter of 2π rad and hence is $\pi/2$ rad. Thus 90° is the same as $\pi/2$ rad.

Thus to convert an angle θ in degrees into radians:

$$\text{number of radians} = \frac{\theta}{360} \times 2\pi$$

and to convert an angle θ in radians to degrees:

$$\text{number of degrees} = \frac{\theta}{2\pi} \times 360$$

Example

Convert the following angles in degrees into angles in radian measure:
(a) 30°, (b) 60°, (c) 90°, (d) 180°, (e) 270°.

(a) Since 360° is 2π radians, then 30° is $\frac{30}{360} \times 2\pi = \frac{\pi}{6} = 0.52$ rad.

(b) Since 360° is 2π radians, then 60° is $\frac{60}{360} \times 2\pi = \frac{\pi}{3} = 1.05$ rad.

(c) Since 360° is 2π radians, then 90° is $\frac{90}{360} \times 2\pi = \frac{\pi}{2} = 1.57$ rad.

(d) Since 360° is 2π radians, then 180° is $\frac{180}{360} \times 2\pi = \pi = 3.14$ rad.

(e) Since 360° is 2π radians, then 270° is $\frac{270}{360} \times 2\pi = \frac{3\pi}{2} = 4.71$ rad.

Example

Convert the following angles in radians into angles in degrees:
(a) 0.80 rad, (b) 1.2 rad.

(a) Since 2π radians is 360°, then 0.80 radians is $\frac{0.8}{2\pi} \times 360 = 45.8°$.

(b) Since 2π radians is 360°, then 1.2 radians is $\frac{1.2}{2\pi} \times 360 = 68.8°$.

Revision

1 Convert the following angles in degrees into angles in radian measure:
   (a) 65°, (b) 125°, (c) 245°, (d) 20°, (e) 119°, (f) 312°.

2 Convert the following angles in radians into angles in degrees:
   (a) 0.3 rad, (b) 1.9 rad, (c) 3.5 rad, (d) $\pi/6$ rad, (e) $3\pi/7$ rad, (f) $\pi/5$ rad.

10.3 Trigonometric ratios

Consider the two right-angled triangles shown in Figure 10.3 with the same angle θ. For such triangles the ratios of corresponding sides are the same. For example, for each triangle the ratio opposite side/hypotenuse has the same value, the ratio opposite side/adjacent side has the same value, etc. The ratios between two sides of a right-angled triangle occur frequently and so are given special names. The ratios are defined as:

1 sine of angle θ, written as $\sin \theta$

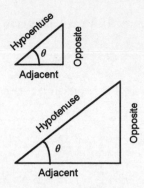

Figure 10.3 *Similar triangles*

$$\sin \theta = \frac{\text{opposite}}{\text{hypotenuse}}$$

2 cosine of angle θ, written as $\cos \theta$

$$\cos \theta = \frac{\text{adjacent}}{\text{hypotenuse}}$$

3 tangent of angle θ, written as $\tan \theta$

$$\tan \theta = \frac{\text{opposite}}{\text{adjacent}}$$

4 cosecant of angle θ, written as $\operatorname{cosec} \theta$

$$\operatorname{cosec} \theta = \frac{\text{hypotenuse}}{\text{opposite}}$$

5 secant of angle θ, written as $\sec \theta$

$$\sec \theta = \frac{\text{hypotenuse}}{\text{adjacent}}$$

6 cotangent of angle θ, written as $\cot \theta$

$$\cot \theta = \frac{\text{adjacent}}{\text{opposite}}$$

Example

For the triangle shown in Figure 10.4, what are the values of (a) $\sin \theta$, (b) $\cos \theta$, (c) $\tan \theta$?

(a) Using ratio 1 in the above list, $\sin \theta = \dfrac{\text{opposite}}{\text{hypotenuse}} = \dfrac{5}{13}$

(b) Using ratio 2 in the above list, $\cos \theta = \dfrac{\text{adjacent}}{\text{hypotenuse}} = \dfrac{12}{13}$

(c) Using ratio 3 in the above list, $\tan \theta = \dfrac{\text{opposite}}{\text{adjacent}} = \dfrac{5}{12}$

Figure 10.4 *Example*

Revision

3 For the triangle shown in Figure 10.5, determine:

(a) $\sin \theta$, (b) $\cos \theta$, (c) $\tan \theta$, (d) $\operatorname{cosec} \theta$, (e) $\sec \theta$, (f) $\cot \theta$.

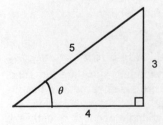

Figure 10.5 *Revision problem 3*

10.3.1 Basic relationships between the ratios

From the definitions of the trigonometric ratios given above it can be seen that:

$$\operatorname{cosec}\theta = \frac{1}{\sin\theta}, \ \sec\theta = \frac{1}{\cos\theta}, \ \cot\theta = \frac{1}{\tan\theta}$$

Also we can write

$$\tan\theta = \frac{\text{opposite}}{\text{adjacent}} = \frac{\text{opposite}}{\text{hypotenuse}} \times \frac{\text{hypotenuse}}{\text{adjacent}}$$

Thus:

$$\tan\theta = \frac{\sin\theta}{\cos\theta}$$

Since the angles in a triangle add up to 180° and one of them in a right angled triangle is 90°, then the angles in such a triangle are always 90°, θ and $(90° - \theta)$. Thus if we consider, for a right-angled triangle, the sine of the angle $(90° - \theta)$ then we have, with the sides labelled as in Figure 10.6:

$$\sin(90° - \theta) = \frac{BC}{AC}$$

But BC/AC is the cosine of angle θ. Thus:

$$\sin(90° - \theta) = \cos\theta$$

If we consider the cosine of the angle $(90° - \theta)$ then:

$$\cos(90° - \theta) = \frac{AB}{AC}$$

But AB/BC is the sine of angle θ. Thus:

$$\cos(90° - \theta) = \sin\theta$$

Figure 10.6 *Angles in a right-angled triangle*

Example

If cos 40° = 0.766, what is sec 40°?

Since sec 40° = 1/cos 40°, then sec 40° = 1/0.766 = 1.305.

Example

If cos 40° = 0.766, what is sin 50°?

Since 40° + 50° = 90°, then sin 50° = cos (90° − 50°) = cos 40°. Thus sin 50° = 0.766.

Revision

4 If sin 20° is 0.342, what are (a) cos 70° and (b) cosec 20°?

5 If cos 45° is 0.707, what are (a) sin 45° and (b) sec 45°?

6 If sin 30° = 0.500 and cos 30° = 0.866, what are (a) tan 30° and (b) cot 30°?

10.3.2 Using a calculator for the trigonometric ratios

The values of the trigonometric ratios for particular angles can be obtained by the use of a calculator. Thus for example to find sin 50°, with my calculator I press the keys for 50 to appear in the display, then the key for sin. The result then appears as 0.7660... Note that the display when carrying out such a calculation displays the symbol deg. indicating that the number 50 is assumed to be in degrees. If the angle is in radians, e.g. sin 0.40, with my calculator I press the mode key until rad appears in the display, then 0.40, then sin. The result then appears as 0.3894...

Sometimes we require to determine the value of an angle that gives a particular trigonometric ratio. For example, we might need to find the angle that gives a sine of 0.6, i.e. we have sin θ = 0.6 and need to find θ. To do this we use the calculator key for sin^{-1}. This symbol indicates that we are carrying out the inverse operation to finding the sine of an angle. The key sequence is thus: press the keys for 0.6, then press the key for sin^{-1}. The result is 36.9°. The keys marked as sin^{-1}, cos^{-1} and tan^{-1} are used to carry out the operations of finding the angles that give particular sines, cosine or tangents.

Example

Using a calculator, determine the values to three decimal places of the following:

(a) sin 25°, (b) cos 78°, (c) tan 35°, (d) cosec 50°, (e) sec 40°,
(f) cot 70°, (g) sin 0.6, (h) cos 0.2, (i) tan 1.3, (j) cosec 1.1, (k) sec 0.5,
(l) cot 0.6.

The following are the key sequences with my calculator.
(a) Press keys for 25, then key for sin. The result is 0.423.
(b) Press keys for 78, then key for cos. The result is 0.208.
(c) Press keys for 35, then key for tan. The result is 0.700.
(d) We can utilise here the fact that cosec 50° = 1/sin 50°. Thus the sequence is: press keys for 50, key for sin, then key for 1/x. The result is 1.305.
(e) We utilise here the fact that sec 40° = 1/cos 40°. Thus the sequence is: press keys for 40, key for cos, then key for 1/x. The result is 1.305.
(f) We utilise here the fact that cot 70° = 1/tan 70°. Thus the sequence is: press keys for 70, key for tan, then key for 1/x. The result is 0.364.

(g) Press key for mode to give rad in the display, keys for 0.6, then key for sin. The result is 0.565.

(h) Press key for mode to give rad in the display, keys for 0.2, then key for cos. The result is 0.980.

(i) Press key for mode to give rad in the display, keys for 1.3, then key for tan. The result is 3.602.

(j) We utilise the fact that cosec 1.1 = 1/sin 1.1. Thus the sequence is: press key for mode to give rad in the display, keys for 1.1, key for sin, then key for 1/x. The result is 1.122.

(k) We utilise the fact that sec 0.5 = 1/cos 0.5. Thus the sequence is: press key for mode to give rad in the display, keys for 0.5, key for cos, then key for 1/x. The result is 1.139.

(l) We utilise the fact that cot 0.6 = 1/tan 0.6. Thus the sequence is: press key for mode to give rad in the display, keys for 0.6, key for tan, then key for 1/x. The result is 1.462.

Example

Use a calculator to determine the following angles: (a) the angle that gives the sine of 0.8, (b) the angle that gives the cosine of 0.3, (c) the angle that gives the tangent of 0.7.

The key sequences are:
(a) Press the keys for 0.8, then the key for $\sin^{-1}$. The result is 53.1°.
(b) Press the keys for 0.3, then the key for $\cos^{-1}$. The result is 72.5°.
(c) Press the keys for 0.7, then the key for $\tan^{-1}$. The result is 35.0°.

Revision

7 Using a calculator, determine the values to three decimal places of the following trigonometric ratios:

(a) sin 55°, (b) cos 15°, (c) tan 60°, (d) cosec 35°, (e) sec 80°, (f) cot 25°, (g) sin 0.9, (h) cos 1.1, (i) tan 0.3, (j) cosec 0.9, (k) sec 0.35, (l) cot 0.45.

8 Use a calculator to determine, to one decimal place, the following angles:

(a) the angle with a sine of 0.85, (b) the angle with a cosine of 0.35, (c) the angle with a tangent of 1.5, (d) the angle with a sine of 0.9, (e) the angle with a sine of 0.21, (f) the angle with a cosecant of 1.9, (g) the angle with a cotangent of 1.2, (h) the angle with a secant of 1.3.

10.3.3 Problems involving the trigonometric ratios

The following are examples of problems involving the trigonometric ratios.

(a)

(b)

(c)

Figure 10.7   *Example*

Figure 10.8   *Example*

Example

Determine the length x, to two decimal places, in the triangles shown in Figure 10.7.

(a) We are given the hypotenuse and asked to obtain the adjacent side. Hence, using the cosine we can write:

$$\frac{x}{15} = \cos 45°$$

and so $x = 15 \cos 45° = 15 \times 0.707 = 10.61$.

(b) We are given the adjacent and asked to obtain the hypotenuse. Hence, using the secant we can write:

$$\frac{x}{3} = \sec 65°$$

$$x = 3 \sec 65° = \frac{3}{\cos 65°} = 7.10$$

Alternatively we could have obtained the answer by writing:

$$\frac{3}{x} = \cos 65°$$

(c) We are given the opposite and asked to obtain the adjacent. Hence, using the cotangent we can write:

$$\frac{x}{4} = \cot 35°$$

$$x = 4 \cot 35° = \frac{4}{\tan 35°} = 5.71$$

Alternatively we could have obtained the answer by writing:

$$\frac{4}{x} = \tan 35°$$

Example

A ladder of length 6 m leans against a wall, making an angle of 70° with the level ground. How high up the wall is the upper end of the ladder?

Figure 10.8 shows the situation. Thus

$$\frac{h}{6} = \sin 70°$$

and so $h = 6 \times \sin 70° = 5.6$ m.

Figure 10.9 *Example*

Example

For the supported beam shown in Figure 10.9, what angle does the supporting stay make with the beam?

For the angle θ we have

$$\tan \theta = \frac{3}{4}$$

We thus need to find the angle which has a tangent of 0.75. We can do this by means of a calculator. The key sequence is: press the keys for 0.75, then press the key for $\tan^{-1}$. The result is 36.9°.

Example

A tower stands on level ground. The angle of elevation of the top of the tower from a point on the ground some distance away is 40°. On moving 30 m towards the tower the elevation becomes 60°. What is the height of the tower?

Figure 10.10 *Example*

Figure 10.10 shows the situation, AB representing the tower. For the triangle ABD we can write:

$$\frac{h}{BD} = \tan 40°$$

Thus BD = h cot 40°. For the triangle ABC we can write:

$$\frac{h}{BC} = \tan 60°$$

Thus BC = h cot 60°. But CD = BD – BC. Thus:

$$30 = h \cot 40° - h \cot 60° = h(\cot 40° - \cot 60°)$$

Hence:

$$h = \frac{30}{\cot 40° - \cot 60°} = \frac{30}{1.192 - 0.577} = 48.8 \text{ m}$$

Revision

9 Solve for x, to two decimal places, in the following right-angled triangles:

(a) Hypotenuse 4, opposite x, angle 46°,
(b) Hypotenuse 10, adjacent x, angle 55°,
(c) Adjacent 10, opposite x, angle 70°,
(d) Adjacent 12, hypotenuse x, angle 58°,
(e) Opposite 4, hypotenuse x, angle 25°,
(f) Opposite 12, adjacent x, angle 67°.

10 The angle of elevation of the top of a flagpole from a point on the ground level with its base and 20 m from it is 40°. What is the height of the flagpole? Note that the angle of elevation is the upward angle from the horizontal.

11 A pendulum of length 200 mm is swinging. Calculate the angle to the vertical it will be when the pendulum bob is displaced horizontally from its vertical position by 50 mm.

12 Figure 10.11 shows a roof truss for a building. Determine the lengths AB, BD and BC.

13 A road up an incline rises 12 m for every 100 m of roadway. What angle does the road make with the horizontal?

14 B is a point 420 m due east of A and C is another point which is 310 m from A in a direction 50° east of north. Determine the distance and bearing of C from B. Note that the bearing of an object from a point of observation is the angle that the line through the two points makes with the north and south line, or east and west line, through the point of observation.

15 When seen from the window of a building 100 m from a tower, the angle of elevation of the top of the tower is 26° and the angle of depression of the bottom of the tower is 10°. What is the height of the tower? Note that the angle of depression is the angle through which a horizontal line must be lowered.

Figure 10.11 *Revision problem 12*

10.4 Ratios for common angles

We can use the Pythagoras theorem to find the values of the sine, cosine and tangent for the commonly encountered angles of 0°, 30°, 45°, 60° and 90°.

Consider a right-angled triangle with the angle θ zero. The opposite size must be zero, with the adjacent side and the hypotenuse both having the same value. Thus:

$$\sin 0° = 0, \quad \cos 0° = 1, \quad \tan 0° = 0$$

Consider a right-angled triangle containing 30° and 60° angles. Since $\sin 30° = 0.5$, then we can draw such a triangle with an opposite side of 1 for the 30° angle and a hypotenuse of 2 (Figure 10.12). The adjacent side must then have a length of:

$$\text{adjacent side} = \sqrt{2^2 - 1^2} = \sqrt{3}$$

Figure 10.12 *Right-angled triangle*

Thus we have:

$$\sin 30° = \frac{1}{2}, \quad \cos 30° = \frac{\sqrt{3}}{2}, \quad \tan 30° = \frac{1}{\sqrt{3}}$$

Figure 10.13 *Right-angled triangle*

For the 60° angle in the triangle we have:

$$\sin 60° = \frac{\sqrt{3}}{2}, \ \cos 60° = \frac{1}{2}, \ \tan 60° = \frac{\sqrt{3}}{1} = \sqrt{3}$$

Consider a right-angled triangle containing two 45° angles (Figure 10.13). Since we have tan 45° = 1 then the opposite and adjacent sides can be 1. The Pythagoras theorem thus gives a hypotenuse of length

$$\text{hypotenuse} = \sqrt{1^2 + 1^2} = \sqrt{2}$$

Thus we have

$$\sin 45° = \frac{1}{\sqrt{2}}, \ \cos 45° = \frac{1}{\sqrt{2}}, \ \tan 45° = 1$$

Consider a right-angled triangle with the angle θ equal to 90°. For this we must have the adjacent side equal to zero and the opposite side and hypotenuse the same value. Thus:

$$\sin 90° = 1, \ \cos 90° = 0, \ \tan 90° \text{ infinite}$$

Table 10.1 summarises the above results for these angles of 0°, 30°, 45°, 60° and 90°.

Table 10.1 *Common angles between 0° and 90°*

θ	$\sin \theta$	$\cos \theta$	$\tan \theta$
0°	0	1	0
30°	$\frac{1}{2}$	$\frac{\sqrt{3}}{2}$	$\frac{1}{\sqrt{3}}$
45°	$\frac{1}{\sqrt{2}}$	$\frac{1}{\sqrt{2}}$	1
60°	$\frac{\sqrt{3}}{2}$	$\frac{1}{2}$	$\sqrt{3}$
90°	1	0	infinity

Example

For the triangle shown in Figure 10.14, using the values given in the above table, determine AB if AC = 12 m.

We can write:

$$\frac{AB}{AC} = \sin 30°$$

Figure 10.14 *Example*

Since sin 30° = 1/2 then AB = ½AC. Thus AB = 6 m.

Example

Show that $\cos 60° \cos 30° + \sin 60° \sin 30° = \dfrac{\sqrt{3}}{2}$.

Using the values given in the above table, then

$$\cos 60° \cos 30° + \sin 60° \sin 30° = \frac{1}{2} \times \frac{\sqrt{3}}{2} + \frac{\sqrt{3}}{2} \times \frac{1}{2} = \frac{\sqrt{3}}{2}$$

Revision

Figure 10.15 *Revision problem 16*

16 For the triangle shown in Figure 10.15, use the values of the ratios given in the above table without working out the square roots.

(a) If AB = 12 m, determine BC.
(b) If BC = 4 m, determine AC.
(c) If AC = 10 m, determine BC.

17 For the triangle shown in Figure 10.16, use the values of the ratios given in the above table without working out the square roots.

(a) If AB = 10 m, determine BC.
(b) If AC = 6 m, determine AB.
(c) If BC = 5 m, determine AB.

Figure 10.16 *Revision problem 17*

18 Show that $\sin^2 30° + \sin^2 45° + \sin^2 60° = 3/2$.

10.4.1 Ratios for angles of any magnitude

In section 10.3, the sine, cosine and tangent trigonometric ratios were defined for acute angles, i.e. angles less than 90°. In this section we consider a more general definition which will enable us to consider the trigonometric ratios for angles greater than 90°. The general definitions are consistent with the definitions for angles less than 90°.

Consider an angle θ formed by a line OP rotating anticlockwise about O from an initial line OX (Figure 10.17). Whatever the size of the angle θ, we will define the trigonometric ratios as:

Figure 10.17 *Angles between 0° and 90°*

$$\sin \theta = \frac{\text{vertical displacement of P}}{\text{OP}}$$

$$\cos \theta = \frac{\text{horizontal displacement of P}}{\text{OP}}$$

$$\tan \theta = \frac{\text{vertical displacement of P}}{\text{horizontal displacement of P}}$$

We consider O to be at the centre of a system of co-ordinates, with distances measured to the right of O being positive, distances to the left negative. Distances measured upwards from O are positive and distances downwards are negative. This is the conventional system of co-ordinates used with graphs with O as the origin.

When the line OP has rotated through an angle greater than 0° and less than 90° and is in the top right-hand quadrant, as in Figure 10.17, then we have for the three basic trigonometric ratios

$$\sin \theta = \frac{QP}{OP}, \quad \cos \theta = \frac{OQ}{OP}, \quad \tan \theta = \frac{QP}{OQ}$$

The displacements QP and OQ are both positive quantities (the radius OP is defined as always being positive) so that, with angles between 0° and 90°, we have with each of the above ratios positive quantities divided by positive quantities and so the sine, cosine and tangent of such angles are all designated as being positive.

Now consider the situation when the line OP is in the top left-hand quadrant (Figure 10.18). To reach this position the line has rotated through more than 90° but less than 180°. Thus, since we have

$$\sin \theta = \frac{QP}{OP}, \quad \cos \theta = \frac{OQ}{OP}, \quad \tan \theta = \frac{QP}{OQ}$$

Figure 10.18 *Angles between 90° and 180°*

then for angles between 90° and 180°, as QP is positive and OQ negative, the sine is positive, the cosine negative and the tangent negative.

For angles in this quadrant, QP/OP could be the sine of θ or, when considering just the triangle OPQ, the sine of the angle (180° − θ). Thus

$$\sin \theta = \sin (180° - \theta)$$

For angles in this quadrant, OQ/OP could be the cosine of θ or, when considering just the triangle OPQ, the cosine of the angle (180° − θ). Thus

$$\cos \theta = -\cos (180° - \theta)$$

For angles in this quadrant, QP/OQ could be the tangent of θ or, when considering just the triangle OPQ, the tangent of the angle (180° − θ). Thus

$$\tan \theta = -\tan (180° - \theta)$$

Now consider the situation when the line OP is in the bottom left-hand quadrant (Figure 10.19). To reach this position the line has rotated through more than 180° but less than 270°. Thus, since we have

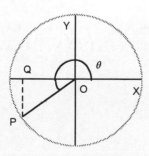

Figure 10.19 *Angles between 180° and 270°*

$$\sin \theta = \frac{QP}{OP}, \quad \cos \theta = \frac{OQ}{OP}, \quad \tan \theta = \frac{QP}{OQ}$$

then for angles between 180° and 270°, as QP is negative and OQ is negative, the sine is negative, the cosine negative and the tangent positive.

For angles in this quadrant, QP/OP could be the sine of θ or, when considering just the triangle OPQ, the sine of the angle $(\theta - 180°)$. Thus:

$$\sin \theta = -\sin (\theta - 180°)$$

For angles in this quadrant, OQ/OP could be the cosine of θ or, when considering just the triangle OPQ, the cosine of the angle $(\theta - 180°)$. Thus:

$$\cos \theta = -\cos (\theta - 180°)$$

For angles in this quadrant, QP/OQ could be the tangent of θ or, when considering just the triangle OPQ, the tangent of the angle $(\theta - 180°)$. Thus:

$$\tan \theta = \tan (\theta - 180°)$$

Now consider the situation when the line OP is in the bottom right-hand quadrant (Figure 10.20). To reach this position the line has rotated through more than 270° but less than 360°. Thus, since we have

$$\sin \theta = \frac{QP}{OP}, \ \cos \theta = \frac{OQ}{OP}, \ \tan \theta = \frac{QP}{OQ}$$

Figure 10.20 *Angles between 270° and 360°*

then for angles between 270° and 360°, as QP is negative and OQ is positive, the sine is negative, the cosine positive and the tangent negative.

For angles in this quadrant, QP/OP could be the sine of θ or, when considering just the triangle OPQ, the sine of the angle $(360° - \theta)$. Thus:

$$\sin \theta = -\sin (360° - \theta)$$

For angles in this quadrant, OQ/OP could be the cosine of θ or, when considering just the triangle OPQ, the cosine of the angle $(360° - \theta)$. Thus:

$$\cos \theta = \cos (360° - \theta)$$

For angles in this quadrant, QP/OQ could be the tangent of θ or, when considering just the triangle OPQ, the tangent of the angle $(360° - \theta)$. Thus:

$$\tan \theta = -\tan (360° - \theta)$$

In summary, we have:

Angles between:

0° and 90° (0 and $\pi/2$ rad)	All the ratios are positive
90° and 180° ($\pi/2$ and π rad)	Only the sine is positive
180° and 270° (π and $3\pi/2$ rad)	Only the tangent is positive
270° and 360° ($3\pi/2$ and 2π rad)	Only the cosine is positive

sin+, cos–, tan–	sin+, cos+, tan+
S	A
T	C
sin–, cos–, tan+	sin–. cos+, tan–

Figure 10.21 *Signs of ratios*

Figure 10.21 summarises the situation with regard to the signs of sines, cosines and tangents of angles between 0° and 360°. A mnemonic that might be used to remember this is, when working anticlockwise round the quadrants,

positively **A**ll **S**ilver **T**ea **C**ups

A = all in the first quadrant, **S** = sine in the second quadrant, **T** = tangent in the third quadrant, **C** = cosine in the fourth quadrant

Example

Determine, to three decimal places, the values of: (a) sin 200°, (b) cos 115°, (c) tan 295°, (d) cos 320°.

If we use a calculator to determine these angles then the signs should automatically come right.
(a) With a calculator the key sequence is: press the keys for 200, then press the key for sin. The result is –0.342.
(b) With a calculator the key sequence is: press the keys for 115, then press the key for cos. The result is –0.423.
(c) With a calculator the key sequence is: press the keys for 295, then press the key for tan. The result is –2.145.
(d) With a calculator the key sequence is: press the keys for 320, then press the key for cos. The result is 0.766.

Alternatively we could have just used the values of the trigonometric ratios for angles between 0° and 90°.
(a) Since sin 200° = –sin (200° – 180°), then sin 200° = –sin 20° and so is –0.342.
(b) Since cos 115° = –cos (180° – 115°), then cos 115° = –cos 65° and so is –0.423.
(c) Since tan 295° = –tan (360° – 295°), then tan 295° = –tan 65° and so is –2.145.
(d) Since cos 320° = cos (360° – 320°), then cos 320° = cos 40° and so is 0.766.

Example

Determine, to three decimal places, the values of (a) sin 1.8, (b) cos 2.1, (c) tan 4.6.

(a) This angle is in radian measure. Thus the key sequence is: press the key for mode to give the display in rad, press the keys for 1.8, then press the key for sin. The result is 0.974.
(b) This angle is in radian measure. Thus the key sequence is: press the key for mode to give the display in rad, press the keys for 2.1, then press the key for cos. The result is –0.505.

(c) This angle is in radian measure. Thus the key sequence is: press the key for mode to give the display in rad, press the keys for 4.6, then press the key for tan. The result is 8.860.

Example

If sin 60° = √(3)/2, determine sin (180° − 60°).

We can use sin θ = sin (180° − θ). Thus sin (180° − 60°) = √(3)/2.

Example

Determine the angles between 0° and 360° having the trigonometric ratios of (a) sin θ = 0.7, (b) cos θ = −0.4, (c) sin θ = −0.3.

(a) Using a calculator the key sequence can be: press the keys for 0.7, then the key for sin⁻¹. The result given by the calculator is 44.4°. But sines are positive in the first and second quadrants (Figure 10.22). Thus (180° − 44.4°) also gives sin θ = 0.7. Hence the angles are 44.4° and 135.6°. You can check that these angles give a sine of 0.7 by using the calculator.

(b) Using a calculator the key sequence can be: press the keys for 0.4, press the key for +/−, then the key for cos⁻¹. The result given by the calculator is 113.6°. But cosines are negative in the second and third quadrants (Figure 10.23). Thus (180° + {180° − 113.6°}) also gives cos θ = −0.4. Hence the angles are 113.6° and 246.4°. You can check that these angles give a cosine of −0.4 by using the calculator.

(c) Using a calculator the key sequence can be: press the keys for 0.3, press the key for +/−, then the key for sin⁻¹. The result given by the calculator is −17.5°. Positive angles are given between the x-axis and the rotating line when it rotates in an anticlockwise direction. A negative angle occurs when the rotation is in clockwise direction. Thus this angle represents (360° − 17.5°) = 342.5°. Sines are negative in the third and fourth quadrants (Figure 10.24). Thus the angle in the third quadrant is (180° + 17.5°). Hence the angles are 197.5° and 342.5°. You can check that these angles give a sine of −0.3 by using the calculator.

Figure 10.22 *Example*

Figure 10.23 *Example*

Figure 10.24 *Example*

Revision

19 Determine, to three decimal places, the values of the following:

(a) sin 125°, (b) cos 145°, (c) tan 170°, (d) sin 200°, (e) cos 220°, (f) tan 260°, (g) sin 290°, (h) cos 300°, (i) tan 340°, (j) sin 345°, (k) sin 2.5, (l) cos 3.1, (m) tan 4.2, (n) sin 3.8.

20 If sin 30° = ½, determine sin (180° − 30°).

21 If cos 45° = √(2)/2, determine cos (180° − 45°).

22 Determine the angles having the following trigonometric ratios:

(a) $\tan \theta = 1.7$, (b) $\sin \theta = 0.8$, (c) $\cos \theta = -0.7$, (d) $\sin \theta = -0.5$,
(e) $\cos \theta = 0.8$.

10.5 The sine rule

Consider the triangle ABC shown in Figure 10.25. This triangle does not contain a right angle. The notation system that is used for the angles and sides of triangles is that the angles are labelled as being A at apex A, B at apex B and C at apex C while the side opposite angle A is labelled as being of length a, that opposite angle B of length b, and that opposite angle C of length c.

If we draw a line from C which is at right angles to AB then we divide the triangle into two right-angled triangles. Suppose this line has a length of h. Then for triangle AHC we can write $\sin A = h/b$ and so $h = b \sin A$. For triangle BHC we can write $\sin B = h/A$ and so $h = a \sin B$. Thus we have $h = b \sin A = a \sin B$, and so:

Figure 10.25 *The sine rule*

$$h = a \sin B = b \sin A$$

and thus:

$$\frac{a}{\sin A} = \frac{b}{\sin B}$$

Similarly, by drawing a line from A at right angles to BC we can show that:

$$\frac{b}{\sin B} = \frac{c}{\sin C}$$

Thus the *sine rule* is stated as:

$$\frac{a}{\sin A} = \frac{b}{\sin B} = \frac{c}{\sin C}$$

This rule may be used with triangles to find the values of unknown sides and angles when one side and any two angles are given, or two sides and an angle not included between those sides are given.

Example

For the triangle shown in Figure 10.26, determine the length of the side AC.

Figure 10.26 *Example*

In the triangle we have angle $A = 35°$, angle $C = 82°$, side $c = 35$ mm and are required to determine b. Since all the angles in a triangle add up to 180° then angle $B = 180° - 35° - 82° = 63°$. Thus, using the sine rule, we have

$$\frac{b}{\sin B} = \frac{c}{\sin C}$$

$$\frac{b}{\sin 63°} = \frac{35}{\sin 82°}$$

Hence:

$$b = \frac{35}{\sin 82°} \times \sin 63° = 31.5 \text{ mm}$$

Thus $AC = 31.5$ mm.

Example

For the triangle shown in Figure 10.27, determine the angle B.

35 mm   45 mm

70°

A   B

Figure 10.27   *Example*

In the triangle we have angle $A = 70°$, side $b = 35$ mm, side $a = 45$ mm and are required to determine angle B. Thus, using the sine rule, we have:

$$\frac{a}{\sin A} = \frac{b}{\sin B}$$

$$\frac{45}{\sin 70°} = \frac{35}{\sin B}$$

Thus:

$$\sin B = \frac{35 \times \sin 70°}{45} = 0.731$$

Hence $B = 47.0°$.

Example

A radio aerial is 20 m high and stands on the roof of a building. The angle of elevation of the top of the aerial is 48° and from the same point the angle of elevation of the base of the aerial is 40°. What is the height of the building?

20 m

a

b

h

48°   40°

Figure 10.28   *Example*

Figure 10.28 shows the situation, with h representing the height of the building. For the triangle with sides of a, b and 20 m we can use the sine rule to write:

$$\frac{b}{\sin(90° - 48°)} = \frac{20}{\sin(48° - 40°)}$$

Thus:

$$b = \frac{20 \times \sin 42°}{\sin 8°} = 96.2 \text{ m}$$

For the right-angled triangle we have $h/b = \sin 40°$ and so the height of the tower $h = 96.2 \sin 40° = 61.8$ m.

Revision

23 Determine the required side or angle in each of the following triangles:

(a) Angle $B = 47°$, angle $C = 78°$, side $a = 340$ mm, determine side c,
(b) Angle $A = 67°$, angle $C = 51°$, side $c = 150$ mm, determine side a,
(c) Angle $A = 80°$, side $a = 200$ mm, side $b = 120$ mm, determine angle B,
(d) Angle $A = 40°$, side $a = 30$ mm, side $c = 37$ mm, determine angle C,
(e) Angle $C = 35°$, side $a = 36$ mm, side $c = 30$ mm, determine side b,
(f) Angle $A = 68°$, angle $C = 30°$, side $c = 26$ mm, determine side b,
(g) Angle $A = 80°$, angle $B = 70°$, side $b = 80$ mm, determine side a,
(h) Angle $B = 55°$, side $a = 70$ mm, side $b = 80$ mm, determine angle A.

24 A roof has a span of 8 m with eaves that slope on one side at 42° to the horizontal and on the other side at 35° to the horizontal. Determine the lengths of the sloping parts of the roof.

25 A tower stands on the top of a slope which rises by 1 m for every 12 m of slope. From a position 120 m down the slope, the angle of elevation of the top of the tower is 18°. Determine the height of the tower.

10.5.1 The ambiguous solution

When given two sides of a triangle and the angle opposite one of the sides, there is the possibility that using the sine rule will lead to two solutions. This is because a positive value for a sine can mean an angle between 0° and 90° or 90° and 180°.

For example, suppose we have angle $A = 25°$, side $a = 28$ mm and side $b = 37$ mm. Using the sine rule we can find angle B. Thus:

$$\frac{28}{\sin 25°} = \frac{37}{\sin B}$$

$$\sin B = \frac{37 \times \sin 25°}{28} = 0.558$$

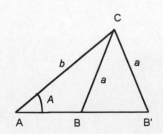

Figure 10.29 *Two solutions*

The angle B which would give a positive sine of 0.558 is either 33.9° or 146.1°. Are both these values feasible? Figure 10.29 shows the possible triangles we could have. With one case we have the triangle ABC, with angle $C = 180° - 25° - 146.1° = 8.9°$, and with the other the triangle AB'C, with angle $C = 180° - 25° - 33.9° = 121.1°$. Thus both the solutions are feasible and there are two possible triangles which fit the supplied data and which can be drawn.

Sometimes, though, there might be two possible values for angle B, one of the answers leads to an impossible situation, such as the sum of the angles $A + B + C$ being greater than 180°, and so one of the answers is then not feasible. Two triangles cannot be drawn on the basis of the data, only one. To illustrate this, consider the determination of the triangle, or triangles, which fit the data given of angle $A = 37°$, side $a = 46$ mm and side $b = 21$ mm. Angle B is given by the sine rule as

$$\frac{46}{\sin 37°} = \frac{21}{\sin B}$$

Thus:

$$\sin B = \frac{21 \times \sin 37°}{46} = 0.275$$

The angles which give this sine are 15.9° and 164.1°. Since we must have $A + B + C = 180°$, then with the first answer for B we have angle C as $180° - 37° - 15.9° = 127.1°$ and with the other answer for B we have angle C as $180° - 37° - 164.1° = -21.1°$. This last answer is not feasible and so we have only one solution, namely angle B is 15.9°.

Revision

26 Determine the required angles in the following triangles:

(a) Angle $A = 25°$, side $a = 18$ mm, side $b = 38$ mm, determine angles B and C,
(b) Angle $A = 42°$, side $a = 160$ mm, side $b = 200$ mm, determine angles B and C.

10.6 The cosine rule

Consider the triangle ABC in Figure 10.30. From apex C a line is drawn at right angles to side AB, intersecting it at point H. Let BH = x, and so we then have AH = $c - x$. Applying the Pythagoras theorem to the triangle ACH gives:

$$b^2 = h^2 + (c - x)^2$$

Applying the Pythagoras theorem to the triangle BCH gives:

$$a^2 = h^2 + x^2$$

Figure 10.30 *The cosine rule*

Subtracting the second equation from the first one gives:

$$b^2 - a^2 = (c - x)^2 - x^2 = c^2 - 2cx + x^2 - x^2$$

Thus:

$$b^2 = a^2 + c^2 - 2cx$$

But, with triangle BCH we have $\cos B = x/a$. Hence $x = a \cos B$ and so

$$b^2 = a^2 + c^2 - 2ac \cos B$$

This is one form of what is termed the *cosine rule*. By repeating the above with lines drawn from A to intersect BC at right angles and from B to intersect AC at right angles, we can also obtain:

$$a^2 = b^2 + c^2 - 2bc \cos A$$

and:

$$c^2 = a^2 + b^2 - 2ab \cos C$$

The cosine rule can be stated as: *the square of a side is equal to the sum of the squares of the other two sides minus twice the product of those sides times the cosine of the angle between them.*

This rule may be used to find the values of unknown sides and angles when two sides and the angle included between those sides, or the three sides are given.

Example

Determine the side a for the triangle given in Figure 10.31.

Using the cosine law we have $a^2 = b^2 + c^2 - 2bc \cos A$, and so

$$a^2 = 12^2 + 18^2 - 2 \times 12 \times 18 \cos 115°$$

Because the cosine of 115° is negative, we have:

$$a^2 = 144 + 324 - (-182.6) = 650.6$$

and so $a = 25.5$.

Figure 10.31   *Example*

Example

Determine the angles in a triangle which has sides of lengths $a = 13$, $b = 15$ and $c = 17$.

Using the cosine rule we can write $a^2 = b^2 + c^2 - 2bc \cos A$, hence:

$$\cos A = \frac{b^2 + c^2 - a^2}{2bc} = \frac{15^2 + 17^2 - 13^2}{2 \times 15 \times 17} = 0.676$$

Hence $A = 47.5°$. Similarly for angle B, $b^2 = a^2 + c^2 - 2ac \cos B$ and so:

$$\cos B = \frac{a^2 + c^2 - b^2}{2ac} = \frac{13^2 + 17^2 - 15^2}{2 \times 13 \times 17} = 0.527$$

Hence $B = 58.2°$. Since $A + B + C = 180°$, then $C = 180° - 47.5° - 58.2°$ and so is 74.3°.

Revision

27 Determine the required angle or length in each of the following triangles:

(a) Angle $A = 60°$, side $b = 14$, side $c = 10$, determine side a,
(b) Side $a = 5$, side $b = 6$, side $c = 7$, determine angle A,
(c) Angle $C = 75°$, side $a = 40$, side $b = 60$, determine side c,
(d) Angle $B = 110°$, side $a = 9.0$, side $c = 11.5$, determine side b,
(e) Side $a = 90$, side $b = 75$, side $c = 65$, determine angle A,
(f) Angle $C = 120°$, side $a = 8$, side $b = 10$, determine side c,
(g) Angle $C = 80°$, side $a = 30$ mm, side $b = 20$ mm, determine side c.
(h) Angle $A = 102°$, side $b = 50$ mm, side $c = 30$ mm, determine side a.

28 An object is supported by two chains connected from the same point on the object to points 3 m apart on the horizontal ceiling. The chain lengths are 1.8 m and 2.1 m. What is the angle between the chains at their connecting point on the object?

29 A parallelogram has sides of length 120 mm and 70 mm and the longer diagonal has a length of 150 mm. What are the angles of the parallelogram?

30 A 600 mm length of wire is bent into the shape of a triangle. If two of the sides have lengths of 220 mm and 180 mm, what are the angles in the triangle?

10.7 The areas of triangles

Figure 10.32 *Area of a triangle*

The area of any triangle is given by

area = ½ base × height

Thus, for the triangle shown in Figure 10.32, the area is ½ch. Since, for triangle ACH, $\sin A = h/b$ we can substitute using $h = b \sin A$ and so write the area as:

area = ½$bc \sin A$

Alternatively we can consider triangle BCH and use $\sin B = h/a$ and so obtain:

area = ½$ac \sin B$

We could alternatively have considered the triangle area in terms of the height from BC as the base and so obtained:

area = ½$ab \sin C$

In general:

Figure 10.33   *Example*

Figure 10.34   *Example*

Figure 10.35   *Revision problem 32*

area = ½ × (product of two sides) × (sine of angle between those sides)

Example

Determine the area of the triangle shown in Figure 10.33.

The data given on the figure is for two sides and the angle between those sides. Hence we can use the equation

$$\text{area} = \tfrac{1}{2}bc \sin A = \tfrac{1}{2} \times 38 \times 30 \times \sin 60° = 493.6 \text{ mm}^2$$

Example

Determine the area of the quadrilateral shown in Figure 10.34.

The quadrilateral can be considered to be made up of two triangles. For the triangle ABD, the area is:

$$\text{area} = \tfrac{1}{2} \times 4 \times 1.5 \times \sin 60° = 2.60 \text{ m}^2$$

For the triangle BCD, the area is

$$\text{area} = \tfrac{1}{2} \times 2.5 \times 3.2 \times \sin 78° = 3.91 \text{ m}^2$$

Thus the total area is 2.60 + 3.91 = 6.51 m³.

Revision

31 Determine the areas of each the following triangles:

(a) Angle A = 45°, side b = 20 mm, side c = 50 mm,
(b) Angle C = 50°, side a = 2 m, side b = 3 m,
(c) Angle B = 110°, side a = 50 mm, side c = 140 mm.
(d) Angle A = 80°, side b = 4 m, side c = 3 m,
(e) Angle C = 38°, side a = 40 mm, side b = 60 mm,
(f) Angle C = 40°, side a = 80 mm, side b = 50 mm,
(g) Angle A = 145°, side b = 30 mm, side c = 40 mm.

32 Determine the area of the quadrilateral shown in Figure 10.35.

10.7.1 Area in terms of the semi-perimeter

If the three sides of a triangle are known but no angle, then an alternative equation can be used to determine the area of the triangle. Let us take the area of a triangle (e.g. that in Figure 10.32) as being given by ½bc sin A. If we square both sides of the equation we obtain:

$$(\text{area})^2 = (\tfrac{1}{2}bc \sin A) \times (\tfrac{1}{2}bc \sin a) = \tfrac{1}{4}b^2c^2 \sin^2 A$$

But $\sin^2 A + \cos^2 A = 1$ (see Chapter 11). Thus we can write

$$(\text{area})^2 = \tfrac{1}{4}b^2c^2(1 - \cos^2 A)$$

We can write this as:

$$(\text{area})^2 = \tfrac{1}{4}b^2c^2(1 - \cos A)(1 + \cos A)$$

Using the cosine law for triangles we can write for $\cos A$:

$$\cos A = \frac{b^2 + c^2 - a^2}{2bc}$$

Hence:

$$(\text{area})^2 = \tfrac{1}{4}b^2c^2\left(1 - \frac{b^2 + c^2 - a^2}{2bc}\right)\left(1 + \frac{b^2 + c^2 - a^2}{2bc}\right)$$

$$= \tfrac{1}{4}b^2c^2\left(\frac{2bc - b^2 - c^2 + a^2}{2bc}\right)\left(\frac{2bc + b^2 + c^2 - a^2}{2bc}\right)$$

$$= \tfrac{1}{16}\left[a^2 - (b-c)^2\right]\left[(b+c)^2 - a^2\right]$$

$$= \tfrac{1}{16}(a - b + c)(a + b - c)(b + c + a)(b + c - a)$$

Let $a + b + c = 2s$, i.e. s is half the length of the perimeter of the triangle. Then subtracting $2b$ from both sides of this equation gives:

$$a - b + c = 2(s - b)$$

Subtracting $2a$ from $a + b + c = 2s$ gives:

$$b + c - 2a = 2(s - a)$$

and if we had subtracted $2c$:

$$a + b - c = 2(s - c)$$

Thus we obtain:

$$(\text{area})^2 = \tfrac{1}{16} \times 2(s - b) \times 2(s - c) \times 2s \times 2(s - a)$$

$$\text{area} = \sqrt{s(s - a)(s - b)(s - c)}$$

Example

Determine the area of a triangle which has sides of $a = 6$ m, $b = 5$ m and $c = 9$ m.

We have $2s = a + b + c = 6 + 5 + 9 = 20$. Hence $s = 10$. Thus:

$$\text{area} = \sqrt{10(10-6)(10-5)(10-9)} = \sqrt{10 \times 4 \times 5 \times 1} = \sqrt{200}$$

Hence the area is 14.1 m^2.

Revision

33 Determine the areas of the following triangles:

(a) Sides $a = 3$ m, $b = 4$ m, $c = 5$ m,
(b) Sides $a = 6$ m, $b = 8$ m, $c = 10$ m,
(c) Sides $a = 7$ m, $b = 10$ m, $c = 12$ m,
(d) Sides $a = 5$ cm, $b = 7$ cm, $c = 8$ cm,
(e) Sides $a = 7$ cm, $b = 8$ cm, $c = 9$ cm,
(f) Sides $a = 5$ cm, $b = 6$ cm, $c = 7$ cm,
(g) Sides $a = 4$ cm, $b = 5$ cm, $c = 7$ cm.

34 Determine the area of a parallelogram which has adjacent sides of lengths 2.4 cm and 3.2 cm and the length of the shortest diagonal is 2.0 cm.

Problems

1 Determine, to three decimal places, the values of the following:

(a) sin 23°, (b) cos 15.1°, (c) tan 19.5°, (d) cosec 34°, (e) sec 59°, (f) cot 57°, (g) sin 0.50, (h) cos 1.21, (i) tan 0.56, (j) cosec 0.85, (k) sec 1.05, (l) cot 1.23, (m) sin 140°, (n) cos 150°, (o) tan 130°, (p) sin 190°, (q) cos 195°, (r) tan 230°, (s) sin 280°, (t) cos 290°, (u) tan 315°, (v) sin 320°, (w) cos 350°, (x) sin 261.2°, (y) sec 301.8°, (z) cot 342.3°.

2 Determine using a calculator the values between 0° and 90°, to one decimal place, of the following angles:

(a) The angle which has a sine of 0.78,
(b) The angle which has a cosine of 0.15,
(c) The angle which has a tangent of 1.9.

3 Solve for x, to two decimal places, in the following right-angled triangles:

(a) Adjacent 10, opposite x, angle 68°,
(b) Adjacent 4, hypotenuse x, angle 55°,
(c) Opposite 6, adjacent x, angle 23°,
(d) Adjacent 12, opposite x, angle 58°,
(e) Hypotenuse 4, adjacent x, angle 70°,
(f) Hypotenuse 2, opposite x, angle 38°.

Figure 10.36 *Problem 6*

Figure 10.37 *Problem 9*

4 What is the height of a flagpole if the angle of elevation of the top of the pole from a point on the level ground 10 m from its base is 70°?

5 A ladder of length 6 m leans against a vertical wall with its base on horizontal ground. If the ladder makes an angle of 70° with the ground, how far is the bottom end from the wall?

6 The roof of a house is as shown in Figure 10.36. What is the length AB if AC = 9.0 m?

7 A and B are two towns 30 km apart. If B is on a bearing 50° east of north from A, how far east is B from A?

8 A guy wire of length 5 m to a pole rising from level ground makes an angle of 62° with the ground. How high above the ground is the wire attached to the pole?

9 In Figure 10.37, AB represents the jib of a crane and BC the tie. If AB is 10.0 m and BC is 4.0 m, what is the length of the tie?

10 Observer A sees an aeroplane as directly overhead. Observer B, who is 300 m due east of A, sees the plane as being at an angle of elevation of 70°. What is the height of the plane?

11 A ramp has to be built with a slope of 15° to enable a heavy machine to be slid up it to a height of 1.3 m. What must be the length of the ramp needed to achieve this?

12 A roof has to span 12 m with one slope pitched at 40° to the horizontal and the other at 50°. What will be the height of the ridge above the eaves?

13 Show that $\dfrac{\tan 60° - \tan 30°}{1 + \tan 60° \tan 30°} = \dfrac{1}{\sqrt{3}}$.

14 Determine the angles between 0° and 360°, to one decimal place, having the trigonometric ratios:

(a) $\sin \theta = 0.35$, (b) $\cos \theta = -0.55$, (c) $\cos \theta = 0.60$, (d) $\tan \theta = 0.10$, (e) $\tan \theta = -1.5$, (f) $\sin \theta = -0.48$, (g) $\cot \theta = -1.24$, (h) $\sec \theta = -2.5$, (i) $\operatorname{cosec} \theta = -2.5$, (j) $\operatorname{cosec} \theta = 1.5$.

15 Determine, to two decimal places, the values of the following:

(a) $\sin \pi/6$, (b) $\tan \pi/2$, (c) $\cos 3\pi/2$, (d) $\sin 5\pi/6$, (e) $\tan 7\pi/4$, (f) $\cos 7\pi/6$, (g) $\tan 7\pi/6$.

16 Determine, using the sine rule, the required sides and angles for each of the following triangles:

(a) Angle $A = 68°$, angle $C = 70°$, side $c = 42$ mm, determine angle B, sides a and b,

(b) Angle $A = 58°$, side $a = 180$ mm, side $b = 150$ mm, determine angles B and C, side c,

(c) Angle $C = 25°$, angle $B = 80°$, side $b = 10$ cm, determine angle A and sides a and c,

(d) Angle $= 10°$, angle $C = 60°$, side $b = 50$ mm, determine angle B and sides a and c,

(e) Angle $B = 50°$, side $b = 30$ mm, side $c = 20$ mm, determine angles A and C and side a,

(f) Angle A $= 70°$, angle $C = 40°$, side $c = 100$ mm, determine angle B and sides a and b,

(g) Angle $A = 62°$, angle $C = 62°$, side $b = 50$ mm, determine angle B and sides a and c,

(h) Angle $A = 53°$, angle $B = 62°$, side $a = 125$ mm, determine angle C and sides b and c,

(i) Angle $A = 80°$, side $a = 165$ mm, side $b = 130$ mm, determine angles B and C and side c,

(j) Angle $A = 30°$, side $a = 50$ mm, side $b = 80$ mm, determine angles B and C and side c.

17 A roof has a span of 9 m with eaves that slope on one side at 44° to the horizontal and on the other side at 36° to the horizontal. Determine the lengths of the sloping parts of the roof.

18 A wall is leaning outwards at an angle to the vertical of 15°. A ladder of length 2 m is placed against the wall with its base on level ground 1 m from the base of the wall. How high up the wall does the ladder reach?

19 A vertical pole of length 6 m stands alongside a road which slopes away from it with a constant slope. The pole casts a shadow which is 14 m long directly downhill when the angle of elevation of the sun is 56°. What is the angle of the road with the horizontal?

20 Determine, using the cosine rule, the required sides and angles for each of the following triangles:

(a) Side $a = 12$, side $b = 63$, side $c = 52$, determine angles A, B and C,

(b) Angle $A = 57.5°$, side $b = 78$, side $c = 97$, determine side a and angles B and C,

(c) Side $a = 9$, side $b = 13$, side $c = 17$, determine angles A, B and C,

(d) Side $a = 7$, side $b = 5$, side $c = 3$, determine angles A, B and C,

(e) Side $a = 4$, side $b = 7$, side $c = 5$, determine angles A, B and C,

(f) Angle $A = 80°$, side $b = 7$, side $c = 8$, determine side a and angles B and C,

(g) Angle $A = 66°$, side $b = 40$ mm, side $c = 50$ mm, determine angles B and C and side a,

(h) Side $a = 40$ mm, side $b = 50$ mm, side $c = 70$ mm, determine angles A, B and C,

(i) Angle $C = 74°$, side $a = 100$ mm, side $b = 50$ mm, determine angles A and B and side c.

21 A plane flies for 50 km in a straight line from its start point before turning through 20° and flying in a straight line on the new bearing for 70 km. What distance is the plane from the start?

22 A parallelogram has adjacent sides of lengths 55 mm and 82 mm with an angle between them of 68°. What is the length of the diagonal through this corner of the parallelogram?

23 A parallelogram has adjacent sides of lengths 30 mm and 50 mm with an angle between them of 144°. What is the length of the diagonal through this corner of the parallelogram?

24 Determine the areas of each of the following triangles:

(a) Angle $A = 85°$, side $b = 40$ mm, side $c = 10$ mm,
(b) Angle $B = 130°$, side $a = 40$ mm, side $c = 60$ mm,
(c) Angle $C = 52°$, side $a = 2$ m, side $b = 5$ m,
(d) Angle $A = 55°$, side $b = 12$ m, side $c = 4$ m,
(e) Side $a = 4$ cm, side $b = 7$ cm, side $c = 9$ cm,
(f) Side $a = 2$ m, side $b = 2$ m, side $c = 2$ m,
(g) Side $a = 5$ cm, side $b = 12$ cm, side $c = 13$ cm.

25 A parallelogram of area 30 cm² has adjacent sides of lengths 6 cm and 9 cm. Determine the angles of the parallelogram.

26 A parallelogram has adjacent sides of lengths 9 cm and 5 cm. If the shortest diagonal has a length of 7 cm, what is the area of the parallelogram?

11 Further trigonometry

11.1 Introduction

This chapter extends the trigonometry in Chapter 10 by considering the use of trigonometric identities in the solution of problems where we have such expressions as $\sin^2 A$, $\cos^2 A$, $\sin (A + B)$, $\cos (A + B)$, $\sin 2A$ and $\cos 2A$. Such types of expressions are frequently encountered in engineering and often need to be converted into other forms to aid the manipulation of equations.

11.2 Trigonometric relationships

In simplifying trigonometric relationships there are a number of useful relationships between trigonometric ratios which can assist.

The *Pythagoras theorem* states that, for a right-angled triangle, the square of the hypotenuse is equal to the sum of the squares of the other two sides. Thus, for the triangle shown in Figure 11.1, we can write

$$AB^2 + BC^2 = AC^2$$

If we divide both sides of the equation by AC^2 we obtain:

$$\left(\frac{AB}{AC}\right)^2 + \left(\frac{BC}{AC}\right)^2 = 1$$

Figure 11.1 *Pythagoras theorem*

But $AB/AC = \sin \theta$ and $BC/AC = \cos \theta$. Hence we have the very commonly used relationship of

$$\sin^2 \theta + \cos^2 \theta = 1$$

$(\sin \theta)^2$ is written as $\sin^2\theta$ and $(\cos \theta)^2$ as $\cos^2\theta$.

11.2.1 Further relationships

We can obtain two further relationships from the above equation. If we divide the $\sin^2 \theta + \cos^2 \theta = 1$ equation by $\cos^2\theta$ we obtain:

$$\left(\frac{\sin \theta}{\cos \theta}\right)^2 + 1 = \left(\frac{1}{\cos \theta}\right)^2$$

and so, since $\sin \theta/\cos \theta = \tan \theta$ and $1/\cos \theta = \sec \theta$, we obtain the relationship:

$$\tan^2\theta + 1 = \sec^2\theta$$

If we divide the $\sin^2\theta + \cos^2\theta = 1$ equation by $\sin^2\theta$ we obtain:

$$1 + \left(\frac{\cos \theta}{\sin \theta}\right)^2 = \left(\frac{1}{\sin \theta}\right)^2$$

and so, since $\cos\theta/\sin\theta = \cot\theta$ and $1/\sin\theta = \operatorname{cosec}\theta$, we obtain:

$$1 + \cot^2\theta = \operatorname{cosec}^2\theta$$

Example

Show that $\dfrac{\sec\theta + \tan\theta}{1 + \sin\theta} = \sec\theta$.

We can write the expression as:

$$\frac{\sec\theta + \tan\theta}{1 + \sin\theta} = \frac{\dfrac{1}{\cos\theta} + \dfrac{\sin\theta}{\cos\theta}}{1 + \sin\theta} = \frac{\dfrac{1 + \sin\theta}{\cos\theta}}{1 + \sin\theta} = \frac{1}{\cos\theta} = \sec\theta$$

Example

Show that $\dfrac{\cos\theta}{1 - \sin\theta} + \dfrac{\cos\theta}{1 + \sin\theta} = 2\sec\theta$.

We can write the equation with a common denominator as:

$$\frac{\cos\theta}{1 - \sin\theta} + \frac{\cos\theta}{1 + \sin\theta} = \frac{\cos\theta(1 + \sin\theta) + \cos\theta(1 - \sin\theta)}{(1 - \sin\theta)(1 + \sin\theta)}$$

Multiplying out the brackets gives:

$$\frac{2\cos\theta}{1 - \sin^2\theta}$$

As $\sin^2\theta + \cos^2\theta = 1$, then this can be written as:

$$\frac{2\cos\theta}{\cos^2\theta} = \frac{2}{\cos\theta} = 2\sec\theta$$

Revision

1 Simplify the following:

  (a) $\dfrac{\operatorname{cosec}\theta}{\cot\theta + \tan\theta}$, (b) $\dfrac{\tan\theta + 1}{\cot\theta + 1}$, (c) $\dfrac{\sin^2\theta\cot\theta}{\cos\theta}$, (d) $\sin\theta\cos\theta\tan\theta$.

2 Show that $\dfrac{\cos\theta\cot\theta\sec^2\theta}{\operatorname{cosec}\theta} = 1$.

3 Show that $\tan\theta + \cot\theta = \sec\theta\operatorname{cosec}\theta$.

4 Show that $1 - 2\sin^2\theta = 2\cos^2\theta - 1$.

5 Show that $\dfrac{1}{1 - \sin^2\theta} = \sec^2\theta$.

11.3 Trigonometric ratios of sums of angles

It is often useful to express the trigonometric ratios of angles such as $A + B$ or $A - B$ in terms of the trigonometric ratios of A and B. Consider the two right-angled triangles OPQ and OQR shown in Figure 11.2:

$$\sin(A + B) = \frac{TR}{OR} = \frac{TS + SR}{OR} = \frac{PQ + SR}{OR} = \frac{PQ}{OQ}\frac{OQ}{OR} + \frac{SR}{QR}\frac{QR}{OR}$$

Hence:

$$\sin(A + B) = \sin A \cos B + \cos A \sin B \qquad [1]$$

If we replace B by $-B$ we obtain:

$$\sin(A - B) = \sin A \cos B - \cos A \sin B \qquad [2]$$

If in equation [1] we replace A by $(\pi/2 - A)$ we obtain:

$$\cos(A + B) = \cos A \cos B - \sin A \sin B \qquad [3]$$

If in equation [3] we replace B by $-B$ we obtain:

$$\cos(A - B) = \cos A \cos B + \sin A \sin B \qquad [4]$$

We can obtain tan $(A + B)$ by dividing sin $(A + B)$ by cos $(A + B)$:

$$\tan(A + B) = \frac{\tan A + \tan B}{1 - \tan A \tan B} \qquad [5]$$

and likewise $\tan(A - B)$ by dividing $\sin(A - B)$ by $\cos(A - B)$:

$$\tan(A - B) = \frac{\tan A - \tan B}{1 + \tan A \tan B} \qquad [6]$$

By adding equations [1] and [3] we obtain:

$$\sin(A + B) + \cos(A + B) = \sin A \cos B + \cos A \sin B$$
$$+ \cos A \cos B - \sin A \sin B$$

$$= 2 \sin A \cos B$$

Thus by adding or subtracting equations from above we can obtain the following group of equations:

$$2 \sin A \cos B = \sin(A + B) + \sin(A - B)$$

$$2 \cos A \cos B = \cos(A + B) + \cos(A - B)$$

$$2 \sin A \sin B = \cos(A - B) - \cos(A + B)$$

$$2 \cos A \sin B = \sin(A + B) - \sin(A - B)$$

Figure 11.2 *Compound angle*

Example

Show that $\dfrac{\sin(A-B)}{\sin A \sin B} = \cot B - \cot A$.

Using $\sin(A-B) = \sin A \cos B - \cos A \sin B$ gives:

$$\frac{\sin A \cos B - \cos A \sin B}{\sin A \sin B} = \frac{\cos B}{\sin B} - \frac{\cos A}{\sin A} = \cot B - \cot A$$

Revision

6 Shows that $\tan(A+B) = \dfrac{\tan A + \tan B}{1 - \tan A \tan B}$.

7 Shows that $\dfrac{\sin(A+B)}{\cos A \cos B} = \tan A + \tan B$.

8 Use the addition of angles expression to simplify $\sin (A + \pi)$.

11.6.2 Trigonometric ratios for double angles

If, for equations [1], [3] and [5], we let $B = A$ we obtain the double-angle equations:

$$\sin 2A = 2 \sin A \cos A$$

$$\cos 2A = \cos^2 A - \sin^2 A = 1 - 2 \sin^2 A = 2 \cos^2 A - 1$$

$$\tan 2A = \frac{2 \tan A}{1 - \tan^2 A}$$

Example

Solve the equation $\cos 2x + 3 \sin x = 2$.

Using expression $\cos 2A = 1 - 2 \sin^2 A$ for $\cos 2x$ gives:

$$1 - 2 \sin^2 x + 3 \sin x = 2$$

This can be rearranged as:

$$2 \sin^2 x - 3 \sin x + 1 = 0$$

$$(2 \sin x - 1)(\sin x - 1) = 0$$

Hence $\sin x = \frac{1}{2}$ or 1. For angles between $0°$ and $90°$, $x = 30°$ or $90°$.

Revision

9 Show that: (a) $\dfrac{1 - \cos 2A}{\sin 2A} = \tan A$,

(b) $\sec 2A + \tan 2A = \dfrac{\cos A + \sin A}{\cos A - \sin A}$,

(c) $\sin 3A = 3 \sin A - 4 \sin^3 A$ (Hint: write $3A$ as $2A + A$.)

10 Solve $2 \cos^2 x + 3 \sin x = 3$, for angles between $0°$ and $90°$.

Problems

1 Show that:

(a) $\sec^2\theta - \dfrac{\sin^2\theta}{\cos^2\theta} = 1$, (b) $(\sin\theta + \cos\theta)^2 + (\sin\theta - \cos\theta)^2 = 2$,

(c) $1 - 2\cos^2\theta = 2\sin^2\theta - 1$, (d) $\dfrac{1 - \tan^2\theta}{1 + \tan^2\theta} = \cos^2\theta - \sin^2\theta$,

(e) $(1 + \tan^2\theta)\cos^2\theta = 1$, (f) $\dfrac{2\tan\theta}{1 + \tan^2\theta} = 2\sin\theta\cos\theta$,

(g) $\dfrac{1}{\sqrt{1 - \sin^2\theta}} = \sec\theta$, (h) $\sin\theta - \sin^3\theta = \dfrac{\sin\theta}{\sec^2\theta}$,

(i) $(1 + \cot\theta)^2 + (1 - \cot\theta)^2 = 2\operatorname{cosec}^2\theta$, (j) $\cot\theta = \sqrt{\dfrac{\cos^2\theta}{1 - \cos^2\theta}}$.

2 The horizontal distance x and the vertical distance y travelled in a time t by a projectile when fired with a velocity v at an angle θ to the horizontal are given by:

$$x = vt\cos\theta \text{ and } y = vt\sin\theta - \tfrac{1}{2}gt^2$$

Substitute for t and so show that:

$$y = x\tan\theta - \frac{gx^2}{2v^2}(\tan^2\theta + 1)$$

3 Show that:

(a) $\tan(A + B) - \tan A = \dfrac{\sin B}{\cos A \cos(A + B)}$,

(b) $\tan A + \cot A = 2\operatorname{cosec} 2A$, (c) $\cot(A + B) = \dfrac{\cot A \cot B - 1}{\cot A + \cot B}$,

(d) $\cos 4A = 8\cos^4 A - 8\cos^2 A + 1$, (e) $\sin 3A + \sin A = 2\sin 2A \cos A$.

4 Use the addition of angles expression to simplify (a) $\cos(A - \pi)$, (b) $\sin(A + \pi/2)$.

12 Vectors

12.1 Introduction

With many engineering problems there are certain quantities for which not only their size but their direction is important. For example, if we want to specify the location of some town in relation to, say, London then we can do this by stating how far away from London it is and in what direction. In the case of forces, both the direction of a force and its size need to be considered if we want to predict the effect of the force on an object. Those quantities for which both the size and direction have to be specified are termed *vectors*. Quantities for which only size is relevant are termed *scalar quantities*. This chapter is an introduction to the problems involving vector quantities, considering how vector quantities can be added or subtracted and how a vector quantity can be replaced by two equivalent vector quantities. The study of vectors offers an example of the usefulness of trigonometry in engineering.

12.1.1 Representation of vectors

To specify a scalar quantity all we need to do is give a single number to represent its size. To specify a vector quantity we need to indicate both size and direction. The term *magnitude* tends to be used for the size of vector quantities. To represent vector quantities on a diagram we use arrows. The length of the arrow is chosen according to some scale to represent the magnitude of the vector and the direction of the arrow, with reference to some reference direction, the direction of the vector. For example, to represent a displacement of 30 km in a north-east direction from A to B we might use the arrow shown in Figure 12.1. The length of the arrow is scaled to represent the magnitude of 30 km and the direction of the arrow is 45° to the horizontal to represent the north-east direction, the horizontal being the reference direction for the east–west direction.

We can indicate that a line from A to B is a vector by writing it as:

$$\overrightarrow{AB}$$

the direction of the arrow indicating that the direction is from A to B. With a direction from B to A we would have used:

$$\overrightarrow{BA}$$

An alternative notation is however more usually used. Vector quantities, such as that in the direction from A to B, are denoted by using a bold letter such as **a**, or when hand-written by underlining the symbol a̲. When we are referring to the vector in the opposite direction, i.e. from B to A, we would use −**a**, or −a̲, with the minus sign being used to indicate that it is in the opposite direction to **a**, or a̲.

Figure 12.1   *A vector*

The term *modulus of a vector* is often used for the magnitude of a vector **a** and is represented by |**a**| or just writing *a*. The direction of a vector with reference to some reference direction is specified by an angle. This angle is often called the *argument of the vector*. For the displacement of 30 km in a north-east direction from A to B represented graphically in Figure 12.1, we have a modulus of 30 km and an argument of 45°.

For two vectors to be equal they must represent the same quantity and have the same magnitude and direction. We cannot say that a force of 2 N is equal to another force of 2 N without knowing that both forces are acting in the same direction. Thus a force of 2 N acting in a northerly direction is the same as another force of 2 N acting in a northerly direction. We can represent these forces as the identical vectors **F** and **F**. A force of 2 N in a northerly direction is however different from a force of 2 N acting in a southerly direction. In this case the magnitudes are the same but the directions are directly opposite. We can represent these forces as the vectors **F** and −**F**.

Example

An object slides down a slope which is at 20° to the horizontal. Draw a diagram to represent the vectors for the vertical force of 30 N acting on the object as a result of gravity and the frictional force acting up the plane and opposing the motion of 10 N.

There is a force of 30 N in a vertical direction and a force of 10 N at an angle of 20° to the horizontal. We can thus represent the two vector quantities by an arrow with a length proportional to 30 N in the vertical direction and an arrow of length proportional to 10 N at 20° to the horizontal. Figure 12.2 shows the result.

Figure 12.2 *Example*

Revision

1 Draw diagrams to represent the following vectors:

(a) An object acted on by a force of 10 N in a horizontal direction and a force of 5 N at an angle of 30° to the horizontal direction and upwards.
(b) A projectile which has a velocity of 2 m/s in the horizontal direction and a velocity of 3 m/s in the vertically upward direction.
(c) An object which has been given a displacement of 200 mm to the right and 400 mm in a direction at right angles to this.

12.2 Addition of vectors

When we add two scalar quantities, say a mass of 2 kg added to a mass of 3 kg, we can write 2 + 3 = 5 and so the addition gives a mass of 5 kg. With vectors we have to consider both the magnitude and direction of each vector in carrying out the addition.

The rule for adding two vectors is called the *triangle rule*. This can be stated as: to add two vectors **a** and **b** we place the tail of the arrow representing one vector at the head of the arrow representing the other and then

Figure 12.3 *Triangle rule*

(a)

(b)

Figure 12.4 *Triangle of forces*

Figure 12.5 *Parallelogram rule*

the line that forms the third side of the triangle represents the vector which is the sum of **a** and **b**. Figure 12.3 shows such a triangle. Note that the directions of **a** and **b** go in one sense round the triangle and the sum goes in the opposite direction. The vector representing the sum of **a** and **b** is termed the *resultant*. We could replace the two vectors **a** and **b** by the vector for the sum and have the same effect.

Note that in engineering you may come across the *triangle of forces* defined as the condition for equilibrium of three forces acting at the same point and in the same plane on some object. This may be stated as: if the three forces are represented in magnitude and direction by arrows F_1, F_2 and F_3, then these arrows when taken in the order of the forces must form a triangle if the three forces are to be in equilibrium (Figure 12.4). This is just the triangle rule. If we consider two forces F_1 and F_2 then the sum of their vectors $F_1 + F_2$ is the third side of the triangle. This vector must be of the same magnitude but in the opposite direction if there is to be no net force acting on the object. Thus $F_1 + F_2$ is $-F_3$.

An alternative and equivalent rule to the triangle rule for determining the sum of two vectors is the *parallelogram rule*. This can be stated as: if we place the tails of the arrows representing the two vectors **a** and **b** together and complete a parallelogram, then the diagonal of that parallelogram drawn from the junction of the two tails represents the sum of the vectors **a** and **b**. Figure 12.5 illustrates this. The triangle rule is just the triangle formed between the diagonal and two adjacent sides of the parallelogram.

Example

The vector **a** has a magnitude of 10 and an argument of 30°. The vector **b** has a magnitude of 5 and an argument of 100°. Determine **a** + **b**.

We can represent **a** by an arrow with a length which represents the magnitude 10 at an angle of 30° to some reference direction. Using the triangle rule then we draw the vector **b** with a length representing 5 at an angle of 100° to the same reference direction and with its tail at the arrow end of **a**. The sum of the two vectors, i.e. the resultant, is then the vector indicated by completing the triangle. Figure 12.6 shows the triangle. We could determine this resultant by using a scale drawing.

Alternatively we could use the cosine rule (see Section 10.6). This rule is: the square of a side is equal to the sum of the squares of the other two sides minus twice the product of those sides times the cosine of the angle between them. Since the angle opposite the resultant of **a** + **b** is $(180° - 100°) + 30° = 110°$, then:

$$(\text{resultant})^2 = 5^2 + 10^2 - 2 \times 5 \times 10 \cos 110° = 25 + 100 + 34.2$$

Hence the $(\text{resultant})^2 = 159.2$ and the resultant has a magnitude of 12.6. The angle this resultant makes with **a** can be determined by means of the sine rule (see Section 10.5). Thus:

$$\frac{12.6}{\sin 110°} = \frac{5}{\sin \theta}$$

Figure 12.6 *Example*

Figure 12.7 *Example*

Figure 12.8 *Example*

Hence

$$\sin \theta = \frac{5 \times \sin 110°}{12.6} = 0.373$$

Hence $\theta = 21.9°$. Thus the resultant is at an angle of $21.9° + 30° = 51.9°$ to the reference axis.

Example

An object is acted on by two forces, of magnitudes 5 N and 4 N, at an angle of 60° to each other. What is the resultant force on the object?

Figure 12.7 shows how these two vectors can be represented by the sides of a parallelogram. The resultant can be determined from a scale drawing from measurements of its length and angle or by calculation. We have a triangle with adjacent sides of 5 and 4 and an angle of $180° - 60° = 120°$ between them. We need to determine the length of the other side. The cosine rule (see Section 10.6) can be used. Thus

$$(\text{resultant})^2 = 5^2 + 4^2 - 2 \times 5 \times 4 \cos 120° = 25 + 16 + 20$$

Hence the resultant is 7.8 N. We can determine the angle θ between the resultant and the 5 N force by the use of the sine rule (see Section 10.5). Thus:

$$\frac{4}{\sin \theta} = \frac{7.8}{\sin 120°}$$

Hence:

$$\sin \theta = \frac{4 \times \sin 120°}{7.8} = 0.444$$

and $\theta = 26.4°$.

Example

Figure 12.8(a) shows an object supported by two wires. If the weight of the object is 25 N, what are the forces in the two wires when the object is in equilibrium?

When the object is in equilibrium then the resultant of the two forces in the wires must be opposite and equal to the weight of 25 N. For the weight we know the magnitude and direction. For the forces in the two wires we only know the directions. Figure 12.8(b) shows the triangle of forces produced by utilising this information. We could obtain the forces in the wires from a scale diagram of the triangle of forces. Alternatively we can use the sine rule (see Section 10.5). The angle between the two wires is, from Figure 12.8(a), $180° - 30° - 60°$. Thus

$$\frac{T_1}{\sin (90° - 60°)} = \frac{25}{\sin (180° - 30° - 60°)}$$

Thus:

$$T_1 = \frac{25 \times \sin 30°}{\sin 90°} = 12.5 \text{ N}$$

For the other string:

$$\frac{T_2}{\sin (90° - 30°)} = \frac{25}{\sin (180° - 30° - 60°)}$$

Thus:

$$T_2 = \frac{25 \sin 60°}{\sin 90°} = 21.7 \text{ N}$$

Example

A tool is given a displacement of 100 mm in the x direction and 60 mm in a direction which is at right angles to the x direction. What is the resultant displacement?

Displacement is a vector quantity and thus the resultant displacement is found by vector means. We can thus use the triangle rule to determine the sum of the two vectors. We draw an arrow with a length scaled to represent 100 mm. The arrow to represent the second displacement is then drawn with its tail starting from the arrow head of the first in a direction at right angles to it and with a length scaled to represent 60 mm. The resultant is the third side of the triangle. Figure 12.9 shows the triangle. We can calculate the resultant by the use of the Pythagoras theorem. Thus:

$$(\text{resultant})^2 = 60^2 + 100^2 = 3600 + 10\ 000$$

Hence the resultant has a magnitude of 116.6 mm. The resultant is at an angle θ where $60/100 = \tan \theta$. Hence $\theta = 31.0°$.

Example

A ship heads on a course due south at 10 km/h. It is driven off course by a current flowing in a north-easterly direction at 6 km/h. What is the resultant velocity?

The ship has two velocities, one in the southerly direction and one in the north-easterly direction. The resultant velocity is the vector sum of the two. If it helps, you can consider the problem in terms of what the resultant displacement of the ship would be if in one hour it has a displacement of 10 km in the southerly direction and 6 km in the north-easterly direction. We can use the triangle rule to determine the result-

Scale 20 mm

60 mm

θ

100 mm

Figure 12.9 *Example*

Figure 12.10 *Example*

ant. Figure 12.10 shows the result. The 10 km/h velocity is represented by an arrow of length scaled to represent 10 km in a direction which represents south. From the end of that arrow an arrow is drawn to represent the 6 km/h velocity. This will be at 45° to the southerly direction in order to give the north-easterly direction. The resultant is then the third side of the triangle. We can determine the resultant from a scale diagram or by calculation using the cosine rule (Section 10.6):

$$(resultant)^2 = 10^2 + 6^2 - 2 \times 10 \times 6 \cos 45° = 100 + 36 - 84.85$$

Hence the resultant has a magnitude of 7.2 km/h. We can use the sine rule (Section 10.5) to determine the angle θ between the resultant and the 10 km/h velocity:

$$\frac{7.2}{\sin 45°} = \frac{6}{\sin \theta}$$

Thus:

$$\sin \theta = \frac{6 \times \sin 45°}{7.2} = 0.589$$

Hence θ = 36.1°. Thus the resultant velocity is 7.2 km/h in a direction which is 36.1° east of south.

Revision

2   If **a** is a vector of magnitude 4 in an easterly direction and **b** a vector of magnitude 3 in a southerly direction, what is the resultant, i.e. the vector sum **a** + **b**?

3   If **a** is a vector of magnitude 5 in an easterly direction and **b** a vector of magnitude 4 in a north-easterly direction, what is the resultant, i.e. the vector sum **a** + **b**?

4   Determine the resultant forces acting on objects subject to the following forces acting at a point on the object:

(a) A force of 3 N acting horizontally and a force of 4 N acting at the same point on the object at an angle of 60° to the horizontal,
(b) A force of 3 N in a westerly direction and a force of 6 N in a northerly direction,
(c) Forces of 4 N and 5 N with an angle of 65° between them,
(d) Forces of 3 N and 8 N with an angle of 50° between them,
(e) Forces of 5 N and 7 N with an angle of 30° between them,
(f) Forces of 9 N and 10 N with an angle of 40° between them,
(g) Forces of 10 N and 12 N with an angle of 105° between them.

5   In a plane structure a particular point is acted on by forces of 1.2 kN and 2.0 kN in the plane, the angle between the forces being 15°. What is the resultant force?

6   An object of weight 30 N is suspended from a horizontal beam by two chains. The two chains are attached to the same point on the object and are at 30° and 40° to the vertical. Determine the tensions in the two chains.

7   Determine the resultant velocity of each of the following pairs of velocities:

(a) 10 km/h due north and 5 km/h due east,
(b) 6 km/h due east and 4 km/h in a north-westerly direction,
(c) 5 km/h due north and 7 km/h in a direction 60° east of south.

8   A boat is required to cross a river 100 m wide to a point on the bank directly opposite the start point. If the boat can have a velocity of 6 km/h and the river is flowing with a current of 3 km/h, in what direction should the boat be steered?

9   A ship is steering a course due south across a current which is in a westerly direction. The resultant velocity of the ship is 18 km/h in a direction 15° west of south. What is the velocity set by the ship?

Figure 12.11   *Polygon of vectors*

12.2.1 Polygon of vectors

The triangle law is used to find the sum of two vectors. Consider now the problem of determining the sum of more than two vectors. Suppose we have vectors **a**, **b**, **c** and **d**. We can use the triangle law to find the sum **e** of vectors **a** and **b**. We can then use the triangle law to find the sum **f** of **e** and **c**. We can then use the triangle law to find the sum **g** of **f** and **d**. Figure 12.11 illustrates the above procedure.

In fact what we have done is take the vectors in sequence and draw the arrows tail to head. The resultant of all the vectors is the vector required to complete the polygon shape and link the head of the last vector to the tail of the start vector.

Example

Determine the force acting on the eyebolt shown in Figure 12.12(a) as a result of the forces applied along the connecting wires.

Starting with the 2.0 kN force at 50° to the vertical, the vector is drawn to represent it. From the head of that vector we draw a vector to represent the next force in sequence as we proceed anticlockwise round the forces, namely the vertical 3.0 kN force. From the head of this vector we draw a vector to represent the next force in the sequence, namely the 4.5 kN force at 20° to the vertical. The resultant force is the

Figure 12.12   *Example*

force completing the polygon. We could calculate this by breaking the polygon down into the triangles indicated by Figure 12.12 and using the cosine law with each triangle in turn. Alternatively we could draw the polygon to scale. With a scale diagram the force is indicated as being about 8.5 kN vertically.

An alternative technique which will be met later in this chapter is to resolve each of the forces into the horizontal and vertical directions. Then the algebraic sum of the horizontal forces and the sum of the vertical forces are obtained. These two forces can then be added, by vector means, to give the resultant.

Revision

10 Determine the resultant force acting on the gusset plate as a result of the forces shown in Figure 12.13.

Figure 12.13

11 Three forces act at a point on an object. One of the forces is 6 N horizontally to the left, another 3 N at 70° anticlockwise to the 6 N force, and the third 4 N at 150° anticlockwise to the 6 N force. Determine the resultant force.

12.3 Subtraction of vectors

Figure 12.14 *Subtracting vectors*

When we subtract two scalar quantities, say taking a mass of 2 kg from one of 5 kg, we can write 5 − 2 = 3 and so the subtraction gives a mass of 3 kg. With vectors we have to consider both the magnitude and direction of each phasor in carrying out the subtraction.

With vectors we define subtraction of vector **b** from vector **a** as being

$$\mathbf{a} - \mathbf{b} = \mathbf{a} + (-\mathbf{b})$$

A negative vector −**b** has the same magnitude but is in the opposite direction to the vector **b**. Thus to subtract **b** from **a**, all we do is reverse the direction of **b** and then add it to **a** by the use of the triangle or parallelogram rule. Figure 12.14 illustrates this process of subtracting vector **b** from vector **a**.

Figure 12.15 *Example*

Example

If vector **a** has a magnitude of 3 in an easterly direction and vector **b** a magnitude of 4 in a northerly direction, determine the value of **a** − **b**.

Figure 12.15 shows the vectors **a** and **b**. To subtract **b** from **a** we reverse the direction of **b** to give −**b** and then add −**b** and **a**. We thus end up with the parallelogram shown in the figure. The vector **a** + (−**b**) has thus a magnitude given by the Pythagoras theorem as:

$$\sqrt{3^2 + (-4)^2} = 5$$

and in a direction which is at an angle θ given by:

$$\tan \theta = \frac{-4}{3}$$

Hence $\theta = -53.1°$.

Revision

12 Determine **a** − **b** if:

(a) vector **a** has a magnitude of 4 in a westerly direction and vector **b** a magnitude of 2 in a northerly direction,
(b) vector **a** has a magnitude of 3 in an easterly direction and vector **b** a magnitude of 5 in a direction 60° south of east,
(c) vector **a** has a magnitude of 7 in an easterly direction and vector **b** a magnitude of 4 in a direction 25° south of west,
(d) vector **a** has a magnitude of 4 in a northerly direction and vector **b** a magnitude of 5 in a direction 25° south of west.

12.3.1 Relative velocity

Consider two objects A and B in motion, one having a velocity v_A and the other a velocity v_B with respect to the fixed ground. What is the velocity of B as perceived by A? The relative velocity of B with respect to A is the velocity that B would appear to have if A was considered to be the fixed point, rather than the fixed ground, from which the velocity of B was measured. To effectively make A at rest we would have to add to it a velocity of $-v_A$. We add this same velocity to B. Thus $v_B + (-v_A)$ is the velocity of B relative to A.

Example

Aircraft A is flying due west at 400 km/h, while aircraft B is flying due north at 300 km/h. What is the velocity of aircraft B relative to aircraft A?

Figure 12.16 *Example*

We need to consider the velocity of aircraft B as perceived by aircraft A if aircraft A is considered to be at rest. Thus we add a velocity of 400 km/h in an easterly direction to that of aircraft A. We then add this same velocity of 400 km/h in an easterly direction to the velocity of aircraft B. Figure 12.16 shows the resulting vector diagram. Using the Pythagoras theorem, then the magnitude of the relative velocity is the length of the diagonal of the parallelogram of vectors, i.e.

$$(\text{relative velocity})^2 = 400^2 + 300^2$$

Hence the relative velocity has a magnitude of 500 km/h. The direction of this velocity is given by $\tan \theta = 400/300$. Hence $\theta = 53.1°$. Thus the relative velocity is 500 km/h in a direction 53.1° east of north.

Revision

13 A passenger on a train travelling due north at 25 m/s sees a second train on a parallel track moving at 40 m/s in the opposite direction. What is the relative velocity of the second train as perceived by the passenger on the first train?

14 Object A is moving at 5 m/s in a westerly direction. Object B is moving at 12 m/s in a south-westerly direction. What is the velocity of B relative to A?

15 A student is walking along a road at 5 km/h. It is raining and the rain is falling vertically with a velocity of 8 km/h. At what angle will the student perceive the rain to be falling?

16 Object A is moving at 5 m/s due west and object B at 12 m/s in a south-easterly direction. What is the velocity of B relative to A?

12.4 Resolution of vectors

Two vectors can be added together to give a single vector called the result-ant. There is a reverse process of taking a single vector **a** and expressing it in terms of two *component vectors* **v** and **h** at right angles to each other. The sum of the components is the original vector, i.e. **a** = **v** + **h**. These components are referred to as the *resolved parts* of the vector.

Consider the vector **a** shown in Figure 12.17. This vector acts along AC, with triangle ABC being a right-angled triangle. If we take the **v** vector as being along BC and the **h** vector along AB, then for the magnitudes of the vectors we have, for the **v** vector

$$\frac{|\mathbf{v}|}{|\mathbf{a}|} = \sin \theta$$

Figure 12.17 *Resolving vectors*

and so $|\mathbf{v}| = |\mathbf{a}| \sin \theta$. For the **h** vector we have

$$\frac{|\mathbf{h}|}{|\mathbf{a}|} = \cos \theta$$

Vertical component

5 N

40°

Horizontal component

Figure 12.18 *Resolving a force*

120 kN

45°

(a)

Vertical component

120 kN

45°

Horizontal comp[onent

(b)

Figure 12.19 *Example*

3.0 kN

30°

2.0 kN

30°

70°

5.0 kN

Figure 12.20 *Example*

and so $|\mathbf{h}| = |\mathbf{a}|\cos\theta$.

In Figure 12.17 the triangle rule was used to resolve a vector into its two components. An alternative, but absolutely equivalent method, would have been to use the parallelogram rule. Thus, suppose we have a force of 5 N at an angle of 40° to the horizontal. Then using the parallelogram rule we can draw the horizontal and vertical components as the sides of the parallelogram, as shown in Figure 12.18. Then we have, using the equation derived above,

horizontal component = 5 cos 40° = 3.8 N

vertical component = 5 sin 40° = 3.2 N

We could therefore replace the 5 N force by the two components of 3.8 N and 3.2 N and the effect would be precisely the same.

Example

Determine the horizontal and vertical components of the 120 kN force acting on the beam shown in Figure 12.19(a).

Figure 12.19(b) shows the parallelogram of vectors which can be used to determine the components:

horizontal component = $|\mathbf{F}|\cos\theta$ = 120 cos 45° = 84.9 kN

vertical component = $|\mathbf{F}|\sin\theta$ = 120 sin 45° = 84.9 kN

Example

Determine the resultant force acting on the bracket shown in Figure 12.20 due to the three forces indicated.

This problem could be solved by drawing the polygon of vectors. However, an alternative method of determining the resultant is to resolve all the forces into their vertical and horizontal components. We can then easily determine the sum of the vertical components and the sum of the vertical components. We then have replaced all the forces by just two components and from these can determine the resultant.

For the 3 kN force we have:

horizontal component = 3.0 cos 60° = 1.5 kN

vertical component = 3.0 sin 60° = 2.6 kN

For the 2.0 kN force we have

horizontal component = 2.0 cos 30° = 1.7 kN

vertical component = 2.0 sin 30° = 1.0 kN

For the 5.0 kN force we have

horizontal component = 5.0 cos 70° = 1.7 kN

vertical component = –5.0 sin 70° = –4.7 kN

The minus sign is because this force is acting downwards and in the opposite direction to the other vertical components which we have taken as being positive. All the horizontal components are in the same direction. Thus

sum of horizontal components = 1.5 + 1.7 + 1.7 = 4.9 kN

sum of vertical components = 2.6 + 1.0 – 4.7 = –1.1 kN

Figure 12.21 shows how we can use the parallelogram rule to find the resultant with these two components. Since the two components are at right angles to each other, the resultant can be calculated using the Pythagoras theorem. Thus:

$$(resultant)^2 = 4.9^2 + 1.1^2$$

Hence the resultant has a magnitude of 5.0 kN. The resultant is at an angle θ downwards from the horizontal given by:

$$\tan \theta = \frac{1.1}{4.9}$$

Thus $\theta = 12.7°$.

4.9 kN

θ

1.1 kN

Figure 12.21 *Example*

Example

An object of weight 30 N rests on an incline which is at 35° to the horizontal. What are the components of the weight acting at right angles to the incline and along the incline?

Figure 12.22(a) describes the situation and Figure 12.22(b) how the triangle rule can be used to determine the components. Thus

component down incline = 30 sin 35° = 17.2 N

component at right angles to incline = 30 cos 35° = 24.6 N

Example

A projectile has an initial velocity of 10 m/s at an angle of 60° to the horizontal. What are the horizontal and vertical components of this velocity?

35°

30 N

(a)

35°

35°

30 N

(b)

Figure 12.22 *Example*

Figure 12.23 *Example*

Figure 12.23 shows how the triangle rule can be used to determine the velocity components. Thus

vertical component = 10 sin 60° = 8.7 m/s

horizontal component = 10 cos 60° = 5.0 m/s

Revision

17 Determine the horizontal and vertical components of the following forces:

(a) 10 N at 40° to the horizontal,
(b) 15 kN at 70° to the horizontal,
(c) 12 N at 20° to the horizontal,
(d) 30 kN at 80° to the horizontal.

18 An object of weight 30 N rests on an incline which is at 20° to the horizontal. What are the components of the weight at right angles to the incline and parallel to the incline?

19 For each of the following systems of forces as a result of considering the components of each force, determine the resultant force:

(a) 2 N in an easterly direction, 3 N at 60° west of north and 2 N due south,
(b) 4 N in a north-easterly direction, 3 N due west and 5 N at 30° south of east,
(c) 2.8 N in a north-easterly direction, 4 N at 60° south of west and 6 N at 30° south of east.

20 An object has a velocity of 40 m/s in a north-easterly direction. What are the components of this velocity in the north and south directions?

21 A rocket starts its motion from the surface of the earth with a velocity of 400 m/s at an angle of 70° to the horizontal. What are the components of the velocity in the vertical and horizontal directions?

22 An object has a displacement of 10 m in a north-easterly direction. What are the components of this displacement in the easterly and northerly directions?

Problems

1 For the vectors **a** of magnitude 2 in a direction due east, **b** of magnitude 3 in a direction due north and **c** of magnitude 4 in a north-westerly direction, determine

(a) **a** + **b**, (b) **b** + **c**, (c) **a** + **c**, (d) **a** − **b**, (e) **b** − **c**, (f) **a** − **c**.

Figure 12.24   *Problem 3*

Figure 12.25   *Problem 4*

Figure 12.26   *Problem 8*

2 Determine the resultants for the following forces acting at the same point on an object:

(a) 2 kN in a northerly direction and 4 kN in a westerly direction,
(b) 10 N in a vertical direction and 20 N in a horizontal direction,
(c) 40 N and 50 N if the angle between the forces is 45°,
(d) 20 N vertically downwards and 10 N in a downward direction at 50° to the vertical,
(e) 10 N and 12 N if the angle between them is 105°,
(f) 20 kN and 12 kN if the angle between them is 120°.

3 An object of weight 30 N is suspended by a string from the ceiling. What horizontal force F must be applied to the object if the string is to become deflected and make an angle of 25° with the vertical (Figure 12.24)?

4 An object of weight 20 N is supported from the ceiling by two cables inclined at 40° and 70° to the ceiling (Figure 12.25). Determine the tensions in the cables.

5 An object of weight 5 N hangs on a vertical string. At what angle to the vertical will the string be when the object is held aside by a force of 3 N in a direction which is 20° above the horizontal?

6 Determine the resultant force acting on an object when two forces of 20 kN and 40 kN are applied to the same point on the object and the angle between the lines of action of the forces is 90°.

7 Three girders in the same plane meet at a point. If there are tensile forces of 70 kN, 80 kN and 90 kN in the girders, what angles will the girders have to be at if there is equilibrium?

8 What is the resultant force acting on the gusset plate shown in Figure 12.26 as a result of the forces shown?

9 Determine the resultant velocity of an object if it is subject to the following velocities:

(a) 4 km/h in a northerly direction and 3 km/h in an easterly direction,
(b) 10 m/s vertically downwards and 4 m/s horizontally,
(c) 20 m/s and 10 m/s if the angle between the velocities is 60°,
(d) 10 km/h in a westerly direction and 20 km/h in a north-easterly direction,
(e) 20 m/s in a southerly direction and 10 m/s in a northerly direction.

10 An object is thrown vertically upwards with a velocity of 10 m/s. If there is a horizontal wind blowing of 5 m/s, what will be the resultant velocity of the object?

11 A pilot sets a course due south with an air speed of 300 km/h. If a 100 km/h wind blows from the south-west, what is the actual velocity of the plane?

12 A cyclist travels 5 km in an easterly direction followed by 7 km in a northerly direction. How far, and in what direction, is the cyclist from the start position?

13 A person walks 6 km in a south-westerly direction and then 4 km in a westerly direction. How far, and in what direction, is the walker from the start position?

14 Determine the resultant force acting on an object if it is acted on by four forces acting in the same plane of 1 N in a westerly direction, 3 N in a south-westerly direction, 6 N in a north-easterly direction and 5 N in a northerly direction.

15 Ship A is moving in a north-easterly direction at 15 km/h. Ship B is moving due west at 8 km/h. To ship A, with what velocity does ship B appear to be moving?

16 Object A moves with a velocity of 4 m/s due north and object B moves with a velocity of 3 m/s due east. What is the velocity of B relative to A?

17 An object of weight 10 N rests on an incline which is at 30° to the horizontal. What are the components of this weight in a direction at right angles to the incline and parallel to the incline?

18 Three forces act in the same plane on the same point on an object. If the forces are 4 N in a direction due north, 7 N in a south-easterly direction and 4 N in a direction 60° south of west, what is the resultant force?

19 A cable exerts a force of 15 kN on a bracket. If the cable is at an angle of 35° to the horizontal, what are the horizontal and vertical components of the force?

20 Forces of 10 N, 12 N and 20 N act in the same plane on an object in the directions west, 30° west of north, and north respectively. Determine the resultant force.

21 Forces of 1 N, 2 N, 3 N, 4 N and 5 N act in the same plane on an object in the directions north, north-east, east, 60° west of south, and due west respectively. Determine the resultant force.

Section E
Graphs

13 Graphs
14 Straight-line motion
15 Linear graphs
16 Polar coordinates
17 Alternating waveforms

The aims of this section are to enable the reader to:

- Plot graphs, choosing suitable scales and clearly marking data points.
- Plot graphs of equations: in particular, linear, exponential and sinusoidal.
- Determine gradients, intercepts and areas from graphs.
- Intepret graphs.
- Recognise the relationship between polar and Cartesian representations of points on a graph; plot polar graphs.
- Reduce non-linear laws to linear form.
- Use and interpret graphs in an engineering context.

This section is about the construction and interpretation of the types of graphs that are commonly encountered in engineering. Particular attention is directed towards linear, exponential and sinusoidal graphs; also, the linearisation of functions in order to obtain linear graphs.

Chapter 13 parallels the development of the mathematics in Section B. This section assumes dexterity with basic algebraic manipulation of equations. Chapter 14 shows the applications of Chapter 13 in a consideration of stright-line motion. Chapter 15 assumes that Chapter 12 and much of the algebra in Section B has ben covered. Chapter 16 assumes a basic knowledge of trigonometry and that the basics of Chapter 13 have been covered. Chapter 17 is an application of graphs in a study of alternating waveforms and their modelling by phasors.

13 Graphs

13.1 Introduction

Graphs are an indispensable element in engineering. They enable trends in experimental data to be more easily seen, and determined, than is possible by just looking at the numbers. Also, equations can be pictorially displayed by means of graphs and the relationships described by the equations more easily comprehended. This chapter is a review of the basic techniques of Cartesian graph drawing and the interpretation of such graphs. Functions which give linear and non-linear graphs are considered.

Chapter 14 looks at the application of graphs to at the engineering context of straight-line motion. Chapter 15 is consideration how relationships can be organised to give linear relationships.

13.2 Cartesian graphs

It is possible to specify the position of a point along a line from some zero point on that line by specifying its distance from the point. In order to specify which side of the zero point the distance is, distances measured to the right of the zero point are given as positive and distances to the left negative (Figure 13.1). Thus, with such a line, a distance specified as +2 is located 2 units to the right of the zero point.

Figure 13.1 *Number line*

If we want to specify the position of a point on a plane then we can use two such number lines at right angles to each other and intersecting at their zero points. The two lines are then called *coordinate axes* and their point of intersection the *origin*. The horizontal axis is called the *x-axis* and the vertical axis the *y-axis* (Figure 13.2). The positive half of the *x*-axis is to the right of the origin and the negative half to the left. The positive half of the *y*-axis is upwards from the origin and the negative half downwards.

Figure 13.2 *Cartesian coordinates*

To specify a point on the plane then we can specify its horizontal displacement and its vertical displacement from the origin, i.e. the intersection of the vertical and horizontal lines through the point with the *x*- and *y*-axes. If these intersections are x_A and y_A for some point A then the point is said to have the *coordinates* of (x_A, y_A). Such coordinates of points are called the *Cartesian coordinates*. The first number in such a pair of numbers is always the *x-coordinate* (or *abscissa*) and the second the *y-coordinate* (or *ordinate*).

Figure 13.3 illustrates this for a number of points. Thus point A is located 2 units in the *x*-direction from the origin and 1 unit in the *y*-direction and so has the coordinates of (2, 1). Point B is located −1 units in the *x*- direction and +2 units in the *y*-direction and so has the coordinates (−1, 2). Point C is located −2 units in the *x*-direction and −1 units in the *y*-direction and so has the coordinates (−2, −1). Point D is located +2 units in the *x*-direction and −2 units in the y direction and so has the coordinates of (2, −2). The sequence of the numbers in the specified coordinate pair is important. Thus (2, 1) and (1, 2) specify two quite different points. The first has an *x-coordinate* of 2 and a *y-coordinate* of 1 while the second has an *x-coordinate* of 1 and a *y-coordinate* of 2.

Figure 13.3 *Examples of coordinates*

Each of the axes in Figure 13.3 has a *scale*. These are the values which indicate how numbers are assigned to distances measured out along the axes. The scales used in Figure 12.3 are *linear scales* in that the number is proportional to the distance from the origin. Chapter 15 shows another type of scale called the *logarithmic scale*.

In selecting the scales to be used for the axes of a graph and in plotting the graph, several points should be taken into account.

1   The scales should be chosen to that the points to be plotted on the graph occupy the full range of the axes used for the graph. There is no point in having a graph with scales from 0 to 100 if all the data points have values between 0 and 10.

2   The scales should not start at zero if starting at zero produces an accumulation of points within a small area of the graph. Thus if all the points have values between 80 and 100 then a scale from 0 to 100 means the points are concentrated in just the end zone of the scale. It is better, in this situation, to have a scale running from 80 to 100.

3   Scales should be chosen so that the location of points between scale marks is made easy. Thus with graph paper subdivided into squares of 10 small squares, it is easy to locate a point of 0.2 if one large square corresponds to 1 but much more difficult if one large square corresponds to, say, 3.

4   The axes should be labelled with the quantities they represent and their units.

5   The data points should be clearly marked, e.g. a large dot or a cross.

6   With experimental data there will be some errors associated with the values being plotted. This will result in some scatter of the points and so the best line should be drawn. This is a smooth line for which there is the same amount of scatter on one side of the line as the other.

Example

The following data is for the voltage across a resistor at different electrical currents. Plot the data as a graph.

Voltage in volts	0	0.5	1.0	1.5	2.0	2.5
Current in amps	0	0.12	0.18	0.30	0.42	0.48

If we take the voltage as being represented by distances measured from the origin along the y-axis then a suitable scale would be one from 0 to 2.5. For the current, represented by distances measured along the x-axis, then a suitable scale would be one from 0 to 0.5. The points to be plotted are (0, 0), (0.1, 0.5), (0.2, 1.0), (0.3, 1.5), (0.4, 2.0) and (0.5, 2.5). Figure 13.4 shows the graph. The best line through the data points is a straight line.

Figure 13.4 *Example*

Figure 13.5 *Example*

Example

The following data is for the length of a metal rod at different temperatures. Plot the data as a graph.

Length in m	20.000	20.004	20.008	20.012	20.016
Temperature in °C	20	40	60	80	100

Starting the scale for the length at 0 would squash all the data into the region of the scale devoted to 20. Thus it is better to start this scale at 20 and have the scale going from 20 to 20.016. For the scale for temperature, we could have the scale starting at 20 and going to 100. Figure 13.5 shows the resulting graph. All the points lie on a straight line.

Revision

1 The following data is for the force applied to stretch a spring and the extension that results. Plot the data as a graph.

Force in N	0	0.2	0.4	0.6	0.8	1.0
Extension in mm	0	5	10	15	20	25

2 The following data is for the distance travelled and the time at which the distance is measured. Plot the data as a graph.

Distance in m	10	15	20	25	30	35
Time in s	10	12	14	16	18	20

13.3 Graphs of equations

Figure 13.6 $y = x + 2$

Consider the equation $y = x + 2$. We can plot a graph of this equation determining the coordinates of a few points, plotting them and then connecting the points with a smooth curve. Thus suppose we take values of x of -2, -1, 0, $+1$, and $+2$. The corresponding y values are:

x	-2	-1	0	$+1$	$+2$
y	0	$+1$	$+2$	$+3$	$+4$

Figure 13.6 shows the points and the resulting graph. It is a straight line.

Consider the equation $y = x^2 - 2$. We can plot this graph by determining the coordinates of a few points, plotting them and then connecting the points with a smooth curve. Thus suppose we take values of x of -2, -1, 0, $+1$, and $+2$. The corresponding y values are:

x	-2	-1	0	$+1$	$+2$
y	2	-1	-2	-1	$+2$

Figure 13.7 $y = 2x^2$

Figure 13.7 shows the points and the resulting graph. In plotting the graph it helps if we determine where the graph line should cut the x-axis, i.e. the values of x when $y = 0$. With $y = 0$ we have $x^2 = 2$ and so $x = \pm \sqrt{2}$ and $+1.4$ and -1.4.

13.3.1 Techniques for graph plotting of equations

The graph of any function can be obtained by simply plotting enough points and drawing a smooth curve through the points. However, in many cases, the amount of labour involved can often be reduced by determining key features.

1 *Is the graph a straight-line graph?*
   A straight-line graph has only y or x to the power 1 and there are no products of x and y or expressions such as trigonometric ratios or exponentials. The equation $y = x + 2$ gives a straight line graph (see Figure 13.6).

2 *Is the graph symmetrical about one of the axes, or both of the axes?*
   If the graph is symmetrical about the y-axis then it contains only even powers of x. If it is symmetrical about the x-axis then it contains only even powers of y. Thus for the graph in Figure 13.7 of $y = x^2 - 2$, we have only even powers of x. Thus the graph is symmetrical about the y-axis.

3 *Is the graph symmetrical about the origin?*
   The graph is symmetrical about the origin if a change in the sign of x causes a change in the sign of y without changing its numerical value. The graph of $y = 1/x$ is symmetrical about the origin because if we change the sign of x we change the sign of y without changing its numerical value. Figure 12.8 shows such a graph.

4 *Where does the graph line cut the axes?*
   The value of where a graph cuts the x-axis is given by putting $y = 0$, the value of where it cuts the y-axis by putting $x = 0$. In the discussion of Figure 13.7 the points where the x-axis was cut were determined.

Figure 13.8 $y = 1/x$

5 *Are there any values of x which make y have an infinite value, or values of y which make x have an infinite value?*
   For $y = 1/x$, when $x = 0$ then $y = 1/0$ which is infinite. Figure 13.8 shows how the graph goes off to infinity as x approaches zero. When $y = 0$ then $x = 1/0$ which is infinite and so the graph goes off to infinity as y approaches zero.

6 *Does the graph have any maximum or minimum values?*
   To determine such points we need to use calculus and this is discussed in section D.

Example

Plot the graph of $y = 2x^2$.

Figure 13.9 *Example*

We have x to the power 2. Thus the graph will not be a straight-line graph. The equation contains only even powers of x and is symmetrical about the y-axis. Thus we need only find values of y for positive values of x since we know that the graph for negative values of x will be just a mirror image of that for the positive values. When $x = 0$ then $y = 0$. If we now take some values of x then we can find the corresponding values of y.

x	0	1	2	3
y	0	2	8	18

Figure 13.9 shows the resulting graph. Graphs of equations of the form $y = ax^2 + bx + c$, where a, b and c are some constant (a must not be zero but b and c can), give cup-shaped graphs called *parabolas*.

Revision

3 Plot graphs of the functions: (a) $y = 5x$, (b) $y = 2x + 3$, (c) $y = x^2$, and (d) $y = x^2 + 4$.

13.4 Graphs of trigonometric functions

Table 13.1 gives values of $y = \sin x$, $y = \cos x$ and $y = \tan x$ as the angle x increase from $0°$ to $360°$ or 0 to 2π radians and Figures 13.10, 13.11 and 13.12 shows the graphs of the ratios. Note that:

1 A particular value of a trigonometric ratio can be given by more than one angle. For example, a sine of 0.71 is given by angles of $45°$ and $135°$.

2 If we consider angles greater than $360°$ the graphs repeat themselves and hence are called *periodic functions*. The sine and cosine curves are continuous. The tan x graph goes towards + infinity and returns from – infinity at $90°$ and $270°$.

2 We have $\sin x$ and $\cos x$ as the same basic graph, one being just the other displaced by $90°$ ($\pi/2$ rad).

Figure 13.10 $y = \sin x$

Table 13.1 *Values of trigonometric ratios*

Angle θ		$\sin \theta$	$\cos \theta$	$\tan \theta$
degrees	radians			
0	0	0	1	0
30	$\pi/6$	0.5	0.87	0.58
45	$\pi/4$	0.71	0.71	1
60	$\pi/3$	0.87	0.50	1.73
90	$\pi/2$	1	0	$\pm$ infinity
120	$2\pi/3$	0.87	-0.50	-1.73
135	$3\pi/4$	0.71	-0.71	-1
150	$5\pi/6$	0.50	-0.87	-0.58
180	π	0	-1	0
210	$7\pi/6$	-0.5	-0.87	0.58
225	$5\pi/4$	-0.71	-0.71	1
240	$4\pi/3$	-0.87	-0.50	1.73
270	$3\pi/2$	-1	0	$\pm$ infinity
300	$5\pi/3$	-0.87	0.5	-1.73
315	$7\pi/4$	-0.71	0.71	-1
330	$11\pi/6$	-0.50	0.87	-0.58
360	2π	0	1	0

Figure 13.11 $y = \cos x$

Figure 13.12 $y = \tan x$

13.4.1 The amplitude term

Consider the graph of $y = 2 \sin x$. This just multiplies the values of $\sin x$ by 2 and thus determines the *amplitude* of the sine wave. Thus we have:

x	0	$\pi/2$	π	$3\pi/2$	2π	$5\pi/2$	3π	$7\pi/2$	etc.
$y = \sin x$	0	+1	0	−1	0	+1	0	−1	
$y = 2 \sin x$	0	+2	0	−2	0	+2	0	−2	

Figure 13.13 shows the graph; compare this with the graph in Figure 13.10 for $y = \sin x$.

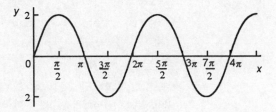

Figure 13.13 $y = 2 \sin x$

13.4.2 The frequency term

Consider the graph of $y = \sin 2x$. For $y = \sin 2x$, the values of x that will give $y = 0$ are the values for which $2x = 0$, π, 2π, etc. (i.e. 0°, 180°, 360°, etc.). We have $\sin 2x = +1$ when $2x = \pi/2$, $5\pi/2$, etc. (i.e. 90°, 450°, etc.). We have $\sin 2x = -1$ when $2x = 3\pi/2$, $7\pi/2$, etc. (i.e. 270°, 630°, etc.). Thus we have:

x	0	$\pi/4$	$\pi/2$	$3\pi/4$	π	$5\pi/4$	$3\pi/2$	etc.
$2x$	0	$\pi/2$	π	$3\pi/2$	2π	$5\pi/2$	3π	
$y = \sin 2x$	0	+1	0	−1	0	+1	0	

Figure 13.14 shows the graph; compare this with the graph in Figure 13.10 for $y = \sin x$.

Figure 13.14 $y = \sin 2x$

13.4.3 The sideways displacement term

Consider the graph of $y = \sin(x + \pi/4)$. The values of x that will give $\sin(x + \pi/4) = 0$ and hence $y = 0$ are the values for which $x + \pi/4 = 0$, π, 2π, 3π, etc. (i.e. 0°, 180°, 360°, 540°, etc.), i.e. $x = -\pi/4$, $3\pi/4$, $7\pi/4$, $11\pi/4$, etc. The values of x that will give $\sin(x + \pi/4) = +1$ and hence $y = +1$ are $x + \pi/4 = \pi/2$, $5\pi/2$, etc. (i.e. 90°, 450°, etc.), i.e. $x = \pi/4$, $9\pi/4$, etc. The values of x that will give $\sin(x + \pi/4) = -1$ and hence $y = -1$ are when $x + \pi/4 = 3\pi/2$, $7\pi/2$, etc. (i.e. 270°, 630°, etc.), i.e. $x = 5\pi/4$, $13\pi/4$, etc. We thus have:

x	0	$\pi/4$	$\pi/2$	$3\pi/4$	π	$5\pi/4$	$3\pi/2$	etc.
$x + \pi/4$		$\pi/4$	$\pi/2$	$3\pi/4$	π	$5\pi/4$	$3\pi/2$	$7\pi/4$
$\sin(x + \pi/4)$		$+0.7$	$+1$	0.7	0	-0.7	-1	-0.7

Figure 13.15(a) shows the graph. The $\sin x$ graph has been displaced sideways to the left by $\pi/4$.

Figure 13.15(b) shows the graph for the equation $y = \sin(x - \pi/4)$. The following table shows the data which can be derived in a similar way to that above:

x	0	$\pi/4$	$\pi/2$	$3\pi/4$	π	$5\pi/4$	$3\pi/2$	etc.
$x - \pi/4$	$-\pi/4$	0	$\pi/4$	$\pi/2$	$3\pi/4$	π	$5\pi/4$	
$\sin(x - \pi/4)$	-0.7	0	$+0.7$	$+1$	$+0.7$	0	-0.7	

The $\sin x$ graph has been displaced to the right by $\pi/4$.

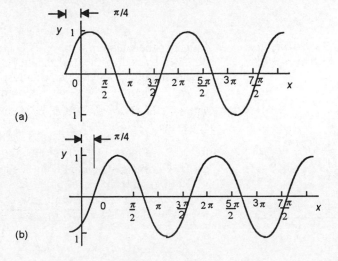

Figure 13.15 *(a)* $y = \sin(x + \pi/4)$, *(b)* $y = \sin(x - \pi/4)$

Example

Plot the graph of $y = 2 \sin 2x$.

For $y = 2 \sin 2x$, the values of x that will give $y = 0$ are the values for which $2x = 0$, π, 2π, etc. (i.e. 0°, 180°, 360°, etc.). Since the sine of an angle oscillates in value between +1 and −1 then the value of y, in this case, will oscillate between +2 and −2. We have $\sin 2x = +1$ when $2x = \pi/2$, $5\pi/2$, etc. (i.e. 90°, 450°, etc.). We have $\sin 2x = -1$ when $2x = 3\pi/2$, $7\pi/2$, etc. (i.e. 270°, 630°, etc.). Thus we have:

x	0	$\pi/4$	$\pi/2$	$3\pi/4$	π	$5\pi/4$	$3\pi/2$	etc.
$2x$	0	$\pi/2$	π	$3\pi/2$	2π	$5\pi/2$	3π	
$y = 2 \sin 2x$	0	+2	0	−2	0	+2	0	

Figure 13.16 *Example*

Figure 13.16 shows the graph.

Revision

4 Plot graphs of the functions: (a) $y = 2 \sin x$, (b) $y = 2 + \sin x$, (c) $y = \cos x$, (d) $y = \cos 2x$, (e) $y = \cos \tfrac{1}{2}x$.

13.5 Graphs of exponentials

See chapter 8 for a discussion and graphs of exponentials.

Revision

5 Plot graphs of the functions: (a) $y = e^x$, (b) $y = 2\,e^x$, (c) $y = e^{2x}$, (d) $y = e^{-x}$, (e) $y = e^{-2x}$, (f) $y = 1 - e^{-x}$.

13.6 Adding and subtracting functions

Consider the equation $y = x + 2$. The value of y at a particular value of x is obtained by adding the 2 to the value of x. We can thus consider the graph of this equation to be formed as a result of adding the graph of $y = x$ to that of $y = 2$. Figure 13.17 shows such an addition. Thus, for example, we can add CD and BD to give AD and so the point A on the line for $y = x + 2$.

Consider the equation $y = x^2 - 2$. We can obtain this graph by adding the graphs for $y = x^2$ and $y = -2$, or by subtracting the graph for $y = 2$ from that for $y = x^2$. Figure 13.18 illustrates this subtraction. We have BC minus AC giving CD and so the point D on the line for $y = x^2 - 2$.

The graph for the equation $y = 2 + \sin x$ (see Revision problem 4(b)) can be obtained by adding the graphs for $y = 2$ and $y = \sin x$, the graph for $y = \sin x + \sin 2x$ by adding the graphs for $y = \sin x$ and $y = \sin 2x$.

Figure 13.17 $y = x + 2$

Example

The distance s travelled by an object when starting at time $t = 0$ with an initial velocity u and having a uniform acceleration a is given by the

Figure 13.18 $y = x^2 + 2$

equation $s = ut + \frac{1}{2}at^2$. Plot a graph showing how the distance varies with time for the first 4 s if $u = 10$ m/s and $a = 4$ m/s^2.

We can consider the required graph to be the sum of the graphs of $s = ut$ and $s = \frac{1}{2}at^2$. Thus, with the values given, we have:

t in s	0	1	2	3	4
$s = 10t$ in m	0	10	20	30	40

t in s	0	1	2	3	4
$s = 2t^2$ in m	0	2	8	18	32

Figure 13.19 shows the two graphs and their sum.

Figure 13.19 $s = 10t + 2t^2$

Revision

6 Plot the graphs of the following equations:

(a) $y = 2x + 1$, (b) $y = x^2 + 2x$, (c) $y = \sin x + \sin 2x$, (d) $y = 2 + e^x$.

13.7 Solving equations graphically

Consider the following pair of simultaneous equations:

$x + y = 3$ and $2x - y = 3$

Solving the equations means finding the values of x and y which simultaneously fit both equations. In graphical terms, this means finding the values of x and y at the points of intersections of their graphs. These are the only points at which the same values of x and y occur for the two graphs. Figure 13.20 shows the graphs of the above equations. There is just one point of intersection, this being when $x = 2$ and $y = 1$. These are then the solutions.

Figure 13.20 *Graphs of simultaneous equations*

Figure 13.21 *Graphs of simultaneous equations*

Figure 13.22 *Solving x + 3 = 0*

We can check that they are by substituting them in the equations. Thus $2 + 1 = 3$ and $2 \times 2 - 1 = 3$.

As a further illustration of this technique of solving simultaneous equations, consider the solution of

$$x - y + 2 = 0 \text{ and } x^2 - y = 0$$

Figure 13.21 shows the graphs of the two equations. There are two points of intersection, these being at $x = 2$ and $y = 4$ and at $x = -1$ and $y = 1$. We can check that these are the solutions by substituting them in the equations. Thus $2 - 4 + 2 = 0$ and $2^2 - 4 = 0$, $-1 - 1 + 2 = 0$ and $(-1)^2 - 1 = 0$.

Revision

7 Solve graphically the following pairs of simultaneous equations:

(a) $2x + y = 4$ and $x + y = 3$, (b) $x + y = 1$ and $2x - y = 5$, (c) $y - x^2 = 0$ and $2y + 3x = 3$, (d) $y - 4x^2 = 0$ and $y - x = 1$.

13.7.1 Solving equations

Suppose we need to solve the equation $x + 3 = 0$. This is fairly straightforward and the answer is easily obtained by algebraic manipulation of the equation (see Chapter 2) as $x = -3$. We could, however, have obtained the result graphically. What we are looking for is where the graph of $y = x + 3$ cuts the x-axis. Then we have the value of $y = 0$. Figure 13.22 shows the graph. What we are essentially doing is finding where the graph lines for $y = x + 3$ and $y = 0$ intersect, i.e. solving the simultaneous equations.

This technique can be used for quadratic, or more complex, equations. Consider the graphical solution of the equation $x^2 + 6x + 5 = 0$. We can do this by plotting the graph of $y = x^2 + 6x + 5$. Taking coordinate values we obtain:

x	−6	−5	−4	−3	−2	−1	0
$y = x^2 + 6x + 5$	+5	0	−3	−4	−3	0	+5

then the graph has the form shown in Figure 13.23. The intercepts with the x-axis occur at −1 and −5. Thus the solutions are $x = -1$ and $x = -5$. What we are essentially doing is finding where the graph lines for $y = x^2 + 6x + 5$ and $y = 0$ intersect, i.e. solving the simultaneous equations.

We could have tackled the solution of this equation in a different way. With the above solution what we have done with $x^2 + 6x + 5 = 0$ is let $y = 0$ and so $x^2 + 6x + 5 = y$ and so have this pair of simultaneous equations to solve. We could, however, have let $y = 6x + 5$ and so $x^2 + y = 0$. We now have a different pair of simultaneous equations to solve. Alternatively we could have let $y = 6x$ and so $x^2 + y + 5 = 0$ and then have to solve this pair of simultaneous equations. The results of such simultaneous equations, i.e.

Figure 13.23 *Solving a quadratic equation*

Figure 13.24 *Example*

Figure 13.27 *Roots*

their intercepts, and hence the solutions of the equation $x^2 + 6x + 5 = 0$, are at $x = -1$ and $x = -5$.

Example

Solve the equation $x^3 - 4x^2 - 2x + 3 = 0$.

For the equation $y = x^3 - 4x^2 - 2x + 3$ then we can obtain the following co-ordinate points:

x	-2	-1	0	1	2	3	4	5
y	-17	0	3	-2	-9	-12	-5	18

Figure 13.24 shows the resulting graph. The intercepts with the x-axis occur at about -1, $+0.75$ and $+4.2$. Thus the solutions are $x = -1$, $x = 0.75$ and $x = 4.2$.

Revision

8   Solve graphically the following equations:

(a) $x - 5 = 0$, (b) $x^2 + 3x + 2 = 0$, (c) $x^2 - 4x + 2 = 0$,
(d) $x^3 - x^2 - 5x = 0$, (e) $x^3 - 1 = 0$, (f) $x^3 - 2x^2 - 5x + 6 = 0$.

13.7.2 Roots of quadratic equations

The quadratic equation has the general form $ax^2 + bx + c = 0$ (see chapter 3), where a, b and c are constants. We can solve such equations graphically by plotting $ax^2 + bx + c = y$ and determining the intercepts of the graph with the $y = 0$ axis. Such graphs are parabolic in shape.

1   $y = x^2$
    Figure 13.25(a) shows the graph of $y = x^2$, i.e. a is 1 but b and c are zero. This graph touches the x-axis at just one point, namely $x = 0$. Thus there is just one value for the roots of this equation.

2   $y = 2x^2$
    Figure 13.26(b) shows the graph of $y = 2x^2$, i.e. a is 2 but b and c are zero. This graph still touches the x-axis at just one point, namely $x = 0$. However, changing the value of a has changed the steepness of the curve.

3   $y = x^2 - 2$
    Figure 13.27(c) shows the graph of $y = x^2 - 2$, i.e. a is 1 and c is -2 but b is 0. Now there are two roots, namely $x = +\sqrt{2}$ and $x = -\sqrt{2}$.

4   $y = x^2 + 2$
    Figure 13.27(d) shows the graph of $y = x^2 + 2$, i.e. $a = 1$ and c is $+2$ but b is 0. There are now no real roots. Changing the value of c shifts the

curve up or down the y-axis. If the curve is shifted above the x-axis then there are no real roots.

5   $y = x^2 - 4x + 3$

Figure 13.27(e) shows the graph of the equation $y = x^2 - 4x + 3$, i.e. a is 1, b is -4 and c is $+3$. The effect of the b term is to shift the graph down the y-axis and that of the c term to shift the graph sideways along the x-axis. Since $x^2 - 4x + 3$ has the factors $(x - 1)(x - 3)$ the intercepts with the x-axis are at $x = +1$ and $x = +3$. There are two roots.

6   $y = x^2 - 4x + 4$

Figure 13.27(f) shows the graph of the equation $y = x^2 - 4x + 4$, i.e. a is 1, b is -4 and c is $+4$. This graph has only one value for its roots. This is because the quadratic has the factors $(x - 2)(x - 2)$.

The graph of $y = ax^2 + bx + c$ cuts the x-axis at two points, i.e. has two real roots, if $b^2 > 4ac$ (see Section 3.4.1). If $b^2 = 4ac$ then the graph just touches the x-axis at just one point. If $b^2 < 4ac$ then the curve will never cut the x-axis and there are no real roots.

Example

At what points, if any, will the graph of $y = x^2 + 5x + 6$ cut the x-axis?

The quadratic equation $x^2 + 5x + 6 = 0$ can be factorised to give $(x + 3)(x + 2) = 0$ and so there will be intercepts at $x = -2$ and $x = -3$.

Example

How will a graph of $y = x^2 + 5$ differ from one of $y = x^2$?

The $y = x^2$ is the basic parabolic shape (as in Figure 13.9 or Figure 13.27(a)). Including a c term of $+5$ shifts the entire graph up the y-axis by 5 units.

Revision

9   What will be the intercepts, if any, with the x-axis of graphs of the following equations:

(a) $y = 3x^2$, (b) $y = x^2 - 4$, (c) $y = x^2 + 2x + 1$, (d) $y = x^2 + 3x + 2$,

(e) $y = x^2 + 3x + 4$.

10   How will the graph of the equation $y = x^2 - 2x + 3$ differ from that of $y = x^2 - 2x + 1$?

13.8 Gradients   Consider the line in Figure 13.28. From point A to point B on the line there is a vertical rise of BC in a horizontal distance of AC. We say that the line

Figure 13.28 *Gradient*

Figure 13.29 *Gradients:*
(a) positive, (b) negative

Figure 13.30 *Straight-line graph*

between the points A and B has a *gradient* (slope) of BC/AC. Since BC is $y_2 - y_1$ and AC is $x_2 - x_1$ then

$$\text{gradient} = \frac{y_2 - y_1}{x_2 - x_1}$$

With a straight line it does not matter what points we consider along the line: the gradient is the same at all positions and so a constant. Note that a horizontal line, i.e. one for which there is no change in y when x increases, has a zero gradient. A line has a positive gradient if when there is an increase in x there is an increase in y. It has a negative gradient if when there is an increase in x there is a decrease in y. Figure 13.29 illustrates such lines.

For a straight-line graph, as in Figure 13.28, if we let the gradient be m:

$$m = \frac{y_2 - y_1}{x_2 - x_1}$$

Hence:

$$y_2 - y_1 = m(x_2 - x_1)$$

We can consider any pair of points on a straight line. Thus suppose we consider A to be where the graph of the straight line intercepts with the y axis (Figure 13.30) and has the value c, i.e. $y_1 = c$ when $x_1 = 0$. If we take the point B to have the coordinates (x, y), then we can write the above equation as:

$$y - c = m(x - 0)$$

Thus, when rearranged, we have the equation:

$$y = mx + c$$

This is the *general equation for a straight line*. All equations of this form give straight-line graphs with a gradient of m and an intercept with the y axis of c.

The gradient measures the rate of change of y compared with x. This is an important concept which is returned to in section D and the discussion of differentiation.

Example

Determine the gradient of the line joining the pair of points (5, 3) and (2, 4).

Using the equation given above for the gradient

$$\text{gradient} = \frac{y_2 - y_1}{x_2 - x_1} = \frac{4 - 3}{2 - 5} = -\frac{1}{3}$$

The minus sign indicates that the value of y is decreasing as x increases.

Example

Determine the gradient of the line given by the equation $y = 2x - 3$.

We could draw the graph of this equation and determine, by taking points on the straight line, the gradient. We could however just consider the coordinates of two points on such a line. Thus, suppose we determine the gradient between the points on the line for which $x = 2$ and $x = 3$. These points have $y = 2 \times 2 - 3 = 1$ and $y = 2 \times 3 - 3 = 3$. The gradient is thus

$$\text{gradient} = \frac{y_2 - y_1}{x_2 - x_1} = \frac{3 - 1}{3 - 2} = 2$$

Note that this is the coefficient of x in the original equation. For an equation of the form $y = mx + c$ the gradient is always m.

Example

For the equation $y = 3x - 4$, what is the gradient of the graph and the value of its intercept with the y-axis?

This equation is of the form $y = mx + c$ and thus the gradient is 3 and the intercept with the y-axis is -4.

Example

Determine the equation of the straight-line graph which has a gradient of 5 and an intercept with the y-axis of $+2$.

The equation is of the form $y = mx + c$ and thus is $y = 5x + 2$.

Revision

11 Determine the gradients of the lines joining the following pairs of points:

(a) (3, 1) and (4, 2), (b) (5, 1) and (10, 2), (c) (–3, 2) and (1, 3),
(d) (5, 3) and (0, 0), (e) (0, 2) and (1, 0).

12 Determine the gradients of the lines given by the following equations and their intercepts with the y-axis:

(a) $y = 5x + 3$, (b) $y = x - 3$, (c) $2y = 4x + 1$, (d) $y = 2 - 2x$.

13 Determine the equations of the straight-line graphs which have:

(a) a gradient of 2 and an intercept with the y-axis of 3,

(b) a gradient of 4 and an intercept with the y-axis of -3,
(c) a gradient of -2 and an intercept with the y-axis of 1,
(d) a gradient of 5 and an intercept with the y-axis of 0.

14 The velocity v of an accelerating object after a time t is given by the equation $v = 4 + 3t$. What is the gradient and the intercept with the y-axis for the straight-line graph of v plotted against t?

13.8.1 Gradient of a curve

Figure 13.31 *Gradient of a curve*

The gradient of a straight line is the same for all points on the line. However, the gradient of a curve is not a constant and varies from point to point along the curve. The gradient of a particular point on a curve is defined as being the gradient of the tangent to the curve at that point. Figure 13.31 illustrates this, the gradient at point P being BC/AC.

Example

Plot the graph of the equation $y = x^2$ for values of x from 0 to 3 and estimate the value of the gradient at $x = 2$.

Figure 13.32 shows the graph. A ruler placed against the curve at the point $x = 2$ can be used to draw the tangent to the curve at that point. This leads to an estimate of the gradient as $8/2 = 4$.

Revision

Figure 13.32 *Example*

15 Plot graphs of the following equations between the values of x indicated and estimate the gradients of the graphs at the indicated points:

(a) $y = x^2 + 1$, x between 0 and 3, gradient at $x = 2$,
(b) $y = x^2 + x$, x between 0 and 3, gradient at $x = 2$,
(c) $y = x^3 + 1$, x between 0 and 3, gradient at $x = 2$.

13.9 Areas under graphs

It is often necessary to estimate the area enclosed between a graph line and the axis of a graph. For example, in estimating the work done when a spring is stretched through a distance x by a force F, the area under the graph line between extensions 0 and x is the work done. With a straight-line graph, the area can be easily calculated as, often, the sum of the areas of simple shapes such as rectangles and triangles. Where the graph is non-linear, methods that can be used include:

1 *Counting squares*
   This involves counting the squares under the graph and then multiplying the number of squares by the value corresponding to their area. See Section 2.4.1 for more details.

Figure 13.33 *Example*

Figure 13.34 *Example*

Figure 13.35 *Example*

2 *The mid-ordinate rule*

The area under the curve is divided into a number of equal width vertical strips and the area is then the sum of the mid-ordinate values of the strips multiplied by their width. See Section 2.4.2 for more details.

3 *The trapezoidal rule*

The area under the curve is divided into a number of equal width vertical strips and the area is then the product of the width of the strips and half the sum of the first and last ordinates plus the sum of the remaining ordinates. See Section 2.4.3 for more details.

Example

Estimate the area under the graph of $y = x^2$ between $x = 0$ and $x = 2$.

Figure 13.33 shows the graph. An estimate of the area under the graph between $x = 0$ and $x = 2$, the shaded area in the graph, is about five graph squares. The square has a vertical height of 1 and a horizontal length of 0.5 and so an area of 0.5 square units. Thus the area is about 2.5 square units.

Example

Estimate the area under the graph of $F = 4x$ between $x = 0$ and $x = 2$.

Figure 13.34 shows the graph. It is a straight-line graph through the origin. Thus the area is that of a triangle and so is half the base × height, i.e. $\frac{1}{2} \times 2 \times 8 = 8$ square units.

Example

Estimate, by the use of the mid-ordinate rule, the area under the graph of $y = x^2 + 2$ between $x = 0$ and $x = 2$.

Figure 13.35 shows the graph with the area divided into just two strips. The mid-ordinate for the first strip occurs at $x = 0.5$. This ordinate has a height, given by the equation $y = x^2 + 2$, of $0.5^2 + 2 = 2.25$. Thus, since the strip has a width of 1, the area of the first strip is $2.25 \times 1 = 2.25$ square units. For the second strip, the mid-ordinate occurs at $x = 1.5$. The height of this ordinate is thus $1.5^2 + 2 = 4.25$. Thus, since the width of the strip is 1, the area is $4.25 \times 1 = 4.25$ square units. The estimate of the total area is thus $2.25 + 4.25 = 6.50$ square units.

We could improve the accuracy of the estimate by using more strips. Thus with strips of width 0.5, and hence four strips, we have:

	Strip 1	Strip 2	Strip 3	Strip 4
Mid-ordinate $x =$	0.25	0.75	1.25	1.75
Height of ordinate	2.06	2.56	3.56	5.06
Area of strip	1.03	1.28	1.78	2.53

The total area is thus $1.03 + 1.28 + 1.78 + 2.53 = 6.62$ square units.

Increasing the number of strips gives yet more accuracy. With a very large number of strips the area is 6.67 square units.

Revision

16 Estimate by the areas under the following graphs:

(a) $y = 2x$ between $x = 0$ and $x = 2$,
(b) $y = x + 2$ between $x = 0$ and $x = 1$,
(c) $y = x^2$ between $x = 0$ and $x = 1$,
(d) $y = 2x^2 + 3$ between $x = 1$ and $x = 2$.

17 Estimate the areas under the graphs of the following equations:

(a) $y = 2x^2$ between $x = 0$ and $x = 2$,
(b) $y = x^2 + 3$ between $x = 0$ and $x = 2$,
(c) $y = x^3$ between $x = 1$ and $x = 3$.

18 Estimate the area under the graph between $x = 0$ and $x = 6$ for the graph which is plotted from the following data:

y	10	18	22	18	10	10	14
x	0	1	2	3	4	5	6

19 Estimate the area under the graph between $t = 0$ and $t = 60$ s for the graph plotted from the velocity–time data given below (note that the area gives the distance travelled in 60 s):

v in m/s	0	14	20	24	27	28	29
t in s	0	10	20	30	40	50	60

Problems

1 Plot graphs of the following equations:

(a) $y = 3x$, (b) $y = x - 1$, (c) $y = x^2 + 1$, (d) $y = x^2 - 1$, (e) $y = 2 \cos 2x$,

(f) $y = \cos x + 1$, (g) $y = 2 \sin \frac{1}{2}x$, (h) $y = 4 + e^{-x}$.

2 Solve, by graphical means, the following equations:

(a) $x + 6 = 0$, (b) $x^2 - 3x + 2 = 0$, (c) $x^2 + 4x + 3 = 0$,

(d) $2x^2 + 3x - 3 = 0$, (e) $x^3 + 2x^2 - 3x = 0$, (f) $x^3 - 3x + 1 = 0$,

(g) $2x^3 - 5x^2 - 10x + 3 = 0$, (h) $3 e^{2x} + 20 = 0$, (i) $\sin \frac{1}{2}x - e^{-x} = 0$.

3   Solve, by graphical means, the following simultaneous equations:

(a) $y + 2x = 3$ and $2y - x = 1$, (b) $y = x^2$ and $y - 2x = 3$,

(c) $y + 3x = 5$ and $2y - 3x = 1$, (d) $y = 3x^2$ and $y - 8 = 0$,

(e) $y = x^2$ and $y = x + 5$, (f) $y = x^2$ and $y = 2x + 1$.

4   What will be the intercepts, if any, with the x-axis of graphs of the following equations:

(a) $y = 5x^2$, (b) $y = x^2 - 9$, (c) $y = x^2 + 3x - 4$, (d) $y = x^2 + 5x + 2$,

(e) $y = x^2 + 4x + 4$, (f) $y = x^2 + 2$, (g) $y = x^2 + x - 6$.

5   Plot graphs, using the first few positive-value points, of the following relationships:

(a) The power P dissipated is a function of the current I passing through a resistor and is given by $P = 2I^2$.
(b) The time T taken for one oscillation of a pendulum is related to the length L of the pendulum by $T = 2\sqrt{L}$.
(c) The velocity v of a moving object is related to the time t for which it is accelerated by $v = 2 + 3t$.
(d) The force F required to stretch a spring to an extension of x is given by $F = 10x$.
(e) The resistance R of a resistor depends on the temperature θ and is given by $R = 10 + 0.01\theta$.
(f) The voltage v across a capacitor varies with time t and is given by $v = 10\,e^{-0.2t}$.

6   Determine the equations of the straight-line graphs which have:

(a) a gradient of 1 and an intercept with the y-axis of 2,
(b) a gradient of 5 and an intercept with the y-axis of -1,
(c) a gradient of -3 and an intercept with the y-axis of 2,
(d) a gradient of -5 and an intercept with the y-axis of -2,
(e) a gradient of 0 and an intercept with the y-axis of 3,
(f) a gradient of 2 and an intercept with the y-axis of -5.

7   Plot graphs of the following equations between the values of x indicated and estimate the gradients of the graphs at the indicated points:

(a) $y = x^2 + 3x$, x between 0 and 3, gradient at $x = 2$,
(b) $y = x^2 - 2x$, x between -1 and $+1$, gradient at $x = 0$,
(c) $y = x^3$, x between 0 and 3, gradient at $x = 1$,
(d) $y = 2x^2 + 1$, x between 0 and 3, gradient at $x = 2$,
(e) $y = x^2 + 2x$, x between 0 and 3, gradient at $x = 2$.

8   Estimate the areas under the graphs of the following equations by using the mid-ordinate or trapezoidal rule:

(a) $y = x^3 + 3$ between $x = 0$ and $x = 2$, with 4 strips,
(b) $y = \sin x$ between $x = 0$ and $x = \pi/2$, with 5 strips,
(c) $y = 3 e^{-x}$ between $x = 0$ and $x = 4$, with 4 strips,
(d) $y = x^3$ between $x = 0$ and $x = 2$, you select the number of strips,
(e) $y = x^2 + 1$ between $x = 0$ and $x = 3$, you select the number of strips,
(f) $y = 2x + 1$ between $x = 1$ and $x = 5$, you select the number of strips.

9   The following are points used to plot a graph:

y	1.00	1.20	1.36	1.50	1.68	1.78	1.86	1.90
x	0	2	4	6	8	10	12	14

Use the mid-ordinate or trapezoidal rule to estimate the area under the graph between $x = 0$ and $x = 14$.

10 The following are points used to plot a graph:

y	1	2	5	10	17	26
x	0	1	2	3	4	5

Use the mid-ordinate or trapezoidal rule to estimate the area under the graph between $x = 0$ and $x = 5$.

11 Use the mid-ordinate or trapezoidal rule to estimate the area under the graph of $y = 20 + x - x^2$ between $x = 0$ and $x = 5$. Use five strips.

12 The following are points used to plot a graph:

y	0	1.43	2.06	2.44	2.69	2.84	2.94	3.00
x	0	1	2	3	4	5	6	7

Use the mid-ordinate or trapezoidal rule to estimate the area under the graph between $x = 0$ and $x = 7$. Use seven strips.

14 Straight-line motion

14.1 Introduction

This chapter illustrates how graphical techniques involving the gradients of graphs and the areas under graphs can be used to solve problems involving motion in a straight line.

14.2 Distance–time graphs

Figure 14.1 *Distance-time graph*

If the distance moved by an object in a straight line, from some reference point on the line, is measured for different times then a distance–time graph can be plotted. Since velocity is the rate at which distance along a straight line changes with time then for the distance-time graph shown in Figure 14.1, where the distance changes from s_1 to s_2 when the time changes from t_1 to t_2, the velocity is

$$\text{velocity} = \frac{s_2 - s_1}{t_2 - t_1}$$

But this is just the gradient of the graph. Thus:

velocity = gradient of distance–time graph

In Figure 14.1 the graph is a straight-line graph. The distance changes by equal amounts in equal intervals of time. There is thus a uniform velocity.

Consider the situation when the graph is not a straight line, as in Figure 14.2. Equal distances are not now covered in equal intervals of time and the velocity is no longer uniform. $(s_2 - s_1)$ is the distance travelled in a time of $(t_2 - t_1)$ and thus the average velocity over that time is:

$$\text{average velocity} = \frac{s_2 - s_1}{t_2 - t_1}$$

Figure 14.2 *Distance-time graph*

The smaller we make the times between A and B then the more the average is taken over a smaller time interval and so the more it approximates to the instantaneous velocity. An infinitesimal small time interval means we have the tangent to the curve. Thus if we want the velocity at an instant of time, say A, then we have to determine the gradient of the tangent to the graph at A. Thus:

instantaneous velocity = gradient of tangent to the distance–time graph at that instant

Example

A car has a constant velocity of 4 m/s for the time from 0 to 3 s and then zero velocity from 3 s to 5 s. Sketch the distance–time graph.

Figure 14.3 *Example*

From 0 to 3 s the gradient of the distance-time graph is 4 m/s. From 3 s to 5 s the velocity is zero and so the gradient is zero. The graph is thus as shown in Figure 14.3.

Example

Figure 14.4 shows an example of a distance–time graph. Describe how the velocity is changing in the motion from A to F.

Figure 14.4 *Example*

From the start position at A to B, the distance increases at a constant rate with time. The gradient is constant over that time and so the velocity is constant and equal to $(4 - 0)/(2 - 0) = 2$ m/s.

From B to C there is no change in the distance from the reference point and so the velocity is zero, i.e. the object has stopped moving. The gradient of the line between B and C is zero.

From C to E the distance increases with time but in a non-uniform manner. The gradient changes with time. Thus the velocity is not constant during that time.

The velocity at an instant of time is the rate at which the distance is changing at that time and so is the gradient of the graph at that time. Thus to determine the velocity at point D we draw a tangent to the graph curve at that instant. So at point D my estimate of the velocity is about $(6.0 - 3.2)/(5 - 3) = 1.4$ m/s.

At point E the distance–time graph shows a maximum. The gradient changes from being positive prior to E to negative after E. At point E the gradient is momentarily zero. Thus the velocity changes from being positive prior to E to zero at E and then negative after E. At E the velocity is zero.

From E to F the gradient is negative and so the velocity is negative. A negative velocity means that the object is going in the opposite direction and so is moving back to its starting point, i.e. the distance is becoming smaller rather than increasing. In this case, the object is back at its starting point after a time of 8 s.

Revision

1 For the distance–time graph shown in Figure 14.5, state whether at points A, B, C and D the object responsible for it is at rest, moving with a constant velocity or moving with a changing velocity.

Figure 14.5 *Revision problem 1*

2 For the following distance–time data, plot the distance–time graph and hence estimate the velocities at times of (a) 2 s, (b) 4 s, (c) 6 s and (d) the average velocity over the time 0 to 6 s.

distance in m	0	2	4	6	6	6	7	9
time in s	0	1	2	3	4	5	6	7

3 A stone falls off the edge of a cliff. The following data shows how the distance the stone has fallen changes with time. Plot the distance–time graph and hence determine the velocities of the stone at (a) 2 s, (b) 3 s, (c) 4 s.

distance in m	0	4.9	19.6	44.1	78.4	122.5
time in s	0	1	2	3	4	5

4 A simple pendulum oscillates back and forth with the pendulum bob having a displacement x from the vertical position which varies with time t and is described by the equation $x = \sin 20t$. Plot the distance–time graph for the pendulum bob for times between $t = 0$ and $t = 0.5$ s. Hence determine the velocities of the bob at times t of (a) 0.05 s, (b) 0.1 s, (c) 0.2 s, (d) 0.3 s.

5 A tool on a machine has a velocity of 5 mm/s for the first 3 s and then a velocity in the opposite direction of 2 mm/s for 5 s. Sketch the distance–time graph.

14.3 Velocity–time graphs

If the velocity of an object is measured at different times then a velocity–time graph can be drawn. Acceleration is the rate at which the velocity changes. Thus, for the graph shown in Figure 14.6, the velocity changes from v_1 to v_2 when the time changes from t_1 to t_2. Thus the acceleration over that time interval is:

$$\text{acceleration} = \frac{v_2 - v_1}{t_2 - t_1}$$

But this is the gradient of the graph. Thus:

$$\text{acceleration} = \text{gradient of the velocity–time graph}$$

Figure 14.6 *Velocity-time graph*

Figure 14.7 *Velocity-time graph*

Figure 14.8 *Distance travelled*

In Figure 14.6 we have a straight line between the points A and B. Thus, since the gradient is the same for all points between A and B we have a uniform acceleration between those points. Consider the situation when the graph is not straight line, as in Figure 14.7. The velocity does not now change by equal amounts in equal intervals of time and the acceleration is no longer uniform. $(v_2 - v_1)$ is the change in velocity in a time of $(t_2 - t_1)$ and thus $(v_2 - v_1)/(t_2 - t_1)$ represents the average acceleration over that time. The smaller we make the times between A and B then the more the average is taken over a smaller time interval and more closely approximates to the instantaneous acceleration. An infinitesimal small time interval means we have the tangent to the curve. Thus if we want the acceleration at an instant of time then we have to determine the gradient of the tangent to the graph at that time, i.e.

instantaneous acceleration = gradient of tangent to the velocity–time graph at that instant

The distance travelled by an object in a particular time interval is given by:

distance = average velocity over the time × the time interval concerned

Thus if the velocity changes from v_1 at time t_1 to v_2 at time t_2, as in Figure 14.8, then the distance travelled between t_1 and t_2 is represented by the area marked on the graph. But this area is equal to the area under the graph line between t_1 and t_2. Thus:

distance travelled between t_1 and t_2 = area under the graph between these times

When the velocity is zero or there is uniform acceleration and so a straight line velocity–time graph, the areas can be easily calculated. However, if the acceleration is not uniform then the techniques outlined in Section 13.9 can be used, i.e. counting squares or the mid-ordinate rule.

Example

For the velocity–time graph shown in Figure 14.9, determine the acceleration at points A, B and C and estimate the distance covered in the 8 s.

Initially the graph is a straight line of constant gradient. The gradient, and hence the acceleration, at A is $2/2 = 1$ m/s^2.

At B the graph is no longer a straight line. The tangent to the curve at B has a gradient, and hence an acceleration, of about $(4.5 - 3.0)/(4 - 2) = 0.75$ m/s^2.

At C the velocity is not changing with time. Thus there is a uniform velocity with zero acceleration.

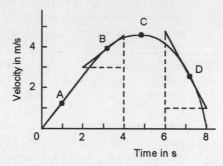

Figure 14.9 *Example*

At D the velocity is decreasing in a non-uniform manner with time. The tangent to the curve at that point has a gradient of about $(0.5 - 4.8)/(8 - 6) = -1.15$ m/s^2.

The distance covered in the 8 s is the area under the graph. Using the mid-ordinate rule with the time interval divided into four strips gives the distance as $1.2 \times 2 + 3.8 \times 2 + 4.6 \times 2 + 2.6 \times 2 = 24.4$ m.

Revision

6 A train, starting from rest, accelerates at a uniform rate for 20 s to reach a velocity 18 m/s. It then travels with this velocity for 160 s before braking results in a uniform retardation to bring the train to rest in 10 s. Sketch the velocity–time graph and hence determine (a) the acceleration during the initial 0 to 20 s time interval, (b) the retardation during the final 10 s, (c) the total distance travelled in the motion.

7 The velocity of a car varies with time and measurements give the following data. Determine (a) the acceleration at 5 s and (b) the distance travelled in the 10 s.

velocity in m/s	0	3.0	5.5	7.5	9.0	10
time in s	0	2	4	6	8	10

8 The velocity v of an object varies with time t and is described by the equation $v = 2t^2 + 3$, with v being in m/s and t in seconds. Sketch the velocity–time graph from $t = 0$ to $t = 5$ s and hence determine the acceleration at $t = 3$ s and the total distance covered in the 5 s.

9 A stone is dropped down a well and takes 4 s to reach the water. Sketch the velocity–time graph for the object, taking the acceleration due to gravity to be 10 m/s^2.

Problems

1 An object moving in a straight line gives the following distance–time data. Plot the distance–time graph and hence determine the velocities at times of (a) 2 s, (b) 3 s, (c) 4 s.

distance in mm	0	10	36	78	136	210
time in s	0	1	2	3	4	5

2 An object moving in a straight line gives the following distance–time data. Plot the distance–time graph and hence determine the velocities at times of (a) 5 s, (b) 8 s, (c) 11 s.

distance in m	0	1	15	31	49	75	110
time in s	0	1	4	6	8	10	12

3 The distance–time data for an object moving along a straight line is described by the equation $s = 4t^2 + 2$, the distance s being in millimetres and the time t in seconds. Plot the distance–time graph and hence determine the velocities when (a) $t = 2$ s, (b) $t = 3$ s, (c) $t = 4$ s.

4 The distance–time data for an object moving along a straight line is described by the equation $s = t^2$, the distance s being in millimetres and the time t in seconds. Plot the distance–time graph and hence determine the velocities when (a) $t = 2$ s, (b) $t = 3$ s, (c) $t = 4$ s.

5 The distance–time data for an object moving along a straight line is described by the equation $s = 4t^2 + 2t$, the distance s being in millimetres and the time t in seconds. Plot the distance–time graph and hence determine the velocities when (a) $t = 2$ s, (b) $t = 3$ s, (c) $t = 4$ s.

6 An object starts from rest and maintains a uniform acceleration of 0.5 m/s² for 4 s. It then moves with a uniform velocity for 3 s before a uniform retardation which brings it to rest in 2 s. Plot the velocity–time graph and hence determine the retardation during the last 2 s and the total distance covered.

7 A stone is thrown vertically upwards. The following are the velocities at different times. Plot a velocity–time graph and hence determine (a) the acceleration at the time $t = 1$ s, (b) the acceleration at $t = 2.5$ s, (c) the distance travelled to the maximum height where the velocity is zero.

velocity in m/s	15	10	5	0	−5	−10	−15
time in s	0	0.5	1.0	1.5	2.0	2.5	3.0

8 A car starts from rest and travels along a straight road. Its velocity changes at a uniform rate from 0 to 50 km/h in 20 s. It then continues for 50 s with uniform velocity before reducing speed at a constant rate to zero in 10 s. Sketch the velocity–time graph and hence determine (a)

the initial acceleration, (b) the final retardation and (c) the total distance travelled.

9 The velocity of an object varies with time. The following are values of the velocities at a number of times. Plot a velocity–time graph and hence estimate the acceleration at a time of 2 s and the total distance travelled in the 5 s.

velocity in m/s	0	1.2	2.4	3.6	3.6	3.6
time in s		0	1	2	3	4	5

10 An object has a maximum acceleration of 4 m/s^2 and a maximum retardation of 8 m/s^2. Determine the shortest time the object will take to move through a distance of 4000 m. Hint: the velocity–time graph will be triangular of area 4000 m.

11 The velocity of an object changes with time. The following are values of the velocities at a number of times. Plot the velocity–time graph and hence determine the acceleration at a time of 4 s and the total distance travelled during the 20 s.

velocity in m/s	0	8	10	16	20	20
time in s		0	4	8	12	16	20

12 The velocity of an object changes with time. The following are values of the velocities at a number of times. Plot the velocity–time graph and hence determine the acceleration at a time of 2 s and the total distance travelled between 1 s and 4 s.

velocity in m/s	25	24	21	16	9	0
time in s		0	1	2	3	4	5

13 State whether the distance–time graphs and the velocity–time graphs will be straight line with a non-zero gradient or zero gradient or curved in the following cases:

(a) The velocity is constant and not zero.
(b) The acceleration is constant and not zero.
(c) The distance covered in each second of the motion is the same.
(d) The distance covered in each successive second doubles.

15 Linear graphs

15.1 Introduction

If two variables x and y give a straight-line graph then the equation relating them is of the form $y = mx + c$, where m is the gradient and c the intercept of the graph line with the y-axis (see Section 13.8). Thus $y = 2x + 3$ is the equation of a straight-line graph with a gradient of 2 and an intercept with the y-axis of +3. The equation $y = -2x + 3$ describes a straight-line graph with a gradient of -2 and an intercept with the y-axis of +3 (see Figure 13.29 for graphs showing the difference between positive and negative gradients).

This chapter is about how, for a set of data points involving two variables which have a linear relationship, we can establish the linear law relating them. It also is about how we can reduce non-linear laws to linear forms and hence determine them.

15.2 Finding linear laws

If when we are given, or determine from an experiment, a set of corresponding values of two variables and plot them on a graph to give a straight line graph, then we know that the variables must be connected by a law of the form $y = mx + c$. To determine the law, all we have to do is find the values of m and c.

For example, suppose we have the following data for two related variables y and x:

y	-3	-1	1	3
x	-1	0	1	2

We will consider two alternative methods of determining m and c.

Figure 15.1 shows the above data plotted as a graph. The gradient of the graph can be found by drawing lines such as AB and BC in order to find by how much the line has risen in some horizontal distance. AB/BC = 4/2 and so $m = +2$. The line intercepts the y-axis at -1 and so $c = -1$. Thus the law is $y = 2x - 1$.

An alternative method is to choose two points, from those given if they can be assumed to fall on the straight-line graph or take values from the straight-line graph, and substitute their values in the equation $y = mx + c$. Thus, if we take the values given of $y = -3$ when $x = -1$ we obtain:

$$-3 = -1m + c$$

If we take another set of values, say $y = 3$ when $x = 2$, then we can also write:

$$3 = 2m + c$$

Figure 15.1   *Linear graph*

We have a pair of simultaneous equations which have to be solved to give m and c:

$$-m + c = -3$$
$$\text{minus} \quad \underline{2m + c = 3}$$
$$-3m \qquad = -6$$

Thus $m = 2$. Substituting this in the first equation gives $-2 + c = -3$ and hence $c = -1$. The equation is thus $y = 2x - 1$.

Example

The following is the data obtained from experimental measurements of the load lifted by a machine and the effort expended. Determine if the relationship between the effort E and the load W is linear and if so the relationship.

E in N	18	27	32	43	51
W in N	40	80	120	160	200

Figure 15.2 *Example*

Figure 15.2 shows the graph. Within the limits of experimental error the results appear to indicate a straight-line relationship. The gradient is AB/BC = 41/200 or about 0.21. The intercept with the E axis is at 10. Thus the relationship is $E = 0.21W + 10$.

Revision

1 Determine, assuming linear, the relationships between the following measured variables:

(a) The volume V and the temperature θ of a gas if we have:

V in m³	19.7	20.5	21.6	22.7	23.1	23.5
θ in °C	20	30	40	50	60	70

(b) The rate of rotation n of the shaft of a motor and the applied voltage V if we have:

n in rev/s	3	10	16	20	26	33
V in volts	40	80	120	160	200	240

(c) The specific latent heat L of steam and the temperature θ if we have:

L in kJ	2340	2300	2280	2250	2220	2160
θ in °C	70	80	90	100	110	120

(d) The tensile strength σ of an alloy at temperature θ if we have:

σ in kPa	83.7	80.2	77.7	75.2	72.7
θ in °C	100	200	300	400	500

15.3 Reduction of laws to linear form

When a graph is a straight line then the relationship between the two variables can easily be stated as being of the form $y = mx + c$ and, as indicated above, we can relatively easily determine the terms m and c from the graph. With a curved graph we can, however, have many possible forms of relationship, e.g. we might have $y = ax^2 + b$, $y = (a/x) + b$, $y = ax^b$, $y = a\,e^{bx}$, etc. To determine the form of the law and the terms in the law in such cases the technique is adopted of changing the variables to put them into a linear form.

The following illustrate how we can do this for a number of common forms of curved graphs. The principle used does, however, apply to other forms.

15.3.1 $y = ax^2 + b$

Consider the equation $y = ax^2 + b$, where a and b are constants. Let $X = x^2$. We can now write $y = aX + b$. This is a linear relationship between y and X and will give a straight-line graph with a gradient of a and an intercept on the y-axis of b. We thus obtain a straight-line graph by plotting y against x^2.

Example

It is believed that the relationship between y and x for the following data is of the form $y = ax^2 + b$. Determine the values of a and b.

y	2.5	4.0	6.5	10.0	14.5
x	1	2	3	4	5

Let $X = x^2$ to give $y = aX + b$.

y	2.5	4.0	6.5	10.0	14.5
$X = x^2$	1	4	9	16	25

Figure 15.3 *Example*

Figure 15.3 shows the graph of y against X. The graph has a gradient of AB/BC = 12.5/25 = 0.5 and an intercept with the y-axis of 2. Thus the relationship is $y = 0.5X + 2$ and so we have $y = 0.5x^2 + 2$.

Revision

2 It is believed that the relationships between y and x for the following data are of the form $y = ax^2 + b$. Determine the values of a and b.

(a)

y	−10	−7	2	17	44
x	0	1	2	3	4

(b)

y	2	3	6	11	18
x	0	1	2	3	4

(c)

y	5	3	−3	−13	−26
x	0	1	2	3	4

15.3.2 $y = (a/x) + b$

Consider the equation $y = a/x + b$, where a and b are constants. Let $X = 1/x$. We can now write $y = aX + b$. This is a linear relationship between y and X and will give a straight-line graph with a gradient of a and an intercept on the y-axis of b. We thus obtain the straight-line graph by plotting y against $1/x$.

Example

It is believed that the relationship between y and x for the following data is of the form $y = (a/x) + b$. Determine the values of a and b.

y	5	4	3.7	3.5	3.4
x	1	2	3	4	5

Let $X = 1/x$ to give $y = aX + b$.

y	5	4	3.7	3.5	3.4
X = 1/x	1	0.50	0.33	0.25	0.20

Figure 15.4 *Example*

Figure 15.4 shows the graph of y against X. The graph has a gradient of AB/BC = $(5 - 3)/1.0 = 2$ and an intercept with the y-axis of 3. Thus the equation for the graph is $y = 2X + 3$ and so we have $y = (2/x) + 3$.

Revision

3 It is believed that the relationships between y and x for the following data are of the form $y = (a/x) + b$. Determine the values of a and b.

(a)

y	2	1.5	1.33	1.25	1.20
x	1	2	3	4	5

(b)

y	1	0	−0.33	−0.50	−0.60
x	1	2	3	4	5

(c)

y	1	1.50	1.67	1.75	1.80
x	1	2	3	4	5

15.3.3 $y = ax^2 + bx$

Consider the equation $y = ax^2 + bx$, where a and b are constants. We can rearrange this equation to give

$$\frac{y}{x} = ax + b$$

Let $Y = y/x$. We can now write $Y = ax + b$. This is a linear relationship between Y and x and will give a straight-line graph with a gradient of a and an intercept on the y-axis of b. We thus obtain the straight-line graph by plotting y/x against x.

Example

It is believed that the relationship between y and x for the following data is of the form $y = ax^2 + bx$. Determine the values of a and b.

y	3	8	15	24	35
x	1	2	3	4	5

Let $Y = y/x$ to give $Y = ax + b$. Then we have

$Y = y/x$	3	4	5	6	7
x	1	2	3	4	5

Figure 15.5 *Example*

Figure 15.5 shows the resulting graph of Y against x. The gradient of the graph is AB/BC = $(7 - 2)/5 = 1$. The intercept with the Y axis is 2. Hence the equation of the graph is $Y = 1x + 2$ and so the required relationship is $y = x^2 + 2x$.

Revision

4 It is believed that the relationships between y and x for the following data are of the form $y = ax^2 + bx$. Determine the values of a and b.

(a)

y	2	10	24	44	70
x	1	2	3	4	5

(b)

y	3	8	15	24	35
x	1	2	3	4	5

(c)

y	0	-2	-6	-12	-20
x	1	2	3	4	5

15.3.4 $y = ax^b$

Consider the equation $y = ax^b$, where a and b are constants. We can put this equation in a linear form by taking logarithms, usually to base 10. Thus

$$\lg y = \lg (ax^b) = \lg x^b + \lg a = b \lg x + \lg a$$

If we let $Y = \lg y$ and $X = \lg X$ then we have $Y = bX + c$, where $c = \lg a$. This is a linear relationship between Y and X and will give a straight-line graph with a gradient of b and an intercept on the y-axis of c. We thus obtain the straight-line graph by plotting $\lg y$ against $\lg x$.

Example

It is believed that the relationship between y and x for the following data is of the form $y = ax^b$. Determine the values of a and b.

y	3.0	4.2	5.2	6.0	6.7
x	1	2	3	4	5

Let $Y = \lg y$ and $X = \lg X$. Then we have

$Y = \lg y$	0.48	0.62	0.72	0.78	0.83
$X = \lg x$	0	0.30	0.48	0.60	0.70

Figure 15.6 *Example*

Figure 15.6 shows the graph of Y plotted against X. The graph has a gradient of AB/BC = $(0.83 - 0.48)/0.70 = 0.5$ and an intercept on the Y-axis of 0.48. Thus the graph has the equation $Y = 0.5X + 0.48$. Thus, since b is the gradient 0.5 and the intercept $c = 0.48 = \lg a$, then the required equation is $y = 3.0x^{0.5}$ or $y = 3\sqrt{x}$.

Revision

5 It is believed that the relationships between y and x for the following data are of the form $y = ax^b$. Determine the values of a and b.

(a)

y	1	8	27	64	125
x	1	2	3	4	5

(b)

y	2	0.50	0.22	0.13	0.08
x	1	2	3	4	5

(c)

y	−2	−2.83	−3.46	−4.00	−4.47
x	1	2	3	4	5

15.3.5 $y = a\,e^{bx}$

Consider the equation $y = a\,e^{bx}$, where a and b are constants. We can put this equation in a linear form by taking logarithms to base e. Thus

$$\ln y = \ln(a\,e^{bx}) = \ln e^{bx} + \ln a = bx + \ln a$$

If we let $Y = \ln y$ then we have $Y = bx + c$, where $c = \ln a$. This is a linear relationship between Y and x and will give a straight-line graph with a gradient of b and an intercept on the y-axis of c. We thus obtain the straight-line graph by plotting $\ln y$ against x.

Example

It is believed that the relationship between y and x for the following data is of the form $y = a\,e^{bx}$. Determine the values of a and b.

y	5.53	6.11	6.75	7.46	8.24
x	1	2	3	4	5

Let $Y = \ln y$.

$Y = \ln y$	1.71	1.81	1.91	2.01	2.11
x	1	2	3	4	5

Figure 15.7 *Example*

Figure 15.7 shows the graph of Y plotted against x. The graph has a gradient AB/BC = $(2.11 - 1.71)/(5 - 1) = 0.1$ and an intercept with the Y-axis of 1.61. Thus the graph has the equation $Y = 0.1x + 1.61$. Hence, we have the relationship $\ln y = 0.1x + 1.61$ and thus the required equation is $y = e^{1.61}\,e^{0.1x} = 5\,e^{0.1x}$.

Revision

6 It is believed that the relationships between y and x for the following data are of the form $y = a\,e^{bx}$. Show that this is the case and determine the values of a and b.

(a)

y	0.67	0.91	1.23	1.66	2.24
x	1	2	3	4	5

(b)

y	2.46	2.01	1.65	1.35	1.10
x	1	2	3	4	5

(c) y -2.21 -2.44 -2.70 -2.98 -3.30

x 1   2   3   4   5

15.4 Log graph paper

When we have relationships involving powers it is often the case that logarithms of the data are taken in order to give a straight-line graph. Thus the data is transformed into log (data) and plotted on the linear axis of the graph. This involves calculating the logarithm of each data point before it can be plotted. A more convenient way is, however, to transform the axis of the graph into a *logarithmic axis* as illustrated in Figure 15.8. This then enables the data values to be directly plotted, the work of calculating the logarithms having been done by the conversion of the axes. The scale markings on a logarithmic axis do not have equal divisions.

The logarithm, to base 10, of 1 is zero and of 10 is 1. The logarithm of 100 is 2. Thus, on a logarithmic scale, the distance between 1 and 10 is the same as between 10 and 100. Each of these distances is referred to as a *cycle*. Figure 15.9 thus shows what an axis would look like with two cycles. Within each cycle we have the same pattern of markings.

Consider the equation $y = 3x^2$. We can convert this into a linear equation by taking logarithms. Thus $\lg y = 2 \lg x + 3$. With conventional, linear, graph paper we would then plot values of $\lg y$ on the y-axis and $\lg x$ on the x-axis. We can, however, have logarithmic axes for both the axes. Such graph paper is then referred to as *log-log paper*. Figure 15.10 shows the form such graph paper takes. We can determine how many cycles are needed by considering the range of values to be plotted. Thus suppose we have x-values ranging from 1 to 50. Then $\lg x$ ranges from $\lg 1 = 0$ to $\lg 50 = 1.7$. Thus for the x-axis we require two cycles, i.e. from 0 to 1 and 1 to 10. For the y-values we have the range extending from 3×1^2 to 3×50^2. Thus $\lg y$ ranges from $\lg 3 = 0.5$ to $\lg 7500 = 3.9$. Thus for the y-axis we require two cycles, i.e. 0 to 1 and 1 to 10. The graph paper required is thus 2 cycle by 2 cycle.

Data   Log of data   Data on
       on linear   log
       axis   axis

Figure 15.8 *Log axis*

Figure 15.8 *Two cycles*

Figure 15.10 *Log-log graph paper*

Figure 15.11   *Log-linear paper*

Consider the equation $y = 3\,e^{2x}$. We can convert this into a linear equation by taking logarithms to base e. Thus $\ln y = 2x + \ln 3$. With conventional, linear, graph paper we would then plot values of $\ln y$ on the Y-axis and x on the x-axis. We can, however, have a logarithmic axis for the y-axis and a linear axis for the x-axis. Such graph paper is referred to as *log-linear paper*. Figure 15.11 shows the form such graph paper takes. The same log-linear paper can be used for logarithms to base e as to base 10. Such logarithms are related by:

$$\ln x = 2.3026\,\lg x$$

Note that the above equation arises because if we take $x = 10^y$ then $\lg x = y$. If we replace the 10 by the identical number $e^{2.3026}$ we obtain $x = (e^{2.3026})^y$ and thus $\ln x = 2.3026y = 2.3026\,\lg x$. Thus $\ln x$ is just the logarithm to base 10 scale multiplied by a constant. We can therefore have the scale markings for the cycles on the log axis representing the values $\ln 1$, $\ln 10$, $\ln 100$, etc.

Example

It is believed that the relationship between y and x for the following data is of the form $y = ax^b$. Show that this is the case and determine, using log-log graph paper, the values of a and b.

y	3.0	4.2	5.2	6.0	6.7
x	1	2	3	4	5

This is a repeat of the example used in section 13.3.3. Taking logarithms to base 10 gives $\lg y = b\,\lg x + \lg a$. The y-axis has to range from $\lg 3 = 0.48$ to $\lg 6.7 = 0.83$ and so just one cycle from 0 to 1 is required. The x-axis has to range from $\lg 1 = 0$ to $\lg 5 = 0.69$ and so one cycle from 0 to 1 is required. Thus 1 cycle × 1 cycle graph paper is required. Figure 15.12 shows the above data plotted on log-log graph paper. The graph is a straight line and so the relationship is of this form.

Figure 15.12 *Example*

The gradient of the graph is:

$$\text{gradient} = \frac{\lg 6.7 - \lg 3.0}{\lg 5 - \lg 1} = \frac{0.83 - 0.48}{0.70 - 0} = 0.5$$

When $x = 1$ then $\lg x = 0$ and so $\lg y = \lg a$. Thus the intercept of the log-log graph with the ordinate for $x = 1$ gives $a = 3$. The equation is thus $y = 3x^{0.5}$.

Example

It is believed that the relationship between y and x for the following data is of the form $y = a\, e^{bx}$. Show that this is the case and determine, using log-linear graph paper, the values of a and b.

y	5.53	6.11	6.75	7.46	8.24
x	1	2	3	4	5

This is a repeat of the example in Section 15.3.5. Taking logarithms to base e gives $\ln y = bx + \ln a$. We thus require log-linear graph paper. The y-axis, which is the ln axis, has to range from $\ln 5.53 = 1.7$ to $\ln 8.24 = 2.1$ and so just one cycle from 1 to 10 is required. Figure 15.13 shows the resulting graph. The graph is straight line and so the relationship is valid. The gradient is

Figure 15.13 *Example*

$$\text{gradient} = \frac{\ln 8.44 - \ln 5.53}{5 - 1} = \frac{2.13 - 1.71}{4} = 0.10$$

The intercept with the y-axis, i.e. $x = 0$ ordinate, is at 5. Thus the required equation is $y = 5\, e^{0.10x}$.

Revision

7 Solve revision problems 5 and 6 using log-log or log-linear graph paper.

8 State the type of logarithmic graph paper, including the number of cycles, that should be used when plotting the following relationships.

(a) $\theta = a\, e^{-bt}$ when t varies from 0 to 30 minutes and θ varies from 200°C to 10°C.
(b) $F = av^b$ when v varies from 1 m/s to 6 m/s and the value of F varies from 10 N to 360 N,
(c) $y = ax^b$ when x varies from 2 to 350 and y varies from 14 to 187.

Problems

1 Determine, assuming linear, the relationships between the following variables:

(a) The load L lifted by a machine for the effort E applied.

E in N	9.5	11.8	14.1	16.3	18.5
L in N	10	15	20	25	30

(b) The resistance R of a wire for different lengths L of that wire.

R in Ω	2.1	4.3	6.3	8.3	10.5
L in m	0.5	1.0	1.5	2.0	2.5

(c) The velocity v of an object at different times t.

v in m/s	4.5	5.1	5.5	5.9	6.5
t in s	5	10	15	20	25

(d) The yield stress σ of a steel sample at different temperatures.

σ in MPa	140	130	122	111	101
θ in °C	200	300	400	500	600

2 Determine, on the basis of the data given, the values of a and b in the following relationships:

(a) $y = ax^2 + b$

y	−1	1	7	17	31
x	0	1	2	3	4

(b) $y = ax^2 + b$

y	3	2	−1	−6	−13
x	0	1	2	3	4

(c) $y = (a/x) + b$

y	4	2.5	2	1.75	1.60
x	1	2	3	4	5

(d) $y = (a/x) + b$

y	−6	−5	−4.7	−4.5	−4.4
x	1	2	3	4	5

(e) $y = ax^2 + bx$

y	3	10	21	36	55
x	1	2	3	4	5

(f) $y = ax^2 + bx$

y	−2	−5	−12	−20	−30
x	1	2	3	4	5

(g) $y = ax^b$

y	0.50	0.71	0.87	1.00	1.12
x	1	2	3	4	5

(h) $y = ax^b$

y	0.6	4.8	16.2	38.4	75.0
x	1	2	3	4	5

(i) $y = a\,e^{bx}$

y	905	819	741	670	607
x	1	2	3	4	5

(j) $y = a\,e^{bx}$

y	13.5	18.2	24.6	33.2	44.8
x	1	2	3	4	5

3 The following are examples of relationships which occur in engineering. Determine what form the variables should take when plotted in order to give straight-line graphs and what the values of the gradient and intercept will have.

(a) The period of oscillation T of a pendulum is related to the length L of the pendulum by the equation:

$$T = 2\pi \sqrt{\frac{L}{g}}$$

where g is a constant.

(b) The distance s travelled by a uniformly accelerating object after a time t is given by the equation:

$$s = ut + \tfrac{1}{2}at^2$$

where u and a are constants.

(c) The e.m.f. e generated by a thermocouple at a temperature θ is given by the equation;

$$e = a\theta + b\theta^2$$

where a and b are constants.

(d) The resistance R of a resistor at a temperature θ is given by the equation:

$$R = R_0 + R_0\alpha\theta$$

where R_0 and α are constants.

(e) The current i in a circuit when a capacitor is being charged varies with time t and is given by the equation:

$$i = I\,e^{t/RC}$$

where I and RC are constants.

(f) The pressure p of a gas and its volume V are related by the equation:

$$pV = k$$

where k is a constant.

(g) The deflection y of the free end of a cantilever due to it own weight of w per unit length is related to its length L by the equation:

$$y = \frac{wL^4}{8EI}$$

where w, E and I are constants.

4 The resistance R of a lamp is measured at a number of voltages V and the following data obtained. Show that the law relating the resistance to the voltage is of the form $R = (a/V) + b$ and determine the values of a and b.

R in Ω	70	62	59	56	55
V in V	60	100	140	200	240

5 The resistance R of wires of a particular material are measured for a range of wire diameters d and the following results obtained. Show that the relationship is of the form $R = (a/d^2) + b$ and determine the values of a and b.

R in Ω	0.25	0.16	0.10	0.06	0.04
d in mm	0.80	1.00	1.25	1.60	2.00

6 The periodic time T of a simple pendulum is measured for a number of different lengths L and the following results obtained. Show that the relationship is of the form $T = aL^b$ and determine the values of a and b.

T in s	1.41	1.55	1.67	1.79	1.90
L in m	0.50	0.60	0.70	0.80	0.90

7 The volume V of a gas is measured at a number of pressures p and the following results obtained. Show that the relationship is of the form $V = ap^b$ and determine the values of a and b.

V in m^3	13.3	11.4	10.0	8.9	8.0
p in 10^5 Pa	1.2	1.4	1.6	1.8	2.0

8 When a gas is compressed adiabatically the pressure p and temperature T are measured and the following results obtained. Show that the relationship is of the form $T = ap^b$ and determine the values of a and b.

p in 10^5 Pa	1.2	1.5	1.8	2.1	2.4
T in K	526	560	589	615	639

9 The cost C per hour of operating a machine depends on the number of items n produced per hour. The following data has been obtained and is anticipated to follow a relationship of the form $C = an^3 + b$. Show that this is the case and determine the values of a and b.

C in £	31	38	67	94	155
n	10	20	30	40	50

10 The following are suggested braking distances s for cars travelling at different speeds v. The relationship between s and v is thought to be of the form $s = av^2 + bv$. Show that this is so and determine the values of a and b.

s in m	5	15	30	50	75
v in m/s	5	10	15	20	25

11 The luminosity I of a lamp depends on the voltage V applied to it. The relationship between I and V is thought to be of the form $I = aV^b$. Use the following results to show that this is the case and determine the values of a and b.

I in candela	3.6	6.4	10.0	14.4	19.6
V in volts	60	80	100	120	140

12 The energy E of electrons emitted from a metal when light of different frequency f is incident on it gives the following results. Show that the data fits the equation $E = hf - \phi$ and determine the constants h and ϕ.

E in electron-volts	0.37	1.36	2.10	2.75	3.90
f in 10^{15} Hz		1.14	1.38	1.56	1.72	2.00

13 How many cycles will be needed for log-log graph paper used to plot the graph of $pV^{1.4} = 2000$ if V ranges from 2 to 12?

14 How many cycles will be needed for log-linear graph paper used to plot the graph of $y = 3\,e^x$ if x ranges from 0 to 3?

15 The following data gives the radius R of the orbit of the planets about the sun and the length T of their year. It is considered that the relationship between T and R might be of the form $T = aR^b$. Show that this is so and determine the values of a and b.

	Mercury	Venus	Earth	Mars	Jupiter
R in 10^6 km	57.9	108.2	149.6	227.9	778.3
T in days	88	225	365	687	4329

	Saturn	Uranus	Neptune	Pluto
R in 10^6 km	1 427	2 870	4 497	5 907
T in days	10 753	30 660	60 150	90 470

16 The following data indicates how the voltage v across a component in an electrical circuit varies with time t. It is considered that the relationship between V and t might be of the form $v = V\,e^{-bt}$. Show that this is so and determine the values of V and b.

v in volts	3.75	1.38	0.51	0.19	0.07
t in s	10	20	30	40	50

17 A hot object cools with time. The following data shows how the temperature θ of the object varies with time t. The relationship between θ and t is expected to be of the form $\theta = a\,e^{-bx}$. Show that this is so and determine the values of a and b.

θ in °C	536	359	241	162	108
t in min	2	4	6	8	10

18 The rate of flow Q of water over a V-shaped notch weir was measured for different heights h of the water above the point of the V and the following data obtained. The relationship between Q and h is thought to be of the form $Q = ah^b$. Show that this is so and determine the values of a and b.

Q in m³/s	0.13	0.26	0.46	2.12	1.07
h in m	0.3	0.4	0.5	0.6	0.7

19 The amplitude A of oscillation of a pendulum decreases with time t and gives the following data. Show that the relationship is of the form $A = a\,e^{bt}$ and determine the values of a and b.

A in mm	268	180	120	81	54
t in s	20	40	60	80	100

20 The tension T and T_0 in the two sides of a belt driving a pulley and in contact with the pulley over an angle of θ is given by the equation $T = T_0\,e^{\mu\theta}$. Determine the values of T_0 and μ for the following data:

T in N	69.5	80.8	91.1	109.1	126.7
θ in radians	1.1	1.6	2.0	2.6	3.1

16 Polar coordinates

16.1 Introduction

Figure 16.1 *Polar coordinates*

The Cartesian coordinate system enables the position of a point on a plane to be specified by stating its x and y coordinates, i.e. the displacement of the point relative to two perpendicular axes. There is another way of specifying the position of a point on a plane and that is to use *polar coordinates*. A reference axis is specified as emanating from some fixed point O, often referred to as the *pole*. A line is then drawn from O to the point being specified. The location of the point is then specified by the angle θ that line makes with the reference axis and the distance along that line to the point, i.e. the radius r of a circle with its centre at O and which would pass through the point concerned (Figure 16.1). The polar coordinates are then (r, θ). The angle θ is a positive angle when measured anticlockwise from the reference axis and negative when measured in a clockwise direction. The angles can be specified in either degrees or radians. This chapter is a consideration of the polar coordinate method of specifying points on a plane and the relationship between polar and Cartesian coordinates.

Such a method of specifying a point on a plane is widely used in electrical and electronic engineering for describing phasors (see Chapter 17). With phasors the important items are the length of a phasor and its angle to a reference axis. Thus the polar coordinate method of specifying it has great advantages. For example, a current phasor of magnitude 2 A with a phase angle of 30° could be specified in polar coordinates as being 2 at an angle of 30° to the reference axis. This can be written as (2, 30°) though, in engineering, it is usually written as $2\angle 30°$ A. The inclusion of the angle sign $\angle$ can remove any doubt as to whether we are dealing with an angle.

Example

Specify the polar coordinates of the points shown in Figure 16.2.

Point A has a radius of 3 and an angle of 30° and so is located at the point $(3, \angle 30°)$. Point B has a radius of 4 and an angle of 110° and so is located at the point $(4, \angle 110°)$.

Figure 16.2 *Example*

Revision

1 The following points are in polar coordinates. Indicate their positions on a polar graph.

(a) $(2, \angle 60°)$, (b) $(3, \angle \pi/4)$, (c) $(2, \angle -45°)$, (d) $(2.5, \angle 4\pi/3)$, (e) $(2, \angle -\pi/2)$.

16.1.1 Conversion from Cartesian to polar coordinates

Figure 16.3 *Coordinates*

Consider a point P (Figure 16.3) which has Cartesian coordinates (x, y). We will take the polar reference axis to coincide with the x-axis. Then radius r can be calculated by the use of the Pythagoras theorem. Thus, we have $r^2 = x^2 + y^2$ and so

$$r = \sqrt{x^2 + y^2}$$

The angle θ is given by:

$$\tan\theta = \frac{y}{x}$$

We can thus write:

$$\theta = \tan^{-1}\frac{y}{x}$$

This reads as 'the angle whose tangent is y/x'.

Example

What are the polar coordinates of a point with Cartesian coordinates (3, 2)?

The radius r is given by:

$$r = \sqrt{3^2 + 2^2} = 3.6$$

and the angle θ by:

$$\theta = \tan^{-1}\frac{2}{3} = 33.7° \text{ or } 0.59 \text{ rad}$$

Thus the polar coordinates are $(3.6, \angle 33.7°)$ or $(3.6, \angle 0.59 \text{ rad})$.

Example

What are the polar coordinates of a point with the Cartesian coordinates of (−2, 3)?

Figure 16.4 *Example*

Figure 16.4 shows the position of the point. The radius r is given by

$$r = \sqrt{(-2)^2 + 3^2} = 3.6$$

As Figure 16.4 indicates, θ is greater than 90° but less than 180°. Thus:

$$\theta = \tan^{-1}\frac{3}{-2} = 123.7° \text{ or } 21.6 \text{ rad}$$

The polar coordinates are thus (3.6, $\angle$123.7°) or (3.6, $\angle$2.16).

Revision

2 Determine the polar coordinates of the points with the following Cartesian coordinates:

(a) (4, 2), (b) (2, −2), (c) (−2, 4), (d) (−2, −2), (e) (2, 0), (f) (0, 2), (g) (−4, 0), (h) (0, −4).

16.1.2 Conversion from polar to Cartesian coordinates

Figure 16.5 *Coordinates*

Consider a point P (Figure 16.5) with the polar coordinates $(r, \angle\theta)$. We have:

$$\cos\theta = \frac{x}{r}$$

and so $x = r\cos\theta$. We can also write:

$$\sin\theta = \frac{y}{r}$$

and so $y = r\sin\theta$.

Example

Determine the Cartesian coordinates of the point with the polar co-ordinates (2, $\angle$30°).

For the x coordinate we have $x = r\cos\theta = 2\cos 30° = 1.7$ and for the y coordinate $y = r\sin\theta = 2.0\sin 30° = 1.0$. Hence the coordinates are (1.7, 1.0).

Revision

3 Determine the Cartesian coordinates of the points with the following polar coordinates:

(a) (4, $\angle$45°), (b) (3, $\angle$60°), (c) (2, $\angle$150°), (d) (3, $\angle$290°), e) (2, $\angle\pi/4$), (f) (4, $\angle\pi/6$), (g) (3, $\angle\pi$).

16.2 Polar graphs

To draw a polar graph we need to know the values of the radius r and angle θ for points that are to fall on the graph line.

Consider the graph of the function which has its coordinates described by the equation:

r = a constant

Figure 16.6 *r = a constant*

Figure 16.7 *θ = a constant*

r is the same for all values of *θ*. Such an equation can only describe a circle with its centre at the pole (Figure 16.6). Only then will all the points on its graph have the same radius.

Now consider the graph of the function which has its coordinates described by the equation:

$$\theta = \text{a constant}$$

θ is the same for all values of *r*. This can only describe a line emanating from the pole (Figure 16.7) so that all the points on its graph are at the same angle.

Consider the graph of the function for which $r = \sin \theta$. Values of *r* and *θ* which fit this equation are:

θ	0°	30°	60°	90°	120°	150°	180°
r	0	0.50	0.87	1.00	0.87	0.50	0

θ	210°	240°	270°	300°	330°	360°
r	−0.50	−0.87	−1.00	−0.87	−0.50	0

Figure 16.8 shows the resulting graph. Plotting the values from 0° to 180° results in a curve which gives a complete circle. When plotting the points from 210° to 360° we are faced with negative values for the radius. A negative value magnitude |*r*| plots as a distance |*r*| in the opposite direction out from the pole as indicated by the angle. Thus we plot for the angle 210° a magnitude of 0.50 in the 30° direction. The values of *r* from 210° to 360° thus end up repeating the circle given by the 0° to 180° values.

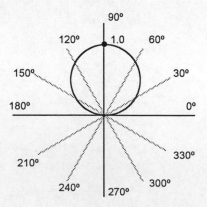

Figure 16.8 $r = \sin \theta$

All functions of the form $r = a \sin \theta$ give circles in the same way. The only effect of the constant *a* is to change the diameter of the circle, the circle being centred on the *y*-axis at $y = a/2$ and with a diameter of *a*. In a

(a)

(b)

Figure 16.9 *(a) r = a sin θ,*
(b) r = a cos θ

similar way, all functions of the form $r = a \cos \theta$ give circles centred on the x-axis at $x = a/2$ and with a diameter a (Figure 16.9).

Consider the graph of the function for which $r = 1 + \cos \theta$. Values of r and θ which fit this equation are:

θ	0°	30°	60°	90°	120°	150°	180°
r	2.0	1.9	1.5	1.0	0.5	0.1	0

θ	210°	240°	270°	300°	330°	360°
r	0.1	0.5	1.0	1.5	1.9	2.0

Figure 15.10 shows the resulting graph. It is called a *cardioid* (i.e. a heart shape).

Figure 16.10 *r = 1 + cos θ*

Revision

4   Plot graphs of the following equations:

(a) $r = 3$, (b) $\theta = \pi/6$, (c) $r = 2(1 + \cos \theta)$, (d) $r = 2(1 - \cos \theta)$.

Problems

1   Determine the polar coordinates of the points with the following Cartesian coordinates:

(a) (0, 3), (b) (−2, 0), (c) (−2, 2), (d) (3, 4), (e) (−3, 4), (f) (4, −2),
(g) (3, 0).

2   Determine the Cartesian coordinates of the points with the following polar coordinates

(a) (2, ∠50°), (b) (3, ∠160°), (c) (3, ∠260°), (d) (4, ∠340°),

(e) $(2, \angle \pi/3)$, (f) $(1, \angle 5\pi/3)$, (g) $(3, \angle \pi)$, (h) $(4, \angle \pi/2)$.

3   Plot polar graphs for the following equations:

(a) $r = 4$, (b) $r = \pi/4$, (c) $r = 2 + \cos \theta$, (d) $r = 2 \sin 2\theta$.

4   A cam has been designed with the following values of radius r at the different angles θ. Sketch the shape of the cam.

θ	0	$\pi/6$	$\pi/3$	$\pi/2$	$2\pi/3$	$5\pi/6$	π
r in mm	10	14	18	22	26	30	34

θ	$7\pi/6$	$4\pi/3$	$3\pi/2$	$5\pi/3$	$11\pi/6$	2π
r in mm	30	26	22	18	14	10

5   Demonstrate by plotting examples of polar graphs that the polar graph of $r = \sin n\theta$ or $r = \cos n\theta$ has n loops when n is an odd positive integer and $2n$ loops when n is an even positive integer.

17 Alternating waveforms

17.1 Introduction

(a)

(b)

(c)

Figure 17.1 *Waveforms*

Same time, the periodic time

Figure 17.2 *Terms*

The term *alternating* voltage or current is used when the polarity or direction of flow of the voltage or current alternates, continually changing with time. Alternating voltages and currents can have many different waveforms. A particularly important form is one which is in the form of a sine graph, the form shown in Figure 17.1(a). This is because electrical power generation in the entire world is virtually all in the form of such sinusoidal voltages. Consequently the mains electrical supply to houses, offices and factories is sinusoidal. The sinusoidal waveform in the figure has a current which oscillates from positive values to negative values to positive values, to ... and so on.

Figure 17.1(b) and (c) show some further examples of waveforms: (b) is a rectangular waveform and (c) a triangular waveform. The rectangular waveform in the figure starts with a positive current which then abruptly switches to a negative current, which then abruptly switches to a positive current, which then ... and so on. The triangular waveform shows a similar form of behaviour.

Alternating waveforms oscillate from positive to negative values in a regular, periodic manner. One complete sequence of such an oscillation is called a *cycle* (Figure 17.2). The time T taken for one complete cycle is called the *periodic time* and the number of cycles occurring per second is called the *frequency f*. Thus $f = 1/T$. The unit of frequency is the hertz (Hz), 1 Hz being 1 cycle per second.

This chapter is a consideration of the mean value and the root-mean-square values of such alternating waveforms and how phasors can be used to simplify the analysis of electrical circuits when sinusoidal voltages and currents occur.

17.2 Average value

Figure 17.3 *Average value*

The *average*, or *mean*, value of a set of numbers is their sum divided by the number of numbers summed. The mean value of some function between specified limits that is described by a graph can be considered to be the mean value of all the ordinates representing the values between these limits. We can give an approximation of this if we divide the area into a number of equal width strips (Figure 17.3); the average value will be the sum of all the mid ordinates y_1, y_2, y_3, etc. of the strips divided by the number n of strips considered. The number of strips n is the range of values $(b - a)$ between the limits divided by the width w of a strip. Thus:

$$\text{average} = \frac{y_1 + y_2 + y_3 + ...}{n} = \frac{(y_1 + y_2 + y_3 + ...)w}{b - a}$$

But wy_1 is the area of the first strip, wy_2 is the area of the second strip, and so on. Thus the average is the sum of the areas of the strips divided by the range of values over which the average is taken. Hence:

Figure 17.4 *Average value*

$$\text{average} = \frac{\text{area under graph}}{b-a}$$

The average value multiplied by the range $(b - a)$ is thus equal to the area under the graph. Thus, the value of the horizontal graph line representing the average value is thus the one for which the area under it is the same as the area under the waveform. Figure 17.4 illustrates this. The area under the average line is that of the shaded rectangle shown in Figure 17.4 and this thus has the same area as the area under the graph line between the specified limits.

17.2.1 Using the mid-ordinate rule

We can use the *mid-ordinate rule* (or perhaps the trapezium rule) to determine the area under a waveform. Thus, as in Figure 17.3, the waveform is divided into a number of equally spaced time interval strips and then a vertical line, an ordinate, is drawn at the middle of each strip and the value of the graph at that instant taken. Then, since the area of a strip is its mid-ordinate value × width of the strip and the widths are the same for each strip:

area under waveform = sum of mid-ordinate values × width of strips

But the range considered is just the number of strips × width of the strips. Hence we can write:

$$\text{average value} = \frac{\text{sum of mid-ordinate values} \times \text{width of strips}}{\text{number of mid-ordinates} \times \text{width of strips}}$$

and so:

$$\text{average value} = \frac{\text{sum of mid-ordinate values}}{\text{number of mid-ordinates}}$$

17.2.2 The average value of waveforms

The average value for an alternating current, or voltage, waveform is thus:

$$\text{average value} = \frac{\text{area under waveform}}{\text{range considered}}$$

Figure 17.5 shows a sinusoidal waveform which has been subdivided into 18 equal width strips. If the instantaneous values of the current for the mid-ordinates of these strips are i_1, i_2, i_3, etc. then the average value over one cycle will be the sum of all these mid-ordinate values divided by the number of values taken. But, because the waveform has a negative half cycle which is just the mirror image of the positive half cycle, then the average over a full cycle must be zero. For every positive value there will be a corresponding negative value.

Figure 17.5 *Obtaining the average value*

Over one half cycle the average value of the sinusoidal waveform is:

$$I_{av} = \frac{i_1 + i_2 + i_3 + \ldots + i_9}{9}$$

When the maximum value of the current is 1, readings taken from the graph for the currents (or obtained using a calculator to give the values of the sine at the mid-ordinate values) are:

i_1	i_2	i_3	i_4	i_5	i_6	i_7	i_8	i_9
0.17	0.50	0.77	0.94	1.00	0.94	0.77	0.50	0.17

The average of these values is:

$$\frac{0.17 + 0.50 + 0.77 + 0.94 + 1.00 + 0.94 + 0.77 + 0.50 + 0.17}{9}$$

The average is thus 0.64. The accuracy of the average value is improved by taking more values for the average. The average value for a sinusoidal waveform of maximum value I_m over half a cycle is then found to be 0.6371. This is the value of $2/\pi$ and so for a current of maximum value I_m we have:

$$I_{av} = \frac{2I_m}{\pi} = 0.637I_m$$

In a similar way we could have derived the average value for a sinusoidal voltage as

$$V_{av} = \frac{2V_m}{\pi} = 0.637V_m$$

Example

Determine the average value of the half cycle of the current waveform shown in Figure 17.6.

Dividing the half cycle into segments of width 1 s by the solid lines then the mid-ordinates are indicated by the fainter lines in the figure. The values of these are

Mid-ordinate in ms	0.5	1.5	2.5	3.5	4.5
Current in A	1.0	3.0	4.0	4.0	2.0

The average value is thus

$$\text{average} = \frac{1.0 + 3.0 + 4.0 + 4.0 + 2.0}{5} = 2.8 \text{ A}$$

Figure 17.6   *Example*

Figure 17.7 *Revision problem 1*

Revision

1 Determine the average values of the segments of waveforms shown in Figure 17.7.

2 Show that for a triangular waveform, of the form shown in Figure 17.7(a), with a maximum value of V_m that over half the cycle shown the average value is $0.5V_m$.

17.3 Root-mean-square value

Since we are frequently concerned with the power developed by a current passing through a circuit component, a useful measure of an alternating current is in terms of the direct current that would give the same power dissipation in a resistor. For an alternating current, the power at an instant of time is i^2R, where i is the current at that instant and R is the resistance. Thus to obtain the power developed by an alternating current over a cycle, we need to find the average power developed over that time. In terms of mid-ordinates we add together all the values of power given at each mid-ordinate of time in the cycle and divide by the number of mid-ordinates considered. Because we are squaring the current values, negative currents give positive values of power. Thus the powers developed in each half cycle add together.

For the sinusoidal waveform shown in Figure 17.5, the average power P_{av} developed in one complete cycle is the sum of the mid-ordinate powers divided by the number of mid-ordinates used and so:

$$P_{av} = \frac{i_1^2 R + i_2^2 R + i_3^2 R + ... + i_{18}^2 R}{18} = \frac{i_1^2 + i_2^2 + i_3^2 + ... + i_{18}^2}{18} R$$

For a direct current I to give the same power as the alternating current, we must have $I^2 R = P_{av}$. Thus the square of the equivalent direct current I^2 is the average (mean) value of the squares of the instantaneous currents during the cycle. This equivalent current is known as the *root-mean-square current* I_{rms}. Hence:

$I_{rms} = \sqrt{}$(mean of the sum of the values of the squares of the alternating current)

In the above we considered the power due to a current through a resistor. We could have considered the power in terms of the voltage developed across the resistor. The power at an instant when the voltage is v is v^2/R. The equivalent direct voltage V to give the same power is thus when V^2/R is equal to the average power developed during the cycle. Thus, as before, we obtain:

$V_{rms} = \sqrt{}$(mean of the sum of the values of the squares of the alternating voltage)

Consider the use of the of the mid-ordinate rule to obtain values of the current, or voltage, for the sinusoidal waveform given in Figure 17.5. When the maximum value is 1, we have for each of the mid-ordinates:

i_1	i_2	i_3	i_4	i_5	i_6	i_7	i_8	i_9
0.17	0.50	0.77	0.94	1.00	0.94	0.77	0.50	0.17

i_{10}	i_{11}	i_{12}	i_{13}	i_{14}	i_{15}	i_{16}	i_{17}	i_{18}
−0.17	−0.50	−0.77	−0.94	−1.00	−0.94	−0.77	−0.50	−0.17

The square of these current values are:

i_1^2	i_2^2	i_3^2	i_4^2	i_5^2	i_6^2	i_7^2	i_8^2	i_9^2
0.03	0.25	0.59	0.88	1.00	0.88	0.59	0.25	0.03

i_{10}^2	i_{11}^2	i_{12}^2	i_{13}^2	i_{14}^2	i_{15}^2	i_{16}^2	i_{17}^2	i_{18}^2
0.03	0.25	0.59	0.88	1.00	0.88	0.59	0.25	0.03

The sum of these values is 9.00 and thus the mean value is 9.00/18 = 0.50. The root-mean-square current is therefore $\sqrt{0.50} = 1/\sqrt{2}$. Thus, with a maximum current of I_m:

Figure 17.8 *Root-mean-square current*

$$I_{rms} = \frac{I_m}{\sqrt{2}}$$

Similarly:

$$V_{rms} = \frac{V_m}{\sqrt{2}}$$

where V_m is the maximum voltage.

We could have arrived at the above results for the sinusoidal waveform by considering the form of the graph produced by plotting the square of the current values, or the voltage values. Figure 17.8 shows the graph. The squares of the positive and negative currents are all positive quantities and so the resulting graph oscillates between a maximum value of I_m^2 and 0. The mean value of the i^2 graph is $I_m^2/2$, the i^2 graph being symmetrical about this value. The mean power is thus $R I_m^2/2$ and so the root-mean-square current is $I_m/\sqrt{2}$. Thus, for a sinusoidal waveform, the root-mean-square current is the maximum current divided by $\sqrt{2}$.

For alternating currents, or voltages, with other waveforms the relationships between the root-mean-square values and the maximum values are different. Figure 17.9 shows some examples of waveforms with their average over half a cycle and root-mean-square values in terms of their maximum values. The ratio of the root-mean-square value to the average value over half a cycle is called the *form factor* and is an indication of the shape of the waveform.

$$\text{form factor} = \frac{\text{rms value}}{\text{average value over half a cycle}}$$

Wave shape	Average	RMS	Form factor
Time	$0.637\ V_m$	$0.707\ V_m$	1.11
Time	$0.319\ V_m$	$0.500\ V_m$	1.57
Time	V_m	V_m	1.00
Time	$0.500\ V_m$	$0.577\ V_m$	1.15

Figure 17.9 *Form factors*

Example

Using the mid-ordinate rule the following current values were obtained for the half cycle of a waveform, the other half cycle being just a mirror image. Estimate (a) the average value over half a cycle, (b) the root-mean-square value, (c) the form factor.

Current values in A 2 2 2 2 2 4 4 4

(a) The average value is:

$$I_{av} = \frac{2+2+2+2+2+4+4+4}{8} = 2.75 \text{ A}$$

(b) The root-mean-square value is:

$$I_{rms} = \sqrt{\frac{4+4+4+4+4+16+16+16}{8}} = 2.92 \text{ A}$$

(c) The form factor is:

$$\text{form factor} = \frac{\text{rms value}}{\text{average value}} = \frac{2.92}{2.75} = 1.06$$

Revision

3 Determine the root-mean-square current for a triangular waveform which gave the following mid-ordinate values over a full cycle:

current in mA   5 15 25 15 5 −5 −15 −25 −15 −5

4 Determine the root-mean-square voltage for an irregular waveform which gave the following mid-ordinate values over a full cycle:

voltage in V   5 10 12 8 2 −5 −10 −12 −8 −2

5 Show that the root-mean-square value of a triangular waveform of the form shown in Figure 17.7(a) and having a maximum value of V_m is given by $V_m/\sqrt{3}$.

6 Determine the average value, the root-mean-square value and the form factor of the waveform giving the current shown in Figure 17.10.

Figure 17.10   *Revision problem 6*

17.4 Phasors

Consider the line OA in Figure 17.11 rotating in an anticlockwise direction about O with a constant angular velocity ω (note that the unit of angular velocity is radians/second). A constant angular velocity means that the line rotates through equal angles in equal intervals of time. The line starts from the horizontal position and rotates through an angle θ in a time t.

Figure 17.11 *Deriving a sinusoidal waveform*

Since AB/OA = sin θ we can write:

$$AB = OA \sin \theta$$

where AB is the vertical height of the line at some instant of time, OA being its length. The maximum value of AB will be OA and occur when $\theta = 90°$. But an angular velocity ω means that in a time t the angle θ covered is ωt. Thus the vertical projection AB of the rotating line will vary with time and is described by the equation

$$AB = OA \sin \omega t$$

If we represent an alternating current i by the perpendicular height AB then its maximum value I_m is represented by OA and we can thus write

$$i = I_m \sin \omega t$$

This is the equation describing the sinusoidal waveform and how its current i at any instant varies with time t. In a similar way we can write for a sinusoidal alternating voltage;

$$v = V_m \sin \omega t$$

Thus we can think of an alternating current and voltage in terms of a model in which the instantaneous value of the current or voltage is represented by the vertical projection of a line rotating in an anticlockwise direction with a constant angular velocity. The term *phasor*, being an abbreviation of the term phase vector, is used for such rotating lines. The length of the phasor can represent the maximum value of the sinusoidal waveform or the root-mean-square value. This is because the maximum value is proportional to the root-mean-square value. The root-mean-square value is the maximum value divided by $\sqrt{2}$. It is often more convenient to use the root-mean-square values but it must be remembered that in an equation such as $v = V_m \sin \omega t$, the V_m must be the maximum value. The line

Figure 17.12 *Phasor*

representing a phasor is drawn with an arrowhead at the end that rotates and is drawn in its position at time $t = 0$, i.e. the phasor represents a frozen view of the rotating line at one instant of time of $t = 0$. Thus for the phasor to represent the current or voltage described by Figure 17.11 we have the phasor shown in Figure 17.12.

It is usual to give angular velocities in units of radians per second, an angular rotation through 360° being a rotation through 2π radians. Since the periodic time T is the time taken for one cycle of a waveform, then T is the time taken for OA to complete one revolution, i.e. 2π radians. Thus:

$$T = \frac{2\pi}{\omega}$$

The frequency f is $1/T$ and so $\omega = 2\pi f$. Because ω is just 2π times the frequency, it is often called the *angular frequency*. We can thus write the above equations as:

$$i = I_m \sin 2\pi ft$$

and:

$$v = V_m \sin 2\pi ft$$

Example

A sinusoidal alternating current has a frequency of 50 Hz and a maximum value of 4.0 A. What is the instantaneous value of the current (a) 1 ms, (b) 2 ms after it is at zero current?

The sinusoidal alternating current has the equation:

$$i = I_m \sin 2\pi ft = 4.0 \sin 2\pi \times 50t = 4.0 \sin 314t$$

(a) When $t = 1$ ms we have:

$$i = 4.0 \sin 314 \times 1 \times 10^{-3} = 4.0 \sin 0.314$$

The 0.314 is in units of radians, not degrees. This can be worked out using a calculator operating in the radian mode, the key sequence being change mode to rad, press the keys for 0.314, then for sin, the $\times$ key, the key for 4 and then the = key. Thus $i = 1.24$ A.
(b) When $t = 2$ ms we have:

$$i = 4.0 \sin 314 \times 2 \times 10^{-3} = 4.0 \sin 0.628$$

The 0.314 is in units of radians, not degrees. This can be worked out using a calculator operating in the radian mode. Thus $i = 2.35$ A.

Example

A sinusoidal alternating current is represented by $i = 10 \sin 500t$, where i is in mA. What is (a) the size of the maximum current, (b) the angular frequency, (c) the frequency and (d) the current after 1 ms from when it is zero?

(a) The maximum current is 10 mA.
(b) The angular frequency is 500 rad/s.
(c) The frequency is given by the equation $\omega = 2\pi f$ and so:

$$f = \frac{\omega}{2\pi} = \frac{500}{2\pi} = 79.6 \text{ Hz}$$

(d) After 0.01 s we have:

$$i = 10 \sin 500 \times 0.01 = 10 \sin 0.50$$

This angle is in radians. Thus $i = 4.79$ mA.

Revision

7 A sinusoidal voltage has a maximum value of 10 V and a frequency of 50 Hz. (a) Write an equation describing how the voltage varies with time. (b) Determine the voltages after times from $t = 0$ of (i) 0.002 s, (ii) 0.006 s and (iii) 0.012s.

8 A sinusoidal current has a maximum value of 50 mA and a frequency of 2 kHz. (a) Write an equation describing how the current varies with time. (b) Determine the currents after times from $t = 0$ of (i) 0.4 ms, (ii) 0.8 ms, (iii) 1.6 ms.

9 For a sinusoidal voltage described by $v = 10 \sin 1000t$ volts, what will be (a) the value of the voltage at time $t = 0$, (b) the maximum value of the voltage, (c) the voltage after 0.2 ms?

10 Complete the following table for the voltage $v = 1 \sin 100t$ volts.

t in ms	0	2	4	6	8	10	12	14	16
v in V									

17.4.1 Phase differences

In Figure 17.11 the rotating line OA was shown as starting from the horizontal position at time $t = 0$ with the phasor being as shown in Figure 17.12. But we could have an alternating voltage or current starting from some value other than 0 at $t = 0$. Figure 17.13 shows such a situation. At the time $t = 0$ the line OA is already at some angle ϕ.

Figure 17.13 *Waveform not zero at* $\theta = 0$, *i.e.* $t = 0$

As the line OA rotates with an angular velocity ω then in a time t the angle swept out is ωt and thus at time t the angle with respect to the horizontal is $\omega t + \phi$. Thus we have

$$i = I_m \sin (\omega t + \phi) \text{ or } v = V_m \sin (\omega t + \phi)$$

Figure 17.14   *Phasor*

The phasor for such alternating currents or voltages is represented by a phasor (Figure 17.14) at an angle ϕ to some reference line, this line being generally taken as being the horizontal. The angle ϕ is termed the *phase angle*. We can use polar coordinates (see Chapter 16) to describe a phasor. Thus $2\angle 40°$ A describes a phasor with a magnitude, represented by its length, of 2 A and with a phase angle of 40°.

In discussing alternating current circuits we often have to consider the relationship between an alternating current through a component and the alternating voltage across it. If we take the alternating voltage as the reference and consider it to be represented by OA for the voltage being horizontal at time $t = 0$, then the current may have some value at that time and so be represented by another line OA at some angle ϕ at $t = 0$. There is said to be a *phase difference* of ϕ between the current and the voltage. If ϕ has a positive value then the current is said to be *leading* the voltage, if a negative value then *lagging* the voltage (Figure 17.15).

Figure 17.15   *(a) Current leading voltage, (b) current lagging voltage*

As with vectors, in books the usual convention is to write a phasor in a bold font as **I** or **V**. When we refer to the magnitude of the phasor we can write $|\mathbf{I}|$ or $|\mathbf{V}|$. However, the magnitude of a phasor is just the maximum value, or sometimes the root-mean-square value, and so is often written as I_m or V_m, or in the case of root-mean-square values just I or V. To specify a phasor all we need to do is specify its magnitude and whether it has a phase angle. Thus, for example, we can specify a current phasor as $20\angle 30°$ mA (r.m.s.). The magnitude, and hence in this case the root-mean-square value, is thus 20 mA and the phase angle is 30°. The phasor is thus an arrow with a length proportional to 20 mA at an angle of 30° to the horizontal.

Example

The instantaneous value i of a sinusoidal alternating current is described by the equation $i = 2 \sin (\omega t + \pi/4)$ amps. Determine the instantaneous current when (a) $\omega t = 0$, (b) $\omega t = \pi/4$ rad, (c) $\omega t = \pi/2$ rad.

(a) When $\omega t = 0$ then:

$i = 2 \sin \pi/4 = 1.41$ A

(b) When $\omega t = \pi/4$ then:

$i = 2 \sin (\pi/4 + \pi/4) = 2 \sin \pi/2 = 2$ A

(c) When $\omega t = \pi/2$ then:

$i = 2 \sin (\pi/2 + \pi/4) = 2 \sin 3\pi/4 = 1.41$ A

Example

A sinusoidal voltage has a maximum value of 10 V and a frequency of 100 Hz. If the voltage has a phase angle of 30°, what will be the instantaneous voltage at times of (a) $t = 0$, (b) $t = 0.5$ ms?

The equation for the sinusoidal voltage will be:

$v = V_m \sin (2\pi f t + \phi)$

The term $2\pi f t$, i.e. ωt, is in radians. Thus, for consistency, we should express ϕ in radians. An angle of 30° is $\pi/6$ radians. Thus:

$v = 10 \sin (2\pi \times 100t + \pi/6)$ volts

It should be noted that it is quite common in engineering to mix the units of radians and degrees in such expressions. Thus you might see:

$v = 10 \sin (2\pi \times 100t + 30°)$ volts

However, when carrying out calculations involving the terms in the bracket there must be consistency of the units.

(a) When $t = 0$ then:

$v = 10 \sin \pi/6 = 5$ V

(b) When $t = 0.5$ ms then:

$v = 10 \sin (2\pi \times 100 \times 0.5 \times 10^{-3} + \pi/6)$ volts

and so $v = 10 \sin 0.838 = 7.43$ V.

Revision

11 Complete the following table for the sinusoidal voltage described by the equation $v = 5 \sin (\omega t + \pi/2)$ volts:

ωt rad	0	$\pi/3$	$\pi/2$	$2\pi/3$	π	$4\pi/3$	$3\pi/2$	$5\pi/3$	2π
v volts									

12 A sinusoidal voltage has a maximum value of 1 V and a frequency of 1 kHz. If the voltage has a phase angle of 60°, what will be the instantaneous voltage at times of (a) $t = 0$, (b) $t = 0.5$ ms?

13 A sinusoidal alternating current has an instantaneous value i at a time t, in seconds, given by $i = 100 \sin (200\pi t - 0.25)$ mA. Determine (a) the maximum current, (b) the frequency, (c) the phase angle.

14 A sinusoidal alternating voltage has an instantaneous value v at a time t, in seconds, given by $v = 12 \sin (100\pi t + 0.5)$ volts. Determine (a) the maximum voltage, (b) the frequency, (c) the phase angle.

15 Draw two phasors to represent the following:

(a) A sinusoidal voltage with a root-mean-square value 2 V leading another voltage with a root-mean-square value of 4 V and the same frequency by 45°.
(b) A sinusoidal current of root-mean-square value of 12 mA lagging another current with a root-mean-square value of 6 mA and the same frequency by 60°.
(c) A sinusoidal voltage with a root-mean-square value 10 V lagging another voltage with a root-mean-square value of 5 V and the same frequency by 30°.

17.5 Adding phasors

Suppose we want to add the voltages across two components in series. If they are alternating voltages we must take account of the possibility that the two voltages may not be in phase, despite having the same frequency since they are supplied by the same source. This means that if we consider the phasors, they will rotate with the same angular velocity but may have different lengths and start with a phase angle between them. Consider one of the voltages to have a magnitude of V_1 and zero phase angle (Figure 17.16(a)). The other voltage we will consider as having a magnitude of V_2 and a phase difference of ϕ from the first voltage (Figure 17.16 (b)). We can obtain the sum of the two by adding the two graphs, point-by-point, to obtain the result shown in Figure 17.16(c). Thus at the instant of time indicated in the figures, the two voltages are v_1 and v_2. Hence the total voltage is $v = v_1 + v_2$. We can repeat this for each instant of time and hence end up with the graph shown in Figure 17.16(c).

Figure 17.16 *Adding two sinusoidal signals of the same frequency*

The following table illustrates this for a number of times when we add the voltages $v_1 = 1 \sin 1000t$ and $v_2 = 1.5 \sin (1000t + \pi/3)$:

t in ms	0	0.5	1.0	1.5	2.0
$v_1 = 1 \sin 1000t$	0	0.48	0.84	1.00	0.91
$v_2 = 1.5 \sin (1000t + \pi/3)$	1.31	1.50	1.34	0.84	0.14
$v = v_1 + v_2$	1.31	1.98	2.18	1.84	1.05

However, exactly the same result is obtained by adding the two phasors by means of the *parallelogram rule* or *triangle rule* of vectors (see Section

12.2). If we place the tails of the arrows representing the two phasors together and complete a parallelogram, then the diagonal of that parallelogram drawn from the junction of the two tails represents the sum of the two phasors. Figure 17.16(c) shows such a parallelogram and the resulting phasor with magnitude V.

If the phase angle between the two phasors of sizes V_1 and V_2 is 90°, as in Figure 17.17, then the resultant can be calculated by the use of the Pythagoras theorem as having a size V of:

$$V^2 = V_1^2 + V_2^2$$

and is at a phase angle ϕ relative to the phasor for V_2 of:

$$\tan\phi = \frac{V_2}{V_1}$$

Figure 17.17   *Adding two phasors*

If the phase angle is some angle other than 90° then we might use the cosine rule (see Section 12.4) to determine the resultant with the sine rule (see Section 10.5) or cosine rule to determine the phase angle. Alternatively we might resolve the phasors into their horizontal and vertical components (see Section 10.4), sum the components in each direction, and then use the Pythagoras theorem to determine the resultant. The following examples illustrate these methods.

Example

Two sinusoidal alternating voltages are described by the equations of $v_1 = 10\sqrt{2}\sin\omega t$ volts and $v_2 = 15\sqrt{2}\sin(\omega t + \pi/2)$ volts. Determine the sum of these voltages.

Figure 17.18   *Example*

Figure 17.18 shows the phasor diagram for the two voltages. The magnitudes of the phasors have been drawn as the root-mean-square values and so are of lengths proportional to 10 V and 15 V. The example could have been worked through with the maximum values and the result would have been the same. The angle between the phasors is $\pi/2$, i.e. 90°. We could determine the sum from a scale drawing or by calculation using the Pythagoras theorem. Thus:

$$(\text{sum})^2 = 10^2 + 15^2$$

Hence the magnitude of the sum of the two voltages is 18.0 V. The phase angle is given by:

$$\tan\phi = \frac{15}{10}$$

Hence $\phi = 56.3°$ or 0.983 rad. Thus the sum is an alternating voltage described by a phasor of root-mean-square magnitude 18.0 V and phase angle 56.3° (or 0.983 rad). This alternating voltage is thus described by $v = 18.0\sqrt{2}\sin(\omega t + 0.983)$ volts.

Example

Two sinusoidal alternating currents are described by the equations of $i_1 = 50\sqrt{2} \sin \omega t$ mA and $i_2 = 100\sqrt{2} \sin (\omega t + \pi/3)$ mA. Determine the total current when these currents both flow through the same branch of a circuit.

Figure 17.19 shows the phasor diagram for the two currents. The magnitudes of the phasors have been drawn as the root-mean-square values and so are of lengths proportional to 50 mA and 100 mA. The angle between the phasors is $\pi/3$ rad, i.e. 60°. We could determine the resultant from a scale drawing. Alternatively we can calculate it by means of the cosine law. Thus:

$$(\text{resultant})^2 = 100^2 + 50^2 - 2 \times 100 \times 50 \cos 120°$$

Thus the resultant has a magnitude, i.e. a root-mean-square value, of 132.3 mA. The phase angle ϕ of the resultant can be determined by the use of the sine rule. Thus:

$$\frac{132.3}{\sin 120°} = \frac{100}{\sin \phi}$$

Hence $\phi = 40.9°$ or 0.714 rad. The total current is thus a phasor with a root-mean-square magnitude of 132.3 mA and a phase angle of 40.9° (or 0.714 rad). This is a current of $i = 132.3\sqrt{2} \sin (\omega t + 0.714)$ mA.

Alternatively we could have resolved the phasors into their horizontal and vertical components. Thus for the $100\angle 60°$ mA phasor we have:

horizontal component $= 100 \cos 60°$

vertical component $= 100 \sin 60°$

For the $50\angle 0°$ mA phasor we have:

horizontal component $= 50$

vertical component $= 0$

Thus the sum of the horizontal components is:

$100 \cos 60° + 50 = 100$ mA

and the sum of the vertical components is:

$100 \sin 60° + 0 = 86.6$ mA

The resultant of these components (Figure 17.20) is given by the Pythagoras theorem as

![Figure 17.19 phasor diagram with 100 mA, 50 mA, Resultant, 60°, 120°]

Figure 17.19 *Example*

Figure 17.20 *Example*

$$(\text{resultant})^2 = 100^2 + 86.6^2$$

and hence the resultant has a magnitude of 132.3 mA. The phase angle is given by:

$$\tan\phi = \frac{86.6}{100}$$

and so $\phi = 40.9°$ or 0.714 rad. The result is thus as obtained earlier by the use of the cosine and sine rules.

Example

The voltage across one component in a circuit is $10 \sin \omega t$ volts and that across another component in series with it $20 \sin (\omega t - \pi/2)$ volts. What is the total voltage across the two components?

Figure 17.21 *Example*

The voltage across two components in series is the sum of the voltages across the two components. Figure 17.21 shows the phasor diagram. For convenience the magnitudes of the phasors have been drawn as the maximum values. The phase difference of $-\pi/2$ means that the voltage across the second component is drawn as lagging that across the first by 90°. Using the Pythagoras theorem, the sum is thus given by

$$(\text{sum})^2 = 10^2 + 20^2$$

Hence the sum phasor has a magnitude of 22.4 V. The phase angle is given by

$$\tan\phi = \frac{-20}{10}$$

and so $\phi = -63.4°$ or -1.11 rad. Thus the voltage is described by the expression $22.4 \sin (\omega t - 1.11)$ V.

Revision

16 The currents in two parallel branches of a circuit are $10 \sin \omega t$ milliamps and $20 \sin (\omega t + \pi/2)$ milliamps. What is the total current entering the parallel arrangement?

17 The voltage across a component in a circuit is $5.0 \sin \omega t$ volts and across another component in series with it $2.0 \sin (\omega t + \pi/6)$ volts. Determine the total voltage across both components.

18 The sinusoidal alternating voltage across a component in a circuit is $50\sqrt{2} \sin (\omega t + 40°)$ volts and across another component in series with it $100\sqrt{2} \sin (\omega t - 30°)$ volts. What is the total voltage across the two components?

19 The currents in two parallel branches of a circuit are $4\sqrt{2} \sin \omega t$ amps and $6\sqrt{2} \sin (\omega t - \pi/3)$ amps. What is the total current entering the parallel arrangement?

17.6 Subtracting phasors

Consider a situation where we know the circuit alternating current entering a junction of two parallel components and the current through one of them. What is the current through the other component? What we need to do is to subtract, at each instant of time, the value of the current through one component from the circuit current. Figure 17.22 illustrates this. Figure 17.22(a) shows the circuit current and Figure 17.22(b) the current through one of the components. Figure 17.22(c) shows the result of subtracting, at each instant of time, the values given by (b) from those given by (a).

Suppose we have the circuit current of Figure 17.22(a) as $1 \sin 1000t$ and the current through the first component as $1.5 \sin (1000t + \pi/3)$, as in Figure 17.22(b). Then the following table illustrates some of the results of a subtraction:

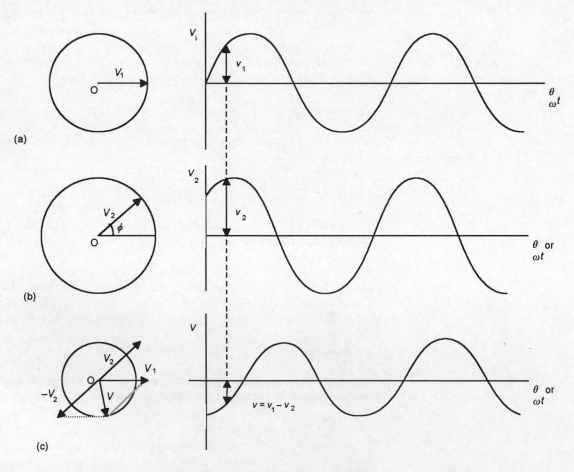

(a)

(b)

(c)

Figure 17.22 *Subtracting two sinusoidal signals of the same frequency*

t in ms	0	0.5	1.0	1.5	2.0
$v = 1 \sin 1000t$	0	0.48	0.84	1.00	0.91
$v_1 = 1.5 \sin (1000t + \pi/3)$	1.31	1.50	1.34	0.84	0.14
$v_2 = v - v_1$	−1.31	−1.02	−0.50	0.16	0.77

However, there is a simpler way of subtracting two sinusoidal alternating quantities and that is by subtracting the phasors. Figure 11.22(c) illustrates this. Subtraction of phasors is carried out in the same way as the subtraction of vectors (see section 12.3). Thus to subtract a phasor I_2 from a phasor I_1 we add the phasor $-I_2$ to phasor I_1. The phasor $-I_2$ is just the phasor I_2 with its direction reversed.

Example

The current flowing into a junction in a circuit is $i = 10 \sin \omega t$ milliamps. If the circuit branches at the junction and there are two currents leaving the junction and one is $i_1 = 6 \sin (\omega t + \pi/4)$ milliamps, what is the other current i_2?

Figure 17.23 *Example*

We can tackle this example by subtracting phasors. Figure 17.23 shows the phasors for the currents i and i_1. To subtract i_1 from i we add the phasor for $-i_1$ to that for i. The phasor for $-i_1$ has the same magnitude as that for i_1 but is in the opposite direction. Thus, using the cosine rule:

$$(\text{difference})^2 = 10^2 + 6^2 - 2 \times 10 \times 6 \cos 45°$$

Hence the difference has a magnitude of 7.15 mA. The phase angle ϕ can be obtained by the use of the sine rule. Thus:

$$\frac{7.15}{\sin 45°} = \frac{6}{\sin \phi}$$

Thus $\phi = 36.4°$ or 0.635 rad. Hence the current i_2 is given by the equation $i_2 = 7.15 \sin (\omega t - 0.635)$ mA.

Alternatively we could have obtained the solution by resolving the phasors for i and $-i_1$ into the horizontal and vertical directions. The sum of these phasor components in the horizontal direction is:

horizontal component $= 10 - 6 \cos 45° = 5.76$ mA

The sum of the vertical components is:

vertical component $= -6 \sin 45° = -4.24$ mA

The minus sign has been included to indicate that this component is directed downwards and not upwards. The magnitude of the difference phasor is thus obtained by the use of the Pythagoras theorem as:

$$(\text{difference})^2 = 5.76^2 + (-4.24)^2$$

Hence the difference has a magnitude of 7.15 mA. The phase angle of this difference phasor is given by:

$$\tan \phi = \frac{-4.24}{5.76}$$

Hence the phase angle is 36.4° or 0.635 rad. The result is thus as obtained by the cosine and sine rules.

Revision

20 If the voltage across a pair of series components is $4 \sin \omega t$ volts and that across one of the components is $2 \sin (\omega t + \pi/4)$ volts, what is the voltage across the other component?

21 The current through a component is equal to the current $5 \sin \omega t$ amps in the circuit minus the current $4 \sin (\omega t - \pi/2)$ amps through a parallel component. Determine the current through the component.

22 The current through a component is equal to the current $120 \sin \omega t$ milliamps in the circuit minus the current $80 \sin (\omega t - \pi/4)$ milliamps through a parallel component. Determine the current through the component.

Problems

1 A sinusoidal alternating current has a maximum value of 2 A. What is the average value over (a) half a cycle, (b) a full cycle?

2 A rectangular shaped alternating voltage has a value of 5 V for half a cycle and −5 V for the other half. What is the average value over (a) half a cycle, (b) a full cycle?

3 Show that the average value for a square waveform, as in Figure 17.7(b), over half a cycle is equal to the maximum value.

4 Show that the average value for a sawtooth waveform over half a cycle, as in Figure 17.5(d), is half the maximum value.

5 A sinusoidal alternating current has a maximum value of 4 V. What is the root-mean-square value?

6 A rectangular shaped alternating current has a maximum value of 4 A for half a cycle and −4 A for the other half. What is the root-mean-square value?

7 A half cycle of a waveform gave the following voltage values for the mid-ordinates, the other half cycle being a mirror image. What are (a) the half cycle average value, (b) the root-mean-square value, and (c) the form factor?

Voltages in V 1 3 5 7 9 9 9 9

8 A half cycle of a waveform gave the following current values for the mid-ordinates, the other half cycle being a mirror image. What are (a) the half cycle average value, (b) the root-mean-square value, and (c) the form factor?

Current in mA 5 15 25 25 15 5

9 Show that the root-mean-square value for a square waveform, as in Figure 17.7(b), over half a cycle is equal to the maximum value.

10 Show that the root-mean-square value for a sawtooth waveform over half a cycle, as in Figure 17.7(d), is the maximum value divided by $\sqrt{3}$.

11 Determine the average value, the root-mean-square value and the form factor for the voltage with the waveform shown in Figure 17.24.

12 Determine the root-mean-square value of a square wave voltage waveform with a maximum value of 1 V.

13 Determine the root-mean-square value of a sinusoidal voltage waveform with a maximum value of 2 V.

14 Determine the root-mean-square value of a sinusoidal current waveform with a maximum value of 10 mA.

15 Determine the maximum value of a sinusoidal current waveform with a root-mean-square value of 100 mA.

16 Determine the maximum value of a square wave voltage waveform with a root-mean-square value of 10 V.

Figure 17.24 *Problem 11*

17 Complete the following tables:

(a)

t in ms	0	2	4	6	8	10
$v = 2 \sin 500t$ in V						
$i = 10 \sin 100t$ in mA						
$v = 5 \sin 1000t$ in V						

(b)

ωt rad	0	$\pi/3$	$\pi/2$	$2\pi/3$	π
$v = 10 \sin \omega t$ in V					
$v = 10 \sin (\omega t + \pi/2)$ in V					
$v = 10 \sin (\omega t + \pi/3)$ in V					

18 A sinusoidal alternating current has an angular frequency of 100 rad/s and a maximum value of 10 A. What is the current 5 ms after it has a zero value?

19 A sinusoidal alternating voltage has a frequency of 2 MHz and a maximum value of 10 V. What is the instantaneous voltage 0.1 μs after it is at a zero value?

20 A sinusoidal alternating voltage in mV is represented by the equation $v = 50 \sin (500t + 20°)$. What is (a) the angular frequency, (b) the frequency, (c) the value of the voltage at $t = 0$?

21 Draw the phasors to represent the following sinusoidal alternating quantities when each has the same frequency:

(a) Two currents with root-mean-square values of 3 A and 5 A, with the second current leading the first by a phase difference of 60°.
(b) Two voltages with root-mean-square values of 1 V and 2 V, with the second voltage lagging the first by a phase difference of 30°.

22 A sinusoidal alternating current has an instantaneous value i at a time t, in seconds, given by $i = 20 \sin (500\pi t - 0.5)$ mA. Determine (a) the maximum current, (b) the frequency, (c) the phase angle.

23 A sinusoidal alternating voltage has an instantaneous value v at a time t, in seconds, given by $v = 5 \sin (1000\pi t + 0.2)$ volts. Determine (a) the maximum voltage, (b) the frequency, (c) the phase angle.

24 The voltage across a component in a circuit is $4.0 \sin \omega t$ volts and across another component in series with it $2.0 \sin (\omega t + \pi/2)$ volts. Determine the total voltage across both components.

25 The voltage across a component in a circuit is $5.0 \sin \omega t$ volts and across another component in series with it $2.0 \sin (\omega t - \pi/6)$ volts. Determine the total voltage across both components.

26 The voltage across a component in a circuit is $4.0 \sin (\omega t + \pi/3)$ volts and across another component in series with it $3.0 \sin (\omega t - \pi/4)$ volts. Determine the total voltage across both components.

27 The currents in two parallel branches of a circuit are $20 \sin \omega t$ milliamps and $30 \sin (\omega t + \pi/3)$ milliamps. What is the total current entering the parallel arrangement?

28 The currents in two parallel branches of a circuit are $30 \sin \omega t$ milliamps and $40 \sin (\omega t + \pi/2)$ milliamps. What is the total current entering the parallel arrangement?

29 The voltage across a component in a circuit is $4.0 \sin (\omega t + \pi/2)$ volts and across another component in series with it $4.0 \sin \omega t$ volts. Determine the total voltage across both components.

30 The voltage across a component in a circuit is $20 \sin \omega t$ volts and across another component in series with it $15 \sin (\omega t - \pi/2)$ volts. Determine the total voltage across both components.

31 The current through a component is equal to the current $20 \sin \omega t$ milliamps in the circuit minus the current $10 \sin (\omega t + \pi/4)$ milliamps through a parallel component. Determine the current through the component.

32 The current through a component is equal to the current $25 \sin \omega t$ milliamps in the circuit minus the current $16 \sin (\omega t - \pi/4)$ milliamps through a parallel component. Determine the current through the component.

33 If the voltage across a pair of series components is $4 \sin \omega t$ volts and that across one of the components is $2 \sin (\omega t - \pi/2)$ volts, what is the voltage across the other component?

34 If the voltage across a pair of series components is $10 \sin \omega t$ volts and that across one of the components is $5 \sin (\omega t + \pi/3)$ volts, what is the voltage across the other component?

35 Determine for the alternating currents $i_1 = 10 \sin (\omega t + \pi/6)$ milliamps and $i_2 = 8 \sin (\omega t - \pi/3)$ milliamps (a) the sum of the two currents and (b) the difference when the second current is subtracted from the first.

36 A circuit contains two parallel branches. The circuit current is $10 \sin \omega t$ amps and the current in one branch is $10 \sin (\omega t - \pi/3)$ amps. What is the current in the other branch?

37 A circuit contains two parallel branches. The root-mean-square current in one branch is 1.2 A in the reference direction. The current in the other branch has a root-mean-square magnitude of 2.0 A and lags the current in the first branch by 30°. What is the circuit current entering the parallel branches?

Section F
Calculus

18 Differentiation
19 Further differentiation
20 Maxima and minima
21 Integration
22 Calculus in action

The aims of this section are to enable the reader to:

- Identify differentiation in terms of the gradient of a graph and rate of change.
- Differentiate basic equations from first principles and use standard rules.
- Differentiate algebraic, trigonometric, exponential and logarithmic equations.
- Differentiate products, quotients and functions of a function.
- Determine second derivatives.
- Determine maxima and minima.
- Identify integration as the inverse of differentiation and also as the area under a graph.
- Integrate algebraic, trigonometric and exponential equations.
- Use integration to determine the area under graphs.
- Solve engineering problems involving the use of differentiation or integration.

This section is an introduction to calculus. Chapter 18 is an introduction to differentiation and Chapter 19 extends this introduction. Chapter 20 shows the application of differentiation to the determination of maxima and minima. Chapter 21 is an introduction to integration. Chapter 22 shows applications of differentiation and integration in the solution of engineering problems involving straight-line motion, work and circuits involving capacitors and inductors.

Dexterity with algebraic equations and graphs is assumed. Chapters 18, 19 and 21 assume that Chapter 17 has been completed. Chapter 22 assumes that Chapters 18 and 21 have been completed.

18 Differentiation

18.1 Introduction

Differentiation is a mathematical technique which can be used to determine the rate at which functions change at some particular point and hence the gradients of the tangents to graphs at a particular point. For example, we might have an equation describing how the distance covered in a straight line by a moving object varies with time. We could plot a graph of distance against time and determine the velocity at some instant as the gradient of the tangent to the curve at that instant. By taking a number of such gradient measurements we could then determine how the velocity varied with time. However, differentiation is a technique which enables the velocity to be obtained from the equation without drawing the graph and tangents. This chapter is an introduction to differentiation.

18.2 Gradients

Consider the problem of determining the gradient of a tangent at a point on a graph. It might, for example, be a distance–time graph for a moving object or the current–time graph for the current in an electrical circuit. Suppose we want to determine the gradient at point A on the curve shown in Figure 18.1. We can select another point B on the curve and join them together and then find the gradient of the line AB. The value of the gradient determined in this way will obviously depend on where we locate the point B. If we let B slide along the curve towards A then the closer B is to A the more the line approximates to the tangent at point A. Thus the line AB_1 is a better approximation to the tangent than the line AB. The method we can use to determine the gradient of the tangent at a point A on a curve is thus:

1 Take another point B on the same curve and determine the gradient of the line joining A and B.

2 Then move B closer and closer to A. In the limit as the distance between A and B becomes infinitesimally small then the gradient of the line becomes the gradient of the tangent at A.

Figure 18.1 *Gradient at A*

Consider A having the coordinates (x_1, y_1). If we select a point B with the coordinates (x_2, y_2) then the gradient of the line AB is:

$$\text{gradient} = \frac{y_2 - y_1}{x_2 - x_1}$$

We thus have the difference in the value of y between points A and B divided by the difference in the value of x between the points. We can write this difference in the value of x as δx and this difference in the value of y as δy. The δ symbol in front of a quantity means 'a small bit of it' or 'an interval of'. Thus the equation can be written as

$$\text{gradient} = \frac{\delta y}{\delta x}$$

An alternative symbol which is often used is Δx, with the Δ symbol being used to indicate that we are referring to a small bit of the quantity x. These forms of notation do *not* mean that we have δ or Δ multiplying x. The δx or Δx should be considered as a single symbol representing a single quantity.

As we move B closer to A then the interval δx is made smaller. The gradient of the line AB then becomes closer to the tangent to the curve at point A. Eventually when the difference in x between A and B, i.e. δx, tends to zero then we have the gradient of the tangent at point A. We can write this as:

$$\lim_{\delta x \to 0} \frac{\delta y}{\delta x} = \frac{dy}{dx}$$

This reads as: the limiting value of $\delta y/\delta x$ as δx tends to a zero value equals dy/dx. A *limit* is a value to which we get closer and closer as we carry out some operation. Thus dy/dx is the value of the gradient of the tangent to the curve at A. Since the tangent is the instantaneous rate of change of y with x at that point then dy/dx is the instantaneous rate of change of y with respect to x. dy/dx is called the *derivative* of y with respect to x. The process of determining the derivative for a function is called *differentiation*. The notation dy/dx should not be considered as d multiplied by y divided by d multiplied by x, but as a single symbol representing the gradient of the tangent and so the rate of change of y with x.

18.2.1 Determining derivatives from first principles

Suppose we have the two variables x and y which are related by the equation $y = x^2$. As x varies then y varies in such a way that the y is always equal to the square of the value of x. Consider a point A on the curve (Figure 18.2) with the coordinates (x, y). Now consider a point B on the curve with coordinates $(x + \delta x, y + \delta y)$. The gradient of the line AB is then:

$$\text{gradient} = \frac{BC}{AC} = \frac{\delta y}{\delta x}$$

Points A and B both lie on the same curve. Thus at point A, we must have:

$$y = x^2$$

At B, we must have:

$$y + \delta y = (x + \delta x)^2 = x^2 + 2x\,\delta x + (\delta x)^2$$

Subtracting these two equations gives:

$$
\begin{aligned}
y + \delta y &= x^2 + 2x\,\delta x + (\delta x)^2 \\
\text{minus} \quad y &= x^2 \\
\hline
\delta y &= \qquad 2x\,\delta x + (\delta x)^2
\end{aligned}
$$

Figure 18.2   *Gradient at A for $y = x^2$*

Therefore:

$$\frac{\delta y}{\delta x} = 2x + \delta x$$

If we want the gradient at $x = 4$ and take a value for δx of 2, i.e. point B is at $x = 4 + 2 = 6$, then the gradient of the line AB is $2 \times 4 + 2 = 10$. If we now move point B closer to A so that we now have $\delta x = 1$ then the gradient of the line AB is $2 \times 4 + 1 = 9$. If we move point B yet closer to A so that $\delta x = 0.5$ then the gradient is the line AB is $2 \times 4 + 0.5 = 8.5$. With $\delta x = 0.1$ we have the gradient of the line AB as $2 \times 4 + 0.1 = 8.1$. As δx tends to the value 0 then the gradient of the line at $x = 4$ tends to the value 8, i.e. $2x$.

In the limit as we make δx smaller and smaller then δx tends to the value 0. Thus, in the above equation, $\delta y/\delta x$ tends to the value $2x$ as δx tends to zero. The gradient of the tangent at point A, coordinates (x, y), is thus given as:

$$\text{gradient} = \lim_{\delta x \to 0} \frac{\delta y}{\delta x} = \frac{dy}{dx} = 2x$$

Thus we have a gradient at $x = 4$ of $2x = 2 \times 4 = 8$.

The procedure adopted above to determine the derivative of y with respect to x when we have $y = x^2$ can be written as:

$$\frac{dy}{dx} = \lim_{\delta x \to 0} \frac{\delta y}{\delta x} = \lim_{\delta x \to 0} \frac{\text{change in } y \text{ when } x \text{ changes by } \delta x}{\delta x}$$

$$= \lim_{\delta x \to 0} \frac{(x + \delta x)^2 - x^2}{\delta x}$$

$$= \lim_{\delta x \to 0} \frac{x^2 + 2x\,\delta x + (\delta x)^2 - x^2}{\delta x}$$

$$= \lim_{\delta x \to 0} (2x + \delta x)$$

$$= 2x$$

We can use this technique with other functions. The following example and the next section give some examples of using this technique.

Example

The variable y varies with another variable x according to the equation $y = 2x + x^2$. Determine the derivative of y with respect to x and hence the value of the gradient of the graph of y against x at $x = 1$.

Using the procedure outlined above:

$$\frac{dy}{dx} = \lim_{\delta x \to 0} \frac{\text{change in } y \text{ when } x \text{ changes by } \delta x}{\delta x}$$

$$= \lim_{\delta x \to 0} \frac{\{2(x+\delta x)+(x+\delta x)^2\} - \{2x+x^2\}}{\delta x}$$

$$= \lim_{\delta x \to 0} (2+2x+\delta x)$$

$$= 2 + 2x$$

At $x = 1$, the gradient of the graph y plotted against x will have the value $2 + 2 \times 1 = 4$.

Revision

1 Differentiate from first principles, i.e. determine dy/dx, for the following equations and hence determine the gradients of their graphs at $x = 0$ and $x = 1$.

(a) $y = 3x^2$, (b) $y = x^2 + x$, (c) $y = 3x + 1$, (d) $y = 1/(2x)$, (e) $y = 4$, (f) $y = 3x$, (g) $y = x/4$.

18.3 Derivative of $y = ax^n$

Consider the equation $y = ax$, where a is some constant. We can find the derivative dy/dx using the procedures outlined above. Thus

$$\frac{dy}{dx} = \lim_{\delta x \to 0} \frac{\text{change in } y \text{ when } x \text{ changes by } \delta x}{\delta x}$$

$$= \lim_{\delta x \to 0} \frac{a(x+\delta x) - ax}{\delta x}$$

$$= \lim_{\delta x \to 0} a$$

$$= a$$

This what we would expect since $y = ax$ describes a straight-line graph with a gradient of a which is the same for all values of x.

Now consider the equation $y = ax^2$, where a is some constant.

$$\frac{dy}{dx} = \lim_{\delta x \to 0} \frac{\text{change in } y \text{ when } x \text{ changes by } \delta x}{\delta x}$$

$$= \lim_{\delta x \to 0} \frac{a(x+\delta x)^2 - ax^2}{\delta x}$$

$$= \lim_{\delta x \to 0} (2ax + a\delta x)$$

$$= 2ax$$

Now consider the equation $y = ax^3$.

$$\frac{dy}{dx} = \lim_{\delta x \to 0} \frac{\text{change in } y \text{ when } x \text{ changes by } \delta x}{\delta x}$$

$$= \lim_{\delta x \to 0} \frac{a(x + \delta x)^3 - ax^3}{\delta x}$$

$$= \lim_{\delta x \to 0} \{3ax^2 + 3ax\,\delta x + a(\delta x)^2\}$$

$$= 3ax^2$$

If we summarise the above, and add further derivatives calculated in the same way, we can establish a pattern.

$$y = ax \qquad \frac{dy}{dx} = a$$

$$y = ax^2 \qquad \frac{dy}{dx} = 2ax$$

$$y = ax^3 \qquad \frac{dy}{dx} = 3ax^2$$

$$y = ax^4 \qquad \frac{dy}{dx} = 4ax^3$$

$$y = ax^5 \qquad \frac{dy}{dx} = 5ax^4$$

The pattern in the above data is that if we have $y = ax^n$, where a and n are constants, then:

$$\frac{dy}{dx} = nax^{n-1}$$

This relationship applies for positive, negative and fractional values of n. Note that if we have just $y = a$, i.e. $n = 0$, then the above equation gives $dy/dx = 0$. The equation $y = a$ represents a horizontal straight line and the gradient of it is 0. Also note that when a function is multiplied by a constant, as for example the a in the above equations, then the derivative is multiplied by the same constant.

Example

Determine the derivative of $y = x^5$ and hence the gradient of its graph at $x = 1$.

Using the equation derived above:

$$\frac{dy}{dx} = nax^{n-1} = 5x^{5-1} = 5x^4$$

At $x = 1$ then $dy/dx = 5$. Hence the gradient at $x = 1$ is 5.

Example

Determine the derivative of $y = 4x^2$ and hence the gradient of its graph at $x = 2$.

Using the equation derived above:

$$\frac{dy}{dx} = nax^{n-1} = 2 \times 4x^{2-1} = 8x$$

At $x = 2$ then $dy/dx = 16$. Hence the gradient at $x = 2$ is 16.

Example

Determine the derivative of $y = 3/x$ and hence the gradient of its graph at $x = 2$.

This equation can be written as $y = 3x^{-1}$. Thus, using equation derived above,

$$\frac{dy}{dx} = nax^{n-1} = -1 \times 3x^{-1-1} = -\frac{3}{x^2}$$

At $x = 2$ then $dy/dx = -3/4$ and so the gradient at $x = 2$ is $-3/4$.

Example

Determine the derivative of $y = 2x^{3/4}$ and hence determine the gradient of its graph at $x = 4$.

Using the equation derived above:

$$\frac{dy}{dx} = nax^{n-1} = \frac{3}{4} \times 2x^{3/4-1} = \frac{3}{2}x^{-1/4}$$

At $x = 4$ then $dy/dx = 1.5 \times 0.71 = 1.1$. Hence the gradient is 1.1.

Example

The volume V of a cube is given by $V = L^3$, where L is the length of a side. Determine the rate of change of volume of a cube with respect to its side length.

Using the equation derived above:

$$\frac{dy}{dx} = nax^{n-1} = 3L^{3-1} = 3L^2$$

The rate of change of the volume with respect to the side length is $3L^2$.

Revision

2 Show from first principles that dy/dx for $y = x^{-1}$ is $-x^{-2}$.

3 Determine, using the rule derived above, the derivatives of:

(a) $y = 2x^3$, (b) $y = x^6$, (c) $y = 2x^{-2}$, (d) $y = x^{3/2}$, (e) $y = 2x^{1/4}$,

(f) $y = 3x^{10}$, (g) $y = 3x^{-1/5}$, (h) $y = 0.1x^4$

4 Determine the gradient of the graph of y against x for $y = 5x^2$ at $x = 1$.

5 Determine the gradient of the graph of y against x for $y = 2x^3$ at $x = 2$.

6 Determine the gradient of the graph of y against x for $y = 2/x$ at $x = 2$.

7 Determine the rate of change of the area of a circle with respect to its radius when the radius is 10 mm.

18.4 Basic rules

The following are some basic rules we can use when differentiating. Further rules are discussed in Chapter 19.

18.4.1 The derivative of a constant

The derivative of a constant is zero. Thus, if a is a constant:

$$\frac{d}{dx}a = 0$$

18.4.2 The constant multiplier rule

The derivative of x^2 is $2x$. The derivative of $5x^2$ is $5 \times 2x$. The derivative of any function which is multiplied by a constant is the derivative of that function multiplied by the same constant. Thus if u is some function and a is a constant:

$$\frac{d}{dx}(au) = a\frac{du}{dx}$$

18.4.3 The sum rule

Consider the problem of determining the derivative of $y = x + x^2$.

$$\frac{dy}{dx} = \lim_{\delta x \to 0} \frac{\text{change in } y \text{ when } x \text{ changes by } \delta x}{\delta x}$$

$$= \lim_{\delta x \to 0} \frac{\{(x+\delta x) + (x+\delta x)^2\} - \{x + x^2\}}{\delta x}$$

$$= \lim_{\delta x \to 0} (1 + 2x + \delta x)$$

$$= 1 + 2x$$

But this is just what we would obtain if we took the sum of the derivatives of the separate terms in the equation $y = x + x^2$. Thus, for $y = x$ we have $dy/dx = 1$ and for $y = x^2$ we have $dy/dx = 2x$. The sum of these two derivatives is $1 + 2x$.

To illustrate it with another example, consider the problem of determining the derivative of $y = 3x^3 - 4x^2$.

$$\frac{dy}{dx} = \lim_{\delta x \to 0} \frac{\text{change in } y \text{ when } x \text{ changes by } \delta x}{\delta x}$$

$$= \lim_{\delta x \to 0} \frac{\{3(x+\delta x)^3 - 4(x+\delta x)^2\} - \{3x^3 - 4x^2\}}{\delta x}$$

$$= \lim_{\delta x \to 0} \{9x^2 + 9x\,\delta x + 3(\delta x)^2 - 8x - 4\,\delta x\}$$

$$= 9x^2 - 8x$$

But this is just what we would obtain if we took the sum of the derivatives of the separate terms in the equation $y = 3x^3 - 4x^2$. Thus, for $y = 3x^3$ we have $dy/dx = 9x^2$ and for $y = -4x^2$ we have $dy/dx = -8x$. The sum of these two derivatives is $9x^2 - 8x$.

Thus in general, if we have two functions u and v, then the derivative of their sum is given by:

$$\frac{d}{dx}(u + v) = \frac{du}{dx} + \frac{dv}{dx}$$

The sum rule can be extended to sums of more than two functions.

Example

Determine the derivative of $y = 2x^2 + x + 4$.

For $y = 2x^2$ we have $dy/dx = 4x$. For $y = x$ we have $dy/dx = 1$. For $y = 4$ we have $dy/dx = 0$. Hence for $y = 2x^2 + x + 4$ we have

$$\frac{dy}{dx} = 4x + 1$$

Example

Determine the derivative of $y = 3x^4 - 2x^3 + 3x$.

For $y = 3x^4$ we have $dy/dx = 12x^3$. For $y = -2x^3$ we have $dy/dx = -6x^2$. For $y = 3x$ we have $dy/dx = 3$. Hence for $y = 3x^4 - 2x^3 + 3x$ we have:

$$\frac{dy}{dx} = 12x^3 - 6x^2 + 3$$

Revision

8 Determine, using the rule for the derivative of a sum, the derivatives of the following:

(a) $y = x^2 - 5$, (b) $y = 2x^4 + 3x^2$, (c) $y = 4x^3 - 2x^2 + 3x$,

(d) $y = 4x^3 - 3x^2 + 4x - 2$, (e) $y = 2x^5 + 3x^3 - 2x$, (f) $y = 4x^5 - 4$,

(g) $y = \frac{2}{x} + \frac{5}{x^2}$, (h) $y = 2x^2 - \frac{3}{x}$.

18.5 Derivatives of trigonometric functions

Consider the determination of how the gradients of the graph of $y = \sin x$ (Figure 18.3(a)) vary with x. Examination of the graphs shows that:

1 As x increases from 0 to $\pi/2$ then the gradient, which is positive, gradually decreases to become zero at $x = \pi/2$.

2 As x increases from $\pi/2$ to π then the gradient, which is now negative, becomes steeper and steeper to reach a maximum value at $x = \pi$.

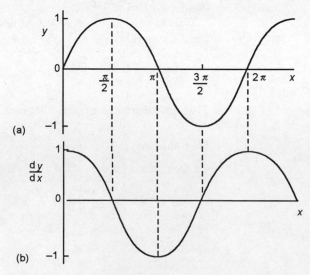

(a)

(b)

Figure 18.3 *Gradients of* $y = \sin x$

3   As x increases from π to $3\pi/2$ then the gradient, which is negative, decreases to become zero at $x = 3\pi/2$.

4   As x increases from $3\pi/2$ to 2π the gradient, which is now positive again, increases to become a maximum at $x = 2\pi$.

Figure 18.3(b) shows the result that is obtained by plotting the gradients against x. The figure shows a cosine curve. Thus, the derivative of $y = \sin x$ is:

$$\frac{d}{dx}(\sin x) = \cos x$$

If we had considered $y = \sin ax$, where a is a constant, then we would have obtained for the derivative

$$\frac{d}{dx}(\sin ax) = a\cos ax$$

18.5.1 Derivative of cosines

Consider $y = \cos x$ (Figure 18.4(a)) and the gradients at various points along the graph.

1   Between $x = 0$ and $x = \pi/2$ the gradient, which is negative, becomes steeper and steeper and reaches a maximum value at $x = \pi/2$.

2   Between $x = \pi/2$ and $x = \pi$ the gradient, which is negative decreases until it becomes zero at $x = \pi$.

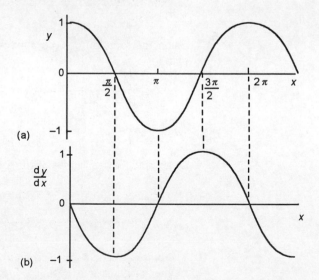

Figure 18.4 *Gradients of $y = \cos x$*

3 Between $x = \pi$ and $x = 3\pi/2$ the gradient, which is positive, increases until it becomes a maximum at $x = 3\pi/2$.

4 Between $x = 3\pi/2$ and $x = 2\pi$ the gradient, which is positive, decreases to become zero at $x = 2\pi$.

Figure 18.4(b) shows how the gradient varies with x. The result is an inverted sine graph. Thus, for $y = \cos x$:

$$\frac{d}{dx}(\cos x) = -\sin x$$

If we had considered $y = \cos ax$, where a is a constant, then we would have obtained for the derivative:

$$\frac{d}{dx}(\cos ax) = -a \sin ax$$

Example

Determine the derivative of the function $y = \sin 3x$.

Using the equation derived above for the derivative of $y = \sin ax$,

$$\frac{dy}{dx} = a \cos ax = 3 \cos 3x$$

Example

The current i in an electrical circuit varies with time t and is given by $i = \cos 100t$. Derive an equation for the rate of change of the current with time.

Using the equation derived above for the derivative of $y = \cos ax$, we obtain:

$$\frac{di}{dt} = -100 \sin 100t$$

Revision

9 Determine the derivatives of the following:

(a) $y = \sin x$, (b) $y = \sin 5x$, (c) $y = \cos 4x$, (d) $y = \cos 3x$,

(e) $y = 2 \sin 3x$, (f) $y = 10 \cos 4x$.

10 The displacement x of an oscillating object varies with time t according to the equation $y = \cos 50t$. Derive an equation indicating how the velocity, i.e. dy/dt, varies with time.

11 The current i, in amperes, in an electrical circuit varies with the time t, in seconds, according to the equation $i = 2 \sin 100t$. Determine the rate of change of current when $t = 0.01$ s.

18.6 Derivatives of exponential functions

Consider the exponential equation $y = e^x$. Consider a small increase in x of δx. The corresponding increase in the value of y is δy where:

$$y + \delta y = e^{x + \delta x} = e^x e^{\delta x}$$

Thus:

$$\frac{dy}{dx} = \lim_{\delta x \to 0} \frac{\text{change in } y \text{ when } x \text{ changes by } \delta x}{\delta x}$$

$$= \lim_{\delta x \to 0} \frac{e^x e^{\delta x} - e^x}{\delta x} = \lim_{\delta x \to 0} \frac{e^x(e^{\delta x} - 1)}{\delta x}$$

Figure 18.5 $\quad y = e^x$

If we let $\delta x = 0.01$ then $(e^{0.01} - 1)/0.01 = 1.005$. If we take yet smaller values of dx then in the limit we obtain the value 1 for this fraction. Thus:

$$\frac{d}{dx} e^x = e^x$$

The derivative of e^x is e^x. Thus the gradient of the graph of $y = e^x$ at a point is equal to the value of y at that point (Figure 18.5). For example, at the point $x = 0$ on the graph the gradient is $y = e^0 = 1$. At $x = 2$ the gradient is $y = e^2 = 7.39$. At $x = -2$ the gradient is $y = e^{-2} = 0.14$.

If we had $y = e^{ax}$ then

$$\frac{d}{dx}(e^{ax}) = a e^{ax}$$

Example

Determine the derivative of $y = e^{3x}$.

Using the equation given above:

$$\frac{dy}{dx} = a e^{ax} = 3 e^{3x}$$

Example

Determine the derivative of $y = e^{-x}$.

Using the equation given above:

$$\frac{dy}{dx} = a e^{ax} = -1 e^{-x}$$

Example

Determine the derivative of $y = 10 - e^{-3x}$.

This is the sum of two terms, hence the derivative is the sum of the derivatives of the two terms. For $y = 10$ we have $dy/dx = 0$ and for $y = -e^{-3x}$ we have $dy/dx = +3\,e^{-3x}$. Thus

$$\frac{dy}{dx} = 3\,e^{-3x}$$

Revision

12 Determine the derivatives of the following:

(a) $y = e^{2x}$, (b) $y = e^{-3x}$, (c) $y = e^{x/2}$, (d) $y = 2\,e^{3x}$, (e) $y = 4 + e^{2x}$,

(f) $y = x + 3\,e^{2x}$, (g) $y = 2\,e^x + 3\,e^{2x}$.

13 The number of radioactive atoms N in a radioactive element changes with time t according to the equation $N = N_0\,e^{-\lambda t}$, where N_0 and λ are constants. Determine the rate of change of N with time (this is called the activity). Note that multiplying an exponential by a constant multiplies the derivative by the same constant.

18.7 Derivatives of logarithmic functions

If $y = \ln x$, then we can write this as:

$$x = e^y$$

Differentiating with respect to y gives:

$$\frac{dx}{dy} = e^y = x$$

Hence, if we invert this we have:

$$\frac{dy}{dx} = \frac{1}{x}$$

$$\frac{d}{dx}(\ln x) = \frac{1}{x}$$

Thus the gradient of a graph of $y = \ln x$ is $1/x$. For example, at $x = 4$ the gradient is $1/4$. At $x = 1$ the gradient is $1/1 = 1$. Figure 18.6(a) shows the graph of $y = \ln x$ and Figure 18.6(b) how the derivative dy/dx varies with x.

(a)

(b)

Figure 18.6 $y = \ln x$

Example

Determine the derivative of $y = \ln 2x$.

We can write this equation as $y = \ln 2 + \ln x$. We can then obtain the derivative of each term and add them together to give the required derivative. For $y = \ln 2$ we have $dy/dx = 0$ and for $y = \ln x$ we have $dy/dx = 1/x$. Thus, the derivative for $y = \ln 2x$ is:

$$\frac{dy}{dx} = \frac{1}{x}$$

Revision

14 Determine the derivative of $y = \ln 3x$.

18.8 The second derivative

Consider a moving object for which we have a relationship between the displacement s of the object and time t of the form:

$$s = ut + \tfrac{1}{2}at^2$$

where u and a are constants. We can plot this equation to give a distance–time graph. If we differentiate this equation we obtain:

$$\frac{ds}{dt} = u + at$$

ds/dt is the gradient of the distance–time graph. It also happens to be the velocity. The gradient varies with time. We could thus plot a velocity–time graph, i.e. a ds/dt graph against t. Then differentiating for a second time, to obtain the gradients to this graph, we obtain the acceleration a.

$$\frac{d}{dt}\left(\frac{ds}{dt}\right) = a$$

The derivative of a derivative is called the *second derivative* and can be written as:

$$\frac{d}{dx}\left(\frac{dy}{dx}\right) \text{ or } \frac{d^2y}{dx^2}$$

The first derivative gives information about how the gradients of the tangents change. The second derivative gives information about the rate of change of the gradient of the tangents.

Example

Determine the second derivative of $y = x^3$.

The first derivative is:

$$\frac{dy}{dx} = 3x^2$$

The second derivative is given by differentiating this equation again:

$$\frac{d^2y}{dx^2} = 6x$$

Example

Determine the second derivative of $y = x^4 + 3x^2$.

The first derivative is

$$\frac{dy}{dx} = 4x^3 + 6x$$

The second derivative is

$$\frac{d^2y}{dx^2} = 12x^2 + 6$$

Revision

15 Determine the second derivatives of the following functions:

(a) $y = x^2$, (b) $y = 2x^2 + x$, (c) $y = x^3 + 2x^2$, (d) $y = \sin x$, (e) $y = e^{2x}$

16 The charge q on the plates of a capacitor is related to time t by the equation $q = CV\,e^{-t/RC}$, where C, V and R are constants. Determine how (a) the current, i.e. dq/dt, changes with time and (b) how the rate of change of current with time, i.e. the second derivative d^2q/dt^2, changes with time.

18.9 Derivatives

The following lists the derivatives of the functions and rules that have been encountered in this chapter. a is a constant, u and v are functions of x.

$$\frac{d}{dx}(ax^n) = nax^{n-1} \qquad\qquad \frac{d}{dx}(a) = 0$$

$$\frac{d}{dx}(au) = a\frac{du}{dx} \qquad\qquad \frac{d}{dx}(u+v) = \frac{du}{dx} + \frac{dv}{dx}$$

$$\frac{d}{dx}(\sin ax) = a\cos ax \qquad\qquad \frac{d}{dx}(\cos ax) = -a\sin ax$$

$$\frac{d}{dx}(e^{ax}) = a\,e^{ax} \qquad\qquad \frac{d}{dx}(\ln x) = \frac{1}{x}$$

Problems

1 Working from first principles, determine the derivatives of: (a) $y = 4/x$, (b) $y = 1/x^2$.

2 Determine, using the rules developed in this chapter, the derivatives of the following functions:

(a) $y = 2x^6$, (b) $y = 5x^{-2}$, (c) $y = x^{7/3}$, (d) $y = 0.2x^3$, (e) $y = 2\sqrt{x}$,

(f) $y = 2x^{-3/2}$, (g) $y = 4x^2 + x$, (h) $y = 2x^3 + 3x^2 + 1$, (i) $y = 3x^3 - 5x^2 + 3x$,

(j) $y = \frac{5}{x^3}$, (k) $y = 2x^3 + \frac{1}{x^2}$, (l) $y = \frac{5}{x^2} - \frac{2}{x}$, (m) $y = \sin 4x$,

(n) $y = \cos 2x$, (o) $y = e^x$, (p) $y = e^{2x}$, (q) $y = e^{-2x/3}$, (r) $y = \ln 5x$,

(s) $y = 3\sqrt{x}$, (t) $y = \sin x + \cos 2x$.

3 Determine the gradient of the graph of y against x for the equation $y = 3x^2$ at $x = 1$.

4 Determine the gradient of the graph of y against x for the equation $y = 2x^2 + 3x$ at $x = 2$.

5 The displacement s of an object starting from rest and moving with a uniform acceleration a is given by $s = \frac{1}{2}at^2$. Derive an equation for the velocity, i.e. the rate of change of distance with time.

6 Determine the gradients of the graph of $y = \cos 3x$ at the points on the graph (a) $x = \pi/2$, (b) $x = \pi/4$, (c) $x = 0$.

7 Determine the gradient of the graph of $y = e^{4x}$ at $x = 1$.

8 Determine the gradient of the graph of $y = 2 + e^{-x/2}$ at $x = 1$.

9 Determine the rate of change of area of a square with respect to the length of a side when the length is 10 mm.

10 The kinetic energy E of a rotating flywheel depends on the angular velocity ω and is given by the equation $E = \frac{1}{2}I\omega^2$. Determine the rate of change of kinetic energy with angular velocity.

11 The energy E stored by a capacitor depends on the potential difference V across it and is given by the equation $E = \frac{1}{2}CV^2$. Determine the rate of change of energy with respect to potential difference.

12 The height h of an object projected vertically upwards with an initial velocity of 12 m/s depends on the time t and is given by the equation $h = 12t - 5t^2$. Determine the rate at which the height changes with time, i.e. the velocity.

13 The atmospheric pressure p depends on the height h above ground level at which the measurement is made and is given by the equation

$p = p_0\, e^{-ch}$, where p_0 and c are constants. Determine the rate of change of pressure with height.

14 The instantaneous value of an alternating current i, in amperes, depends on the time t, in seconds, and is given by $i = 10 \sin 120t$. Determine the rate of change of the current with time when $t = 0.01$ s.

15 The instantaneous value of an alternating voltage v, in volts, depends on the time t, in seconds, and is given by $v = 12 \cos 100t$. Determine the rate of change of the voltage with time when $t = 0.02$ s.

16 The e.m.f. E, in millivolts, produced by a particular thermocouple is given by $E = a\theta^2$, where θ is the temperature in °C. If the thermocouple produces 200 mV at 100°C, determine the rate of change of the voltage with temperature at 120°C.

17 For a graph of $y = 2x^3 - 6x^2 - 15x + 5$ determine the points at which the gradient is +3.

18 The volume v, in cubic metres, of water in a reservoir varies with time t, in minutes, after a valve is opened according to $v = 10^4 - 10t^2 - 6t^3$. Determine the rate of change of the water volume at the time t of 2 min.

19 Determine the second derivatives of the following:

(a) $y = 3x^2$, (b) $y = e^{2x}$, (c) $y = \cos 2x$, (d) $y = 3x^2 + 4x$.

19 Further differentiation

This chapter is an extension of Chapter 18 and considers further rules that can be used in differentiating functions. The rules considered are those used in the differentiating of products, of quotients and of functions of a function. These enable functions such as $x \sin x$, $\sin x / \cos x$ and $\sin (ax + b)$ to be differentiated.

19.2 The product rule

Consider the equation $y = x \sin x$; how do we obtain dy/dx? The right-hand side of the equation can be considered to be the product of two terms which are functions of x, namely x and $\sin x$. So what we want is a rule for determining the derivative of a product of two functions. Suppose we have u and v which are two functions of x, i.e.

$$y = uv$$

Since u and v are both functions of x, when x increases to $x + \delta x$ then u increases to $u + \delta u$ and v to $v + \delta v$. Thus:

$$y + \delta y = (u + \delta u)(v + \delta v) = uv + v\delta u + u\delta v + \delta u \delta v$$

Subtracting $y = uv$ from this equation gives:

$$\delta y = v\delta u + u\delta v + \delta u \delta v$$

Dividing throughout by δx:

$$\frac{\delta y}{\delta x} = v\frac{\delta u}{\delta x} + u\frac{\delta v}{\delta x} + \frac{\delta u}{\delta x}\delta v$$

In the limit as $\delta x \to 0$, then $\delta v \to 0$ and so the last term in the above equation tends to become 0 and we have:

$$\frac{dy}{dx} = v\frac{du}{dx} + u\frac{dv}{dx}$$

Hence we can write a product rule as:

$$\frac{d}{dx}(uv) = v\frac{du}{dx} + u\frac{dv}{dx}$$

Example

Determine the derivative of $y = x \sin x$.

Let u be x and v be $\sin x$; the product rule then gives:

$$\frac{dy}{dx} = \sin x \frac{d}{dx}(x) + x\frac{d}{dx}(\sin x) = \sin x + x \cos x$$

Example

Determine the derivative of $y = x^2\,e^{3x}$.

Let u be x^2 and v be e^{3x}; the product rule then gives:

$$\frac{dy}{dx} = e^{3x}\frac{d}{dx}(x^2) + x^2\frac{d}{dx}(e^{3x}) = 2x\,e^{3x} + 3x^2\,e^{3x}$$

Example

Determine the derivative of $y = x^5(x^2 + 1)$.

Let u be x^5 and v be $x^2 + 1$; the product rule then gives:

$$\frac{dy}{dx} = (x^2 + 1)\frac{d}{dx}(x^5) + x^5\frac{d}{dx}(x^2 + 1) = 5x^4(x^2 + 1) + x^5(2x)$$

Revision

1   Determine the derivatives of the following:

(a) $y = (x^2 + 1)(x^3 + 3)$, (b) $y = x\,e^{4x}$, (c) $y = \sin x \cos 2x$,

(d) $y = e^{-2x}(x + 5)$, (e) $y = x^3(3x^2 + 2x + 1)$.

19.3 The quotient rule

Consider the equation $y = \sin x/\cos x$; this is the quotient of two terms which are functions of x, namely $\sin x$ and $\cos x$. So what we want is a rule for determining the derivative of a quotient of two functions. Suppose we have u and v which are two functions of x, i.e.

$$y = \frac{u}{v}$$

Since u and v are both functions of x, when x increases to $x + \delta x$ then u increases to $u + \delta u$ and v to $v + \delta v$. Thus:

$$y + \delta y = \frac{u + \delta u}{v + \delta v}$$

Subtracting $y = u/v$ from both sides of the above equation gives:

$$\delta y = \frac{u + \delta u}{v + \delta v} - \frac{u}{v} = \frac{(u + \delta u)v - u(v + \delta v)}{v(v + \delta v)} = \frac{v\delta u - u\delta v}{v(v + \delta v)}$$

Dividing both sides of the equation by δx gives:

$$\frac{\delta y}{\delta x} = \frac{v\frac{\delta u}{\delta x} - u\frac{\delta v}{\delta x}}{v(v+\delta v)}$$

In the limit as $\delta x \to 0$ we have $\delta v \to 0$ and the equation becomes:

$$\frac{dy}{dx} = \frac{v\frac{du}{dx} - u\frac{dv}{dx}}{v^2}$$

Hence we can write the quotient rule as:

$$\frac{d}{dx}\left(\frac{u}{v}\right) = \frac{v\frac{du}{dx} - u\frac{dv}{dx}}{v^2}$$

Example

Determine the derivative of $y = x^2/(x^3 + 2)$.

Let u be x^2 and v be $x^3 + 2$; the quotient rule then gives:

$$\frac{dy}{dx} = \frac{(x^3 + 2)\frac{d}{dx}(x^3) - x^2\frac{d}{dx}(x^3 + 2)}{(x^3 + 2)^2}$$

$$= \frac{3x^2(x^3 + 2) - 3x^4}{(x^3 + 2)^2}$$

Example

Determine the derivative of $y = \sin x/\cos x$.

Let u be $\sin x$ and v be $\cos x$; the quotient rule then gives:

$$\frac{dy}{dx} = \frac{\cos x\frac{d}{dx}(\sin x) - \sin x\frac{d}{dx}(\cos x)}{\cos^2 x} = \frac{\cos^2 x + \sin^2 x}{\cos^2 x} = \frac{1}{\cos^2 x}$$

Example

Determine the derivative of $y = x\,e^x/\cos x$.

This requires the use of both the quotient and product rules. First we use the quotient rule with u as $x\,e^x$ and v as $\cos x$:

$$\frac{dy}{dx} = \frac{\cos x\frac{d}{dx}(x\,e^x) - x\,e^x\frac{d}{dx}(\cos x)}{\cos^2 x}$$

We need to use the product rule to determine the derivative of $x\,e^x$.

$$\frac{d}{dx}(x\,e^x) = e^x\frac{d}{dx}(x) + x\frac{d}{dx}(e^x) = e^x + x\,e^x$$

Hence:

$$\frac{dy}{dx} = \frac{\cos x(e^x + x\,e^x) - x\,e^x(-\sin x)}{\cos^2 x}$$

$$\frac{dy}{dx} = \frac{e^x(\cos x + x\cos x + x\sin x)}{\cos^2 x}$$

Revision

2   Determine the derivatives of:

(a) $y = \dfrac{2x+1}{x^2-2}$, (b) $y = \dfrac{1}{x^3+3x^2}$, (c) $y = \dfrac{1-x}{(1+x)^2}$, (d) $y = \dfrac{1-\cos x}{\sin x}$.

19.4 Function of a function

With some expressions we can regard them as having a function within a function. For example, $\sin(ax + b)$ can be considered to be the function $ax + b$ within the sine function. As another example, $(x^2 + 1)^5$ can be considered to be the function $x^2 + 1$ within the power to the 5 function. We can obtain the derivative of such expressions by substituting u for the inner function. For example, with $y = (x^2 + 1)^5$ we write it as $y = u^5$ where $u = x^2 + 1$. With $y = \sin(ax + b)$ we write it as $y = \sin u$ where $u = ax + b$. If x changes increases by δx then there will be corresponding increases of δy in y and δu in u. We can write:

$$\frac{\delta y}{\delta x} = \frac{\delta y}{\delta x} \times \frac{\delta u}{\delta u} = \frac{\delta y}{\delta u} \times \frac{\delta u}{\delta x}$$

As $\delta x \to 0$ then $\delta u \to 0$ and so the above equation becomes:

$$\frac{dy}{dx} = \frac{dy}{du} \times \frac{du}{dx}$$

This is knows as the *function of a function rule* or the *chain rule*.

Example

Determine the derivative of $y = (x^2 + 1)^5$.

Let $u = x^2 + 1$ and so $y = u^5$. Using the function of a function rule:

$$\frac{dy}{dx} = \frac{dy}{du} \times \frac{du}{dx} = \frac{d}{du}(u^5) \times \frac{d}{dx}(x^2 + 1) = 5u^4 \times 2x$$

Since $u = x^2 + 1$, then $dy/dx = 5(x^2 + 1)^4 \times 2x$.

Example

Determine the derivative of $y = \sin(ax + b)$.

Let $u = ax + b$ and so $y = \sin u$. Using the function of a function rule:

$$\frac{dy}{dx} = \frac{dy}{du} \times \frac{du}{dx} = \frac{d}{du}(\sin u) \times \frac{d}{dx}(ax + b) = \cos u \times a$$

Since $u = ax + b$ then $dy/dx = a\cos(ax + b)$.

Example

Determine the derivative of $y = \sqrt{(x^3 - 5)}$.

Let $u = x^3 - 5$ and so $y = u^{1/2}$. Using the function of a function rule:

$$\frac{dy}{dx} = \frac{dy}{du} \times \frac{du}{dx} = \frac{d}{du}(u^{1/2}) \times \frac{d}{dx}(x^3 - 5) = \tfrac{1}{2}u^{-1/2} \times 3x^2$$

Since $u = x^2 - 5$ then:

$$\frac{dy}{dx} = \frac{3x^2}{2\sqrt{x^2 - 5}}$$

Revision

3 Determine the derivatives of:

(a) $y = (2x + 1)^3$, (b) $y = \sqrt{(x^2 + 1)}$, (c) $y = \sin x^3$, (d) $y = \sin(5x + 2)$,

(e) $y = \dfrac{1}{(6x^3 - x)^4}$, (f) $y = (x + \sin x)^2$.

Problems

1 Determine the derivatives of:

(a) $y = x^3 \sin x$, (b) $y = x^2(3x + 5)$, (c) $y = x\,e^{4x}$, (d) $y = (3x + 1)/(x + 5)$,

(e) $y = 4x/(1 - 3x)$, (f) $y = \sin(4x - 5)$, (g) $y = (\cos x)^5$,

(h) $y = (x^2 - 6x + 1)^3$, (i) $y = (x + 3)^4$, (j) $y = (x^2 + 2x^3 + 3)^8$,

(k) $y = \sqrt{(x^3 + 5)}$, (l) $y = 1/\sin x$, (m) $y = 3x^2 \sin 2x$, (n) $y = \sin 3x/\cos 3x$,

(o) $y = 1/\tan 2x$, (p) $y = \sqrt{(3x^2 + 4x + 1)}$, (q) $y = e^{3x} \sin 4x$,

(r) $y = 5\,e^{2x + 3}$.

20 Maxima and minima

20.1 Introduction

In a graph of y against x, where y is a variable that depends on the value of x, the gradient of the graph is given by dy/dx. If dy/dx is positive then y is increasing as x increases. If dy/dx is negative then y is decreasing as x increases. However, if $dy/dx = 0$ then y is neither increasing or decreasing. Points where this occurs are called *stationary points* or *critical points* or *turning points*. At such points we must have the tangent to the graph parallel to the x-axis. In Figure 20.1 these are the points A, B, C and D.

The turning point at A is termed a *maximum*, that at C a *minimum*. At the maximum the value of y is greater than the neighbouring values either side of A. It does not mean that it is necessarily greater than any other value of y for other points on the curve. At the minimum the value of y is smaller than the neighbouring values either side of C. It does not mean that it is necessarily smaller than any other value of y for other points on the curve. Points B and D are termed *points of inflexion*. At such points y is higher on one side of the point than the other.

There are many situations in engineering where we are concerned with establishing maximum or minimum values. Thus with a projectile we might need to determine the maximum height reached. In determining how large a batch of a product to manufacture we might need to establish the number which will have the minimum cost. With an electrical circuit we might need to establish the condition for which maximum power is generated in a load resistance. In these situations we have effectively a graph of one variable against another and have to establish the point on the graph at which there a maximum or a minimum. This chapter is about determining such points.

Figure 20.1 *Turning points*

20.2 Turning points

Consider a graph of y against x when the values of y depend in some way on the values of x. Points on the graph at which $dy/dx = 0$ can be:

1 *A local maximum*
   The term *local* is used because the value of y is only necessarily a maximum for points in the locality and there could be higher values of y elsewhere on the graph. Figure 20.2 shows such a maximum. At the maximum point A we have zero gradient for the tangent, i.e. $dy/dx = 0$. Consider two points P and Q close to A, with P having a value of x less than that at A and Q having a value greater than that at A. The gradient of the tangent at P is positive, the gradient of the tangent at Q is negative. Thus for a maximum we have the gradient changing from being positive prior to the turning point to negative after it.

2 *A local minimum*
   The term *local* is used because the value of y is only necessarily a minimum for points in the locality and there could be lower values of y elsewhere on the graph. Figure 20.3 shows such a minimum. At the

Figure 20.2 *A maximum*

Figure 20.3 *A minimum*

(a)

(b)

Figure 20.4 *Points of inflexion*

minimum point C we have zero gradient for the tangent, i.e. $dy/dx = 0$. Consider two points R and S close to A, with R having a value of x less than that at C and S having a value greater than that at C. The gradient of the tangent at R is negative, the gradient of the tangent at S is positive. Thus for a minimum we have the gradient changing from being negative prior to the turning point to positive after it.

3 *A point of inflexion*
Consider points of inflexion, as illustrated in Figure 20.4. At such points $dy/dx = 0$. However, in neither of the graphs is there a local maximum or minimum. In Figure 20.4(a), the gradient at a point T prior to the point is negative and the gradient at a point U after the point is also negative. In Figure 20.4(b), the gradient at a point V prior to the point is positive and the gradient at a point W after the point is also positive. For a point of inflexion the sign of the gradient prior to the point is the same as that after the point.

Thus, if we consider the gradients in the vicinity of maxima, minima and points of inflexion, then in moving from points before to after the turning point:

1 At a *maximum* the gradient changes from being positive to negative.

2 At a *minimum* the gradient changes from being negative to positive.

3 At a *point of inflexion* the sign of the gradient does not change.

The gradient at a point on a graph is given by dy/dx. We can thus determine whether a turning point is a maximum, a minimum or a point of inflexion by considering how the value of dy/dx changes for a value of x smaller than the turning point value compared to that for a value of x greater than the turning point value.

There is an alternative method we can use to distinguish between maximum and minimum points. We need to establish how the gradient changes in going from points before to after turning points. Consider, for a maximum, a graph of the gradients plotted against x (Figure 20.5(a)). The gradients prior to the maximum are positive and decrease in value to become zero at the maximum. They then become negative and as x increases become more and more negative. The second derivative d^2y/dx^2 measures the rate of change of dy/dx with x, i.e. the gradient of the dy/dx graph. The gradient of the gradient graph is negative before, at and after the maximum point. Hence at a maximum d^2y/dx^2 is negative.

Consider a minimum (Figure 20.5(b)). The gradients prior to the minimum are negative and become less negative until they become zero at the minimum. As x increases beyond the minimum the gradients become positive, increasing in value as x increases. The second derivative d^2y/dx^2 measures the rate of change of dy/dx with x, i.e. the gradient of the dy/dx graph. The gradient of the gradient graph is positive before, at and after the minimum point. Hence at a minimum d^2y/dx^2 is positive. Thus we can distinguish between maximum and minima in that:

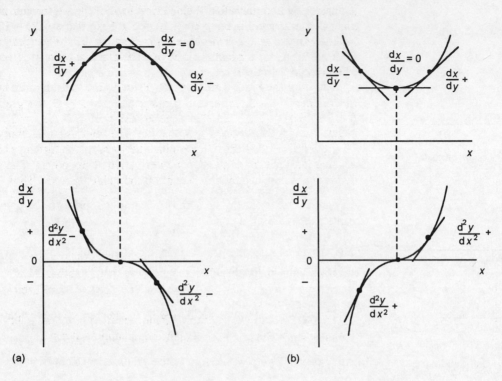

Figure 18.5 *(a) A maximum, (b) a minimum*

1 At a maximum the second derivative is negative.

2 At a minimum the second derivative is positive.

Example

Determine, and identify the form of, the turning points on a graph of the equation $y = 2x^3 - 3x^2 - 12x$.

Differentiating the equation gives

$$\frac{dy}{dx} = 6x^2 - 6x - 12$$

Thus the gradient of the graph is zero when $6x^2 - 6x - 12 = 0$. We can rewrite this as:

$$6(x^2 - x - 2) = 6(x + 1)(x - 2) = 0$$

The gradient is zero, and hence there are turning points, at $x = -1$ and $x = 2$.

To establish the form of these turning points consider the gradients just prior to and just after them. Prior to the $x = -1$ turning point at $x = -2$, the gradient is $6x^2 - 6x - 12 = 6 \times (-2)^2 - 6 \times (-2) - 12 = 12$. After the point at $x = 0$ we have a gradient of -12. Thus the gradient prior to the $x = -1$ point is positive and after the point it is negative. The point is thus a maximum.

Consider the $x = 2$ turning point. Prior to the turning point at $x = 0$ the gradient is -12. After the turning point at $x = 3$ the gradient is $6 \times 3^2 - 6 \times 3 - 12 = 24$. Thus the gradient prior to the $x = 2$ point is negative and after the point it is positive. The point is thus a minimum.

Alternatively we could determine the form of the turning points by considering the sign of the second derivative at the points. The second derivative is obtained by differentiating the $\mathrm{d}y/\mathrm{d}x$ equation. Thus

$$\frac{\mathrm{d}^2 y}{\mathrm{d}x^2} = 12x - 6$$

At $x = -1$ then the second derivative is $12 \times (-1) - 6 = -18$. The negative value indicates that the point is a maximum. At $x = 2$ the second derivative is $12 \times 2 - 6 = 18$. The positive value indicates that the point is a maximum.

Figure 20.6 shows a graph of the equation $y = 2x^3 - 3x^2 - 12x$, showing the maximum at $x = -1$ and the minimum at $x = 2$.

Figure 20.6   *Example*

Example

Determine, and identify the form of, the turning points on a graph of the equation $y = x^3$.

Differentiating the equation gives:

$$\frac{\mathrm{d}y}{\mathrm{d}x} = 3x^2$$

Thus $\mathrm{d}y/\mathrm{d}x = 0$ when $x = 0$. The turning point is thus at $x = 0$.

Consider the value of the gradient just prior to the turning point. At $x = -1$ then the gradient is $3 \times (-1)^2 = 3$. After the turning point at $x = 1$ then the gradient is $3 \times 1^2 = 3$. The gradient is thus positive prior to the turning point and positive after it. The turning point is thus an inflexion point. Figure 20.7 shows the graph.

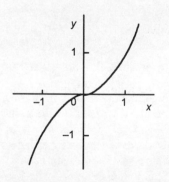

Figure 20.7   *Example*

Example

The displacement y in metres of an object is related to the time t in seconds by the equation $y = 5 + 4t - t^2$. Determine the maximum displacement.

Differentiating the equation gives:

$$\frac{dy}{dt} = 4 - 2t$$

dy/dt is 0 when $t = 2$ s. There is thus a turning point at the displacement $y = 5 + 4 \times 2 - 2^2 = 9$ m.

We need to check that this is a maximum displacement. The gradient prior to the turning point at $t = 1$ has the value $4 - 2 = 2$. After the turning point at $t = 3$ it has the value $4 - 6 = -2$. The gradient changes from a positive value prior to the turning point to a negative value afterwards. It is thus a maximum.

Alternatively we could have established this by determining the second derivative. Differentiating $4 - 2t$ gives $d^2y/dx^2 = -2$. Thus, the turning point is a maximum.

Example

If the sum of two numbers is 40, determine the values which will give the minimum value for the sum of their squares.

Let the two numbers be x and y. Then we must have $x + y = 40$. We have to find the minimum value of S when we have $S = x^2 + y^2$. We need an equation which expresses the sum in terms of just one variable. Thus substituting from the previous equation gives

$$S = x^2 + (40 - x)^2 = x^2 + 1600 - 80x + x^2$$

$$= 2x^2 - 80x + 1600$$

Differentiating this equation, then

$$\frac{dS}{dx} = 4x - 80$$

The value of x to give a zero value for dS/dx is when $4x - 80 = 0$ and so when $x = 20$.

We can check that this is the value giving a minimum by considering the values of dS/dx prior to and after the point. Thus prior to the point at $x = 19$ we have $dS/dx = 4 \times 19 - 80 = -4$. After the point at $x = 21$ we have $dS/dx = 4 \times 21 - 80 = 4$. Thus dS/dx changes from being negative to positive. The turning point is thus a minimum. Alternatively we could check that this is a minimum by obtaining the second derivative. Differentiating $4x - 80$ gives $d^2S/dx^2 = 4$. Since this is positive then we have a minimum. Thus the two numbers which will give the required minimum are 20 and 20.

Example

Determine the maximum area of a rectangle with a perimeter of 32 cm.

If the width of the rectangle is w and its length L then the area $A = wL$. But the perimeter has a length of 32 cm. Thus $2w + 2L = 32$. We need an equation expressing the area in terms of just one variable. Thus, if we eliminate w from the area equation by substitution:

$$A = L(16 - L) = 16L - L^2$$

Hence:

$$\frac{dA}{dL} = 16 - 2L$$

dA/dL is zero when $16 - 2L = 0$ and so when $L = 8$.

We can check that this gives a maximum area by considering values of the gradient at values of L below and above 8. At $L = 7$ the gradient is 2 and at $L = 9$ it is -2. It is thus a maximum. Alternatively we could have considered the second derivative. Since $d^2A/dL^2 = -2$ and so is negative, we have a maximum.

For a maximum area we must therefore have $L = 8$ cm, and after substituting this value in $2w + 2L = 32$, $w = 8$ cm.

Revision

1 Determine, and identify the form of, the turning points on graphs of the following equations:

(a) $y = x^2 - 6x + 4$, (b) $y = 2x^3 - 3x^2 - 12x + 8$, (c) $y = 3x^2 - x^3$,

(d) $y = 2x^3 + 3x^2 - 36x + 7$, (e) $y = x^2 + 3x$, (f) $y = 5x - 10x^2$,

(g) $y = \sin x$.

2 The rate r at which a chemical reaction proceeds is related to the quantity x of chemical present by the equation $r = k(a - x)(b + x)$. Determine the maximum rate.

3 A water tank is to be made of sheet metal. It is to have a square base and be open at the top. What is the least area of metal that can be used if the tank is to have a volume of 32 m³?

4 The bending moment M of a uniform beam of length L at a distance x from one end is given by the equation $M = \frac{1}{2}wLx - \frac{1}{2}wx^2$, where w is the weight per unit length of beam. Determine the point at which the bending moment is a maximum.

5 Determine the two numbers whose sum is 40 and for which the product is a maximum.

6 A stone is thrown vertically upwards. Its height h, in metres, depends on the time t, in seconds, and is given by the equation $h = 40t - 5t^2$. After what time is the height a maximum?

7 A cylindrical container with closed ends is to have a volume of 0.25 m^3. What radius and height will it have if the area of sheet metal used in its construction is to be a minimum?

8 A wire of length 60 cm is to be bent to form a rectangle. What should be the length of its sides if the area is to be a maximum?

9 Determine the dimensions an open box can have for maximum volume if constructed with a square base from a sheet of metal of area 100 cm^2.

10 The power P supplied to an electrical circuit is given by $P = IE - I^2R$, where I is the current taken by the circuit, E the e.m.f. of the supply and r its internal resistance. Determine the condition for maximum power transfer.

11 An electrical voltage v, in volts, depends on the time t, in seconds, and is given by $v = 30 \cos 500t + 15 \sin 500t$. Determine the maximum value of the voltage.

12 The e.m.f. E produced by a thermocouple is related to the temperature θ by $E = a\theta + b\theta^2$. Determine the temperature at which the e.m.f. is a maximum.

Problems

1 Determine, and identify the form of, the turning points on graphs of the following equations:

(a) $y = 4x^2 - 8x + 3$, (b) $y = 2x^2 + 6x + 10$, (c) $y = x^2$, (d) $y = x^3 - x$,

(e) $y = x^3 + x^2$, (f) $y = 2x^3 - 3x^2 - 36x - 15$, (g) $y = x^2 - (16/x)$,

(h) $y = 2x + (8/x) + 3$, (i) $y = 3 - x^3$, (j) $y = x^2 - 2x$,

(k) $y = 3x^2 - 6x + 3$, (l) $y = x^3 - 3x + 7$, (m) $y = 4x^3 + 9x^2 - 12x + 7$,

(n) $y = x^3 - 3x^2 - 24x + 5$.

2 The profit P gained by a company from the sales of product X depends on the number n sold, and is given by $P = 40n - 200 - \frac{1}{2}n^2$. Determine the number that have to be sold to maximise the profit.

3 When one number is added to twice a second number then the result is 24. What should the numbers be if their product is to be a maximum?

4 The production costs C per day for steel tube depends on the length L, in metres, produced per day and is given by the equation:

$$C = 30L + \frac{300\,000}{L}$$

Determine the length that has to be produced per day to keep the costs to a minimum.

5 The speed v, in metres per second, of a car when in first gear is related to the time t, in seconds, by $v = 5 + 3t - t^2$. Determine the maximum speed of the car.

6 An open rectangular tray is constructed from a rectangular sheet measuring 200 mm by 300 mm by cutting equal size squares from each corner and then folding up the sides to form the tray. What should be the side lengths of the squares cut from each corner if the volume of the tray is to be a maximum?

7 A open cylindrical tank is to have a volume of 2 m³. What should be the dimensions of the cylinder if the area of the sheet used to form it is to be a minimum?

8 A rocket moves vertically upwards with its distance h, in metres, from the ground related to the time t by $h = 500t - 5t^2$. Determine the greatest height reached.

9 An open tank is to have a square base and vertical sides. It is to be lined with plastic sheet. Show that the minimum area of plastic sheet will be needed when the depth of the tank is twice the side length of its base.

10 The cost C of producing n units of a product is given by a setting-up cost of £500 which applies regardless of how many items are produced and a cost which depends on the number of units produced and is given, in £, by $0.1n + 0.001n^2$. Determine the production level which will minimise the average cost per item, i.e. the value of C/n, produced.

11 A closed 1 litre (1000 cm³) container is to be made in the form of a cylinder from sheet metal. What dimensions will give the cylinder requiring the smallest area of sheet?

12 The variation of the alternating current i in a circuit with time t is described by the equation $i = 2 \cos t + 2 \sin t$. What is the size of the maximum current?

21 Integration

21.1 Introduction

Integration can be considered to be the mathematical process which is the reverse of the process of differentiation. It also turns out to be a process for finding an area under a graph. The term *integral* is used for the outcome of integration. This chapter can be considered to be a basic introduction to integration.

As an illustration of the application of integration in engineering, consider the situation where the velocity v of an object varies with time t, say $v = 2t$ (Figure 21.1) Since velocity is the rate of change of distance x with time we can write this as:

$$\frac{dx}{dt} = 2t$$

Thus we know how the gradient of the distance–time graph varies with time. Integration is the method we can use to determine from this how the distance varies with time. We thus start out with the gradients and find the graph responsible for them, the reverse of the process used with differentiation where we start with a graph and find the gradients.

But we could have considered the same problem in a different way. Suppose the velocity variation with time was represented as a velocity–time graph. The distance travelled in a particular time can be obtained from the graph by determining the area under the graph between those times. Integration can be used to determine such areas.

21.2 Integration as the reverse of differentiation

Suppose we have an equation $y = x^2$. When this equation is differentiated we obtain the derivative of $dy/dx = 2x$.

> *Differentiation is the determination of the relationship for the gradient of a graph. We can define integration as the mathematical process which reverses differentiation, i.e. given the gradient relationship, finding the equation which was responsible for it.*

Thus, in this case, when given the gradient as $2x$ we need to find the equation which on being differentiated gives $2x$. Thus integrating $2x$ should give us x^2. However, the derivative of $x^2 + 1$ is also $2x$, likewise the derivative of $x^2 + 2$, the derivative of $x^2 + 3$, and so on. Figure 21.1 shows part of the family of graphs which all have the gradients given by $2x$. Thus, for each of the graphs, at a particular value of x, such as $x = 1$, they all give the same gradient of $2x = 2$. Thus in the integration of $2x$ we are not sure whether there is a constant term or not, or what value it might have. Hence a constant C has to be added to the result. Thus the outcome of the integration of $2x$ has to be written as being $x^2 + C$. The integral, which has to have a constant added to it, is referred to as an *indefinite integral*.

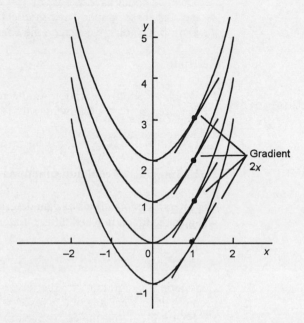

Figure 21.1 dy/dx = 2x

To indicate the process of integration a special symbol:

$\int$ with dx

is used. This sign indicates that integration is to be carried out and the dx that x is the variable we are integrating with respect to. Thus the integration referred to above can be written as

$$\int 2x \, dx = x^2 + C$$

The above thus reads as 'the integral of 2x with respect to x is x^2 plus a constant'.

Example

By considering integration as the reverse of differentiation, what is the integral of 3x^2?

We have to find what function when differentiated gives $3x^2$. If we differentiate x^3 we obtain $3x^2$. If we differentiate $x^3 + C$, where C is some constant, we also obtain $3x^2$. Thus the integral of $3x^2$ is $x^3 + C$.

$$\int 3x^2 \, dx = x^3 + C$$

Thus we could have $y = x^3$, or $y = x^3 + 1$, or $y = x^3 + 2$, or $y = x^3 - 1$, etc. All these graphs have gradient equal to $3x^2$. Because we do not know which one, we include the constant C.

Revision

1   By considering integration as the reverse of differentiation, determine the integrals of the following with respect to x:

(a) $4x$, (b) $6x^2$, (c) $x + 2$, (d) $\sin x$, (e) e^x

21.2.1 Integrals of common functions

The integrals of functions can be determined by considering what equation will give the function when differentiated. For example, consider:

$$\int x^n \, dx$$

Considering integration as the inverse of differentiation, the question becomes as to what function gives x^n when differentiated. The derivative of x^{n+1} is $(n + 1)x^n$. Thus the derivative of $x^{n+1}/(n + 1)$ is x^n. Hence:

$$\int x^n \, dx = \frac{x^{n+1}}{n+1} + C$$

This is true for positive, negative and fractional values of n other than $n = -1$, i.e. the integral of x^{-1}. For the integral of x^{-1}, i.e.

$$\int \frac{1}{x} \, dx$$

then since the derivative of $\ln x$ is $1/x$:

$$\int \frac{1}{x} \, dx = \ln x + C$$

This only applies if x is positive, i.e. $x > 0$. If x is negative, i.e. $x < 0$, then the integral of $1/x$ in such a situation is *not* $\ln x$. This is because we cannot have the logarithm of a negative number as a real quantity.

Consider the integral of the exponential function e^x, i.e.

$$\int e^x \, dx$$

The derivative of e^x is e^x. Thus:

$$\int e^x \, dx = e^x + C$$

Consider the integral of trigonometric functions. For $\sin x$, i.e.

$$\int \sin x \, dx$$

we have, since the derivative of $\cos x$ is $-\sin x$:

$$\int \sin x \, dx = -\cos x + C$$

For the integral of $\cos x$, i.e.

$$\int \cos x \, dx$$

since the derivative of $\sin x$ is $\cos x$, we have:

$$\int \cos x \, dx = \sin x + C$$

The above, with the inclusion of a constant a are listed below in Table 21.1.

Table 21.1 *Common integrals*

Function being integrated	Outcome of the integration
$\int ax^n \, dx$	$\frac{ax^n}{n+1} + C$, n not -1
$\int a \, dx$	$ax + C$
$\int \frac{1}{x} \, dx$	$\ln x + C$
$\int e^{ax} \, dx$	$\frac{1}{a} e^{ax} + C$
$\int \sin ax \, dx$	$-\frac{1}{a} \cos ax + C$
$\int \cos ax \, dx$	$\frac{1}{a} \sin ax + C$

Example

Evaluate the integrals:

(a) $\int x^4 \, dx$, (b) $\int x^{1/2} \, dx$, (c) $\int x^{-4} \, dx$, (d) $\int x^{-1} \, dx$,

(e) $\int \cos 4x \, dx$, (f) $\int e^{2x} \, dx$, (g) $\int 5 \, dx$.

(a) Using $\int x^n \, dx = \frac{x^{n+1}}{n+1} + C$:

$$\int x^4 \, dx = \frac{x^{4+1}}{4+1} + C = \frac{x^5}{5} + C$$

(b) Using $\int x^n \, dx = \frac{x^{n+1}}{n+1} + C$:

$$\int x^{1/2} \, dx = \frac{x^{1/2+1}}{1/2+1} + C = \frac{x^{3/2}}{3/2} + C = \frac{2}{3} x^{3/2} + C$$

(c) Using $\int x^n \, dx = \frac{x^{n+1}}{n+1} + C$:

$$\int x^{-4} \, dx = \frac{x^{-4+1}}{-4+1} + C = \frac{x^{-3}}{-3} + C = -\frac{1}{3}x^{-3} + C$$

(d) Using the relationship giving in the table:

$$\int x^{-1} \, dx = \ln x + C$$

(e) Using $\int \cos ax \, dx = \frac{1}{a} \sin x + C$:

$$\int \cos 4x \, dx = \frac{1}{4} \sin 4x + C$$

(f) Using $\int e^{ax} \, dx = \frac{1}{a} e^x + C$:

$$\int e^{2x} \, dx = \frac{1}{2} e^{2x} + C$$

(g) Using $\int a \, dx = ax + C$:

$$\int 5 \, dx = \int 5x^0 \, dx = 5x + C$$

The above is just a particular version of the standard integral:

$$\int x^n \, dx = \frac{x^{n+1}}{n+1} + C$$

with $n = 0$ since $x^0 = 1$.

Revision

2   Determine the integrals:

(a) $\int x^3 \, dx$, (b) $\int 2x^3 \, dx$, (c) $\int x^{-7} \, dx$, (d) $\int x^{-1/2} \, dx$, (e) $\int x^{1/3} \, dx$,

(f) $\int \sin 3x \, dx$, (g) $\int \cos 2x \, dx$, (h) $\int e^{4x} \, dx$, (i) $\int e^{-5x} \, dx$.

21.2.2 Integral of a sum

The derivative of, for example, $x^2 + x$ is the derivative of x^2 plus the derivative of x, i.e. it is $2x + 1$. The integral of $2x + 1$ is thus $x^2 + x + C$. Thus the integral of the sum of a number of functions is the sum of their separate integrals.

Example

Determine the integral $\int (x^3 + 2x + 4) \, dx$.

We can write this as:

$$\int (x^3 + 2x + 4)\, dx = \int x^3\, dx + \int 2x\, dx + \int 4\, dx$$

Hence the integral is:

$$\frac{x^4}{4} + P + x^2 + Q + 4x + R$$

where P, Q and R are constants. We can combine these constants into a single constant C. Hence the integral is:

$$\frac{x^4}{4} + x^2 + 4x + C$$

Revision

3   Determine the following integrals:

(a) $\int (5x + 3)\, dx$, (b) $\int (4x^3 + 3x^2 + 2)\, dx$, (c) $\int (\sin x + \cos 2x)\, dx$,

(d) $\int (2 + e^{-3x})\, dx$, (e) $\int \left(\frac{1}{x} + 3x\right) dx$, (f) $\int (3x + \sin 4x)\, dx$.

21.2.3 Finding the constant

The solution given by the above integration is a general solution and includes a constant. As was indicated in Figure 21.1 the integration of $2x$ gives $y = x^2 + C$. This solution indicates a family of possible equations which could give $dy/dx = 2x$. We can, however, find a *particular solution* if we are supplied with information giving specific coordinate values which have to fall on the graph curve. Thus, in this case, we might be given the condition that the solution must pass through the point $(1, 1)$. This means that the condition of $y = 1$ when $x = 1$ must fit the equation $y = x^2 + C$. This can only be the case when $C = 0$. Hence the solution is $y = x^2$.

Example

Determine the equation of a graph if it has to pass through the point $(0, 2)$ and has a gradient given by:

$$\frac{dy}{dx} = 3x + 2$$

To obtain the general solution, i.e. the family of curves which fit the above gradient equation, we integrate. Thus:

$$y = \int (3x + 2)\, dx = \tfrac{3}{2}x^2 + 2x + C$$

The particular curve we require must pass through the point (0, 2). Thus we must have $y = 2$ when $x = 0$. Putting this data into the equation gives $2 = 0 + 0 + C$. Hence $C = 2$ and so the particular solution is:

$$y = \tfrac{3}{2}x^2 + 2x + 2$$

Revision

4 Determine the equations of the following curves, given the equation for the gradient of a curve and a point through which the curve must pass:

(a) $\dfrac{dy}{dx} = 2x + 3$, curve passes through (2, 3),

(b) $\dfrac{dy}{dx} = 9x^2 + 2x - 2$, curve passes through (1, 3),

(c) $\dfrac{dy}{dx} = 4x - 9$, curve passes through (2, -5),

(d) $\dfrac{dy}{dx} = 2 - \dfrac{1}{x^2}$, curve passes through (1, 3).

5 The velocity v, in m/s, of an object varies with time t, in seconds, and is given by $v = 10 + 2t$. As $v = dx/dt$, determine how the distance x varies with time if at $t = 0$ the distance is 2 m.

6 The velocity v, in m/s, of an object varies with time t, in seconds, and is given by $v = 2t + 5$. As $v = dx/dt$, determine how the distance x varies with time if at $t = 0$ the distance is 0.

21.3 Integration as the area under a graph

Figure 21.2 *Velocity-time graph*

Consider a moving object and its graph of velocity v against time t. The distance travelled between times of t_1 and t_2 is the area under the velocity–time graph between those times. If we divide the area into a number of equal width strips then we can represent this area under the velocity–time graph as being the sum of the areas of these equal width strip areas, as illustrated in Figure 21.2. If t is the value of the time at the centre of a strip of width δt and v the velocity at this time, then a strip has an area of $v\,\delta t$. Thus the area under the graph between the times t_1 and t_2 is equal to the sum of the areas of all such strips between the times t_1 and t_2,

distance travelled = area under graph between t_1 and t_2

= sum of the areas of all the strips
between t_1 and t_2

We can write this summation as:

$$x = \sum_{t=t_1}^{t=t_2} v\,\delta t$$

The Σ sign is used to indicate that we are carrying out a summation of a number of terms. The limits between which this summation is to be carried out are indicated by the information given below and above the sign. If we make δt very small, i.e. let δt tend to 0, then we denote it by dt. The sum is then the sum of a series of very narrow strips and is written as:

$$x = \lim_{\delta t \to 0} \sum_{t=t_1}^{t=t_2} v\,\delta t = \int_{t_1}^{t_2} v\,\mathrm{d}t$$

The integral sign is an "S" for summation and the t_1 and t_2 are said to be the limits of the range of the variable t. Here x is the *integral* of the v with time t between the limits t_1 and t_2. The process of obtaining x in this way is termed *integration*. Because the integration is between specific limits it is referred to as a *definite integral*.

21.3.1 Integration as reverse of differentiation and area under a graph

The definitions of integration in terms of the reverse of differentiation and as the area under a graph describe the same concept. Suppose we increase the area under a graph of y plotted against x by one strip. Then the increase in the area δA is the area of this strip. Thus:

increase in area $\delta A = y\,\delta x$

So we can write:

$$\frac{\delta A}{\delta x} = y$$

In the limit as δx tends to 0 then we can write dA/dx and so

$$\frac{\mathrm{d}A}{\mathrm{d}x} = y$$

With integration defined as the inverse of differentiation then the integration of the above equation gives the area A, i.e.

$$A = \int y\,\mathrm{d}x$$

This is an indefinite integral, which is the same as that given by the definition for integration as the area under a graph when limits are imposed. An *indefinite integral* has no limits and the result has a constant of integration. Integration between specific limits gives a *definite integral*.

21.3.2 Areas under graphs

Consider the integration of y with respect to x when we have $y = 2x$. This has no specified limits and so is an indefinite integral, with the solution (as the function which differentiated would give $2x$):

$$\int 2x \, dx = x^2 + C$$

Figure 21.3 $y = 2x$

Now consider the area under the graph of $y = 2x$ between the limits of $x = 1$ and $x = 3$ (Figure 21.3). We can write this as the definite integral:

$$\int_1^3 2x \, dx$$

We can consider this as representing the area between $x = 3$ and $x = 0$ minus the area between $x = 1$ and $x = 0$, i.e.

$$\int_0^3 2x \, dx - \int_0^1 2x \, dx$$

The integral of $2x$ is $x^2 + C$. Thus we have the area as being that given by $x^2 + C$ when we have $x = 3$ minus that given by $x^2 + C$ when $x = 1$, i.e.

$$(9 + C) - (1 + C) = 8 \text{ square units}$$

Such a calculation is more usually written as

$$\int_1^3 2x \, dx = [x^2 + C]_1^3$$

The square brackets round the $x^2 + C$ are used to indicate that we have to impose the limits of 3 and 1 on it. Thus the integral is the value of $x^2 + C$ when $x = 3$ minus the value of $x^2 + C$ when $x = 1$.

$$\int_1^3 2x \, dx = (9 + C) - (1 + C) = 8 \text{ square units}$$

The constant term C vanishes when we have a definite integral.

Note that an area below the x-axis is negative. If the area is required when part of it is below the x-axis then the parts below and above the x-axis must be found separately and then added, disregarding the sign of the area.

Example

Determine the area between a graph of $y = x + 1$ and the x-axis between $x = -2$ and $x = 4$.

Figure 21.4 *Example*

Figure 21.4 shows the graph. The area required is that between the values of x of -2 and 4. We can break this area down into a number of elements. The area under the graph between $x = 0$ and $x = 4$ is that of a rectangle 4×1 plus a triangle $\frac{1}{2}(4 \times 4)$ and so is $+12$ square units. The area between $x = -1$ and 0 is that of a triangle $\frac{1}{2}(1 \times 1) = 0.5$ square units and the area between $x = -1$ and $x = -2$ is a triangular area below the axis and so is negative and given by $\frac{1}{2}(1 \times 1) = -0.5$ square units. Hence the total area under the graph is $+12 + 0.5 - 0.5 = 12$ square units.

Alternatively we can consider this area as the integral:

$$\text{area} = \int_{-2}^{4} (x + 1)\,dx = \left[\frac{x^2}{2} + x + C \right]_{-2}^{4}$$

and so:

$$\text{area} = (8 + 4 + C) - (2 - 2 + C) = 12 \text{ square units}$$

Example

Determine the value of the integral $\int_{-2}^{4} e^{2x}\,dx$.

We can consider that this integral represents the area under the graph between $x = -2$ and $x = 4$ of e^{2x} plotted against x.

$$\int_{-2}^{4} e^{2x}\,dx = \left[\tfrac{1}{2} e^{2x} + C \right]_{-2}^{4} = \tfrac{1}{2} e^8 - \tfrac{1}{2} e^{-4} = 1490.479 - 0.009$$

The answer is thus 1490.470.

Example

Determine the value of the integral $\int_{0}^{\pi/3} \cos 2x\,dx$.

We can consider that this integral represents the area under the graph between $x = 0$ and $x = \pi/3$ of $\cos 2x$ plotted against x.

$$\int_{0}^{\pi/3} \cos 2x\,dx = \left[\tfrac{1}{2} \sin 2x + C \right]_{0}^{\pi/3} = \tfrac{1}{2} \sin 2\pi/3 - \tfrac{1}{2} \sin 0 = 0.433$$

Revision

7 Determine, by integration, the area under the following graphs between the specified limits:

(a) $y = x + 5$ between $x = -2$ and $x = 2$,

(b) $y = 4x + 1$ between $x = 1$ and $x = 2$,

(c) $y = 3x^2 + 2x$ between $x = 0$ and $x = 2$,

(d) $y = 6x^2 + 4x + 2$ between $x = -1$ and $x = 1$,

(e) $y = \sin x$ between $x = 0$ and $x = \pi/2$,

(f) $y = \cos x$ between $x = 0$ and $x = \pi/4$,

(g) $y = e^{-x}$ between $x = 0$ and $x = 2$,

(h) $y = 3 \sin 2x$ between $x = 0$ and $x = \pi/3$,

(i) $y = 8 - 2x^2$ between $x = -2$ and $x = 2$,

(j) $y = x^2 - x + 2$ between $x = -1$ and $x = 2$.

8   The work done w by a force is the area under a graph of force F against displacement x in the direction of the force. Determine the work done when $F = 2x$ and x changes from 0 to 2, i.e. the value of the integral:

$$\int_0^2 2x\, dx$$

9   The work done w by a force is the area under a graph of force F against displacement x in the direction of the force. Determine the work done when $F = 2x^2 + 3x$ and x changes from 2 to 5, i.e. the value of:

$$\int_2^5 (2x^2 + 3x)\, dx$$

Problems

1   Determine the following integrals:

(a) $\int x^5\, dx$, (b) $\int x^{-3}\, dx$, (c) $\int \sqrt{x}\, dx$, (d) $\int \frac{1}{\sqrt{x}}\, dx$, (e) $\int 2x\, dx$,

(f) $\int e^{-3x}\, dx$, (g) $\int e^{x/2}\, dx$, (h) $\int \cos x\, dx$, (i) $\int \sin 4x\, dx$,

(j) $\int \sin 2x\, dx$, (k) $\int \cos 2x\, dx$, (l) $\int (2x + x^2)\, dx$,

(m) $\int (x^3 + 2x + 4)\, dx$, (n) $\int (3 + e^{-2x})\, dx$, (o) $\int (\sin 2x + \cos 3x)\, dx$.

2   Determine the equations of the following curves, given the equation for the gradient of a curve and a point through which the curve must pass:

(a) $\frac{dy}{dx} = 3x^2$, the curve passes through (1, 3),

(b) $\frac{dy}{dx} = 4x - 2$, the curve passes through (2, 3),

(c) $\frac{dy}{dx} = 6x^2 + 10x - 5$, the curve passes through (-3, 5),

(d) $\frac{dy}{dx} = 2$, the curve passes through (2, 3).

3   The current i, in milliamps, through a capacitor varies with time t, in seconds, according to the equation $i = 3 \sin 100t$. If $i = dq/dt$, determine how the charge q on the capacitor will vary with time if $q = 0$ at $t = 0$.

4   The velocity v of an object varies with time t according to the equation $v = 10t + 1$. Determine how the displacement x varies with time if $x = 0$ when $t = 0$.

5   A moving iron ammeter has a current sensitivity given by:

$$\frac{d\theta}{dI} = cI$$

where c is a constant. Determine how the angular deflection θ of the pointer depends on the current I if $\theta = 0$ at $I = 0$.

6   Determine the values of the following integrals:

(a) $\int_1^4 x^2 \, dx$, (b) $\int_0^{\pi/2} \sin 2x \, dx$, (c) $\int_0^4 x \, dx$, (d) $\int_{\pi/6}^{\pi/3} \cos x \, dx$,

(e) $\int_0^3 \frac{1}{x} \, dx$, (f) $\int_{-1}^2 (x^3 - 3x^2 - 2x + 2) \, dx$, (g) $\int_1^3 (x^2 + 4) \, dx$.

7   Determine the areas under the following graphs with the x-axis between the limits given:

(a) $y = x^2 - 3x$ between $x = 3$ and $x = 5$,

(b) $y = x^3$ between $x = 0$ and $x = 1$,

(c) $y = 4$ between $x = 1$ and $x = 3$,

(d) $y = 3x^2 + 1$ between $x = 1$ and $x = 4$,

(e) $y = x^3 - 3x + 2$ between $x = 1$ and $x = 3$,

(f) $y = 4x^3$ between $x = -2$ and $x = 2$,

(g) $y = \cos x$ between $x = 0$ and $x = \pi/4$,

(h) $y = 3x^2 + 2x + 4$ between $x = -1$ and $x = 3$,

(i) $y = 4 \cos x$ between $x = 0$ and $x = \pi/2$,

(j) $y = 4 e^{3x}$ between $x = 0$ and $x = 1$.

22 Calculus in action

22.1 Introduction

There are many situations in engineering where we are concerned with the rate at which some quantity changes with time or the area under a graph. This chapter is a look at some of these situations: namely straight-line motion, work, root-mean-square values for alternating waveforms and the behaviour of capacitors and inductors in electrical circuits.

22.2 Straight-line motion

Figure 22.1 *Distance-time graph*

The *velocity* of a particle is the rate at which its distance in a straight line from a fixed point changes with time. Thus if we have the distance–time graph shown in Figure 22.1, then when the displacement changes by δx in a time δt the average velocity over that time interval is $\delta x/\delta t$. As δt tends to 0 then $\delta x/\delta t$ tends to a limiting value dx/dt which is the gradient of the tangent to the distance–time graph at a time t, point A on the graph, and so the instantaneous velocity v at the time t. Thus, since we have:

$$\lim_{\delta t \to 0} \frac{\delta x}{\delta t} = \frac{dx}{dt}$$

then the velocity at an instant of time is

$$v = \frac{dx}{dt}$$

dx/dt will be positive if x increases as t increases and negative if x decreases as t increases.

The *acceleration* of a particle is the rate at which the velocity changes. Thus if we have the velocity–time graph shown in Figure 22.2, then when the velocity changes by δv in a time δt the average acceleration over that time interval is $\delta v/\delta t$. As δt tends to 0 then $\delta v/\delta t$ tends to a limiting value dv/dt which is the gradient of the tangent to the velocity–time graph at time t, point A on the graph, and so the acceleration a at that time t. Thus, since we have:

$$\lim_{\delta t \to 0} \frac{\delta v}{\delta t} = \frac{dv}{dt}$$

then the acceleration at an instant of time is:

$$a = \frac{dv}{dt}$$

dv/dt will be positive if v increases as t increases and negative if v decreases as t increases. The term *retardation* is used for a negative acceleration.

We can rewrite this equation for acceleration as a second derivative. Since the velocity $v = dx/dt$ and the acceleration a is the rate of change of velocity with time, then we can write:

Figure 22.2 *Velocity-time graph*

$$a = \frac{d}{dt}\left(\frac{dx}{dt}\right) = \frac{d^2x}{dt^2}$$

Thus velocity is the first derivative of displacement with time and acceleration the first derivative of velocity with time or the second derivative of displacement with time.

Example

The distance x covered by an object moving in a straight line varies with the time t according to the equation $x = 2t^2$. Determine (a) how the velocity varies with time, (b) how the acceleration varies with time, and (c) the velocity and acceleration after a time of 2 s if, in the above equation, the distance is in metres and the time in seconds.

(a) Using $v = dx/dt$ then we can obtain the velocity from the distance–time equation by differentiation. Thus:

$$v = \frac{dx}{dt} = 4t$$

(b) Using $a = dv/dt$ then we can obtain the acceleration from the velocity–time equation by differentiation. Thus:

$$a = \frac{dv}{dt} = 4$$

(c) The velocity $v = 4t$ and so $v = 4 \times 2 = 8$ m/s. The acceleration $a = 4$. Thus it does not vary with time and is constant at 4 m/s².

Example

The displacement x of a particle from some fixed point when oscillating with simple harmonic motion is given by the equation $x = A \cos \omega t$. Determine how (a) the velocity v and (b) the acceleration a vary with time t.

(a) Using $v = dx/dt$ then we can obtain the velocity from the distance–time equation by differentiation. Thus:

$$v = \frac{dx}{dt} = -A\omega \sin \omega t$$

(b) Using $a = dv/dt$ then we can obtain the acceleration from the velocity–time equation by differentiation. Thus:

$$a = \frac{dv}{dt} = -A\omega^2 \cos \omega t$$

Since $x = A \cos \omega t$ then we can rewrite this equation as $a = -\omega^2 x$.

Example

The distance x covered by an object moving in a straight line varies with the time t according to the equation $x = 2t - t^3$. Determine (a) how the velocity varies with time, (b) how the acceleration varies with time, and (c) the velocity and acceleration after a time of 2 s if, in the above equation, the distance is in metres and the time in seconds.

(a) Using $v = dx/dt$ then we can obtain the velocity from the distance–time equation by differentiation. Thus:

$$v = \frac{dx}{dt} = 2 - 3t^2$$

(b) Using $a = dv/dt$ then we can obtain the acceleration from the velocity–time equation by differentiation. Thus:

$$a = \frac{dv}{dt} = -6t$$

(c) The velocity is $v = 2 - 3t^2 = 2 - 12 = -10$ m/s. The negative sign means that it is moving in the opposite direction to that in which we measure distance outwards from the fixed reference point, i.e. it is moving back towards the point and not away from it. The acceleration $a = -6t = -6 \times 2 = -12$ m/s^2. The negative sign means that there is a retardation and the object is slowing down.

Revision

1 The distance x covered by an object moving in a straight line varies with the time t according to the equation $x = 10t - t^2$. Determine (a) how the velocity varies with time, (b) how the acceleration varies with time, and (c) the velocity and acceleration after a time of 2 s if, in the above equation, the distance is in metres and the time in seconds.

2 The distance x covered by an object moving in a straight line varies with the time t according to the equation $x = \sqrt{t}$. Determine (a) how the velocity and (b) the acceleration vary with time.

3 When an object is thrown vertically upwards, its height h varies with time t and is given by the equation $h = 4t - 5t^2$. Determine the velocity after 0.2 s.

4 Determine the velocity and the acceleration after 2 s for an object for which the displacement x, in metres, is related to the time t, in seconds, by $x = 10 + 20t - 2t^2$.

5 Determine the equations for the velocity and acceleration of a body after a time t if its displacement x varies with time t and is given by the equation $x = \sin 2t + \cos 3t$.

6   A ball rolls from rest down an inclined plane. If the distance travelled down the plane x, in metres, is related to time t, in seconds, by the equation $x = 3t^2$ determine (a) how the velocity varies with time, (b) how the acceleration varies with time and (c) the velocity and the acceleration after the ball has rolled 3 m.

22.2.1 Reversing differentiation

As illustrated in the previous section of this chapter, we can obtain the velocity–time relationship for an object in motion by differentiating the distance–time relationship. Integration can be considered the reverse of differentiation. Thus, given the equation relating the velocity and time for an object in motion, e.g.

$$v = \frac{dx}{dt} = 4t$$

then we can find the distance–time relationship by integration. In this case:

$$x = \int v \, dt = \int 4t \, dt$$

If we want to find the distance travelled between two times, t_1 and t_2, then the integral is written as:

$$x = \int_{t_1}^{t_2} v \, dt = \int_{t_1}^{t_2} 4t \, dt$$

Figure 22.3   *Velocity–time graph*

This represents the area under a velocity–time graph in that time interval (Figure 22.3).

We can obtain the acceleration–time relationship for an object in motion by differentiating the velocity–time relationship. Since integration is the reverse of differentiation, given an equation relating the acceleration and time for an object in motion, e.g.

$$a = \frac{dv}{dt} = 4$$

then we can find the velocity–time relationship by integration. In this case:

$$v = \int a \, dt = \int 4 \, dt$$

If we want the velocity between two times, t_1 and t_2, then the integral is written as:

$$v = \int_{t_1}^{t_2} a \, dt = \int_{t_1}^{t_2} 4 \, dt$$

Example

The acceleration a, in m/s^2, of an object depends on the time t, in seconds, and is given by the equation $a = t + 2$. Determine the equation describing how the velocity varies with time if the velocity is 2 m/s at the time $t = 0$.

We can write the acceleration as dv/dt and so we have

$$\frac{dv}{dt} = t + 2$$

Integrating this equation gives

$$v = \int (t + 2)\, dt = \tfrac{1}{2}t^2 + 2t + C$$

We are given the information that $v = 2$ m/s at $t = 0$. Thus $C = 2$. Hence the equation is:

$$v = \tfrac{1}{2}t^2 + 2t + 2$$

Example

The velocity v, in m/s, of an object is related to the time t, in seconds, by the equation $v = t^2 + 2t + 1$. Determine the distance travelled by the object between $t = 2$ s and $t = 3$ s.

The velocity v can be written as dx/dt and so we have:

$$v = \frac{dx}{dt} = t^2 + 2t + 1$$

The distance–time relationship can be obtained by integrating this equation. Since we require the distance travelled between two times then the integral becomes

$$x = \int_{t_1}^{t_2} v\, dt = \int_2^3 (t^2 + 2t + 1)\, dt = \left[\tfrac{1}{3}t^3 + t^2 + t + C \right]_2^3$$

Thus $x = (9 + 9 + 3 + C) - (\tfrac{8}{3} + 4 + 2 + C) = 12.3$ m.

Example

The acceleration a, in m/s^2, of a moving object varies with time t, in seconds, and is given by the equation $a = 4t + 1$. Determine (a) the velocity–time equation and (b) the distance–time equation, given that $v = 2$ m/s at $t = 0$ and $x = 5$ m at $t = 0$.

(a) The acceleration a can be written as $a = dv/dt$ and so:

$$a = \frac{dv}{dt} = 4t + 1$$

Integrating this gives:

$$v = \int (4t+1)\,dt = 2t^2 + t + C$$

We have $v = 2$ m/s at $t = 0$ and so $2 = 0 + 0 + C$. Hence the equation is:

$$v = 2t^2 + t + 2$$

(b) The velocity v can be written as dx/dt and so we have:

$$v = \frac{dx}{dt} = 2t^2 + t + 2$$

Thus integrating this gives:

$$x = \int (2t^2 + t + 2)\,dt = \tfrac{2}{3}t^3 + \tfrac{1}{2}t^2 + 2t + C$$

We have $x = 5$ m at $t = 0$ and so $5 = 0 + 0 + 0 + C$. Thus the equation is:

$$x = \tfrac{2}{3}t^3 + \tfrac{1}{2}t^2 + 2t + 5$$

Revision

7   The velocity v, in m/s, of a moving object varies with time t, in seconds, and is given by the equation $v = 8t - t^2$. Determine the distance travelled in the time from $t = 0$ to $t = 3$ s.

8   The velocity v, in m/s, of a moving object varies with time t, in seconds, and is given by the equation $v = 2t + 5$. Determine the distance travelled in the time from $t = 0$ to $t = 4$ s.

9   The velocity v, in m/s, of a moving object varies with time t, in seconds, and is given by the equation $v = 5t + 3t^2$. Determine the distance travelled in the time from $t = 1$ to $t = 2$ s.

10   The velocity v, in m/s, of a moving object varies with time t, in seconds, and is given by the equation $v = 4t - 2t^2$. If at time $t = 0$ the displacement is 1 m, what will be the displacement at 4 s?

11   The acceleration a, in m^2/s, of a moving object varies with time t, in seconds, and is given by the equation $a = 2t + 1$. Determine the velocity–time relationship given that the velocity is 10 m/s at $t = 2$ s.

12   The acceleration a, in m/s^2, of a moving object varies with time t, in seconds, and is given by the equation $a = 6t$. Determine (a) the velocity–time equation and (b) the distance–time equation, given that $v = 4$ m/s at $t = 0$ and $x = 10$ m at $t = 0$.

22.3 Work

Figure 22.4 *Force-displacement graph*

Work is the term given to the process by which energy is transferred by the point of application of a force moving through a distance. With a constant force F and a displacement x in the direction of the force, then the work done W on a body, i.e. the energy transferred to it, is given by:

$$W = Fx$$

Consider the graph shown in Figure 22.4 for the force applied to an object and how it varies with the displacement of that object. For a small displacement δx we can consider the force to be effectively constant at F. Thus the work done for that displacement is $F\,\delta x$. This is the area of the strip under the force–distance graph. If we want the work done in changing the displacement from x_1 to x_2 then we need to determine the sum of all such strips between these displacements, i.e. the total area under the graph between the ordinates for x_1 and x_2. Thus:

$$\text{work done} = \sum_{x=x_1}^{x=x_2} F\,\delta x$$

If we make the strips tend towards zero thickness then the above summation becomes the integral, i.e.

$$\text{work done} = \int_{x_1}^{x_2} F\,dx$$

Example

Determine the work done in stretching a spring when a force F is applied and causes a displacement change in its point of application, i.e. an extension, from 0 to x if $F = kx$, where k is a constant.

Figure 22.5 *Example*

Figure 22.5 shows the force–distance graph. The work done is the area under the graph between 0 and x. This is the area of a triangle and so the work done is $\frac{1}{2}Fx$. Since $F = kx$ we can write this as $\frac{1}{2}kx^2$.

We could have solved this problem by integration. Thus:

$$\text{work done} = \int_0^x F\,dx = \int_0^x kx\,dx = \left[\tfrac{1}{2}kx^2 + C\right]_0^x = \tfrac{1}{2}kx$$

Revision

13 Determine the work done in stretching a spring when a force F is applied and causes a displacement change in its point of application, i.e. an extension, from 0 to x if $F = kx^2$, where k is a constant.

14 An object falling under gravity from rest in air experiences a resistive force of $0.6s$, where s is the distance it has fallen. Determine the work done by the falling object against this resistive force when it falls through a distance of 10 m.

15 A horizontal force F, in newtons, acts on an object and causes the object to move from its starting position at $s = 0$ through a distance of 3 m in the direction of the force. If $F = 2s + 3$, determine the work done.

22.3.1 Work done compressing gases

Figure 22.6 *Compressing a gas*

Consider work being done as a result of a piston being moved, as in Figure 22.6, to reduce the volume of a gas. The work done in moving the piston through a small distance δx when the force is F is $F \, \delta x$. But the gas is subject to a pressure as a result of the force. Since pressure is force per unit area, then if the force acts over an area A the pressure $p = F/A$. Thus we can write for the work done:

$$\text{work done} = F \, \delta x = pA \, \delta x$$

But $A \, \delta x$ is the change in volume δV of the gas. Hence:

$$\text{work done} = p \, \delta V$$

The total work done in changing the volume of a gas from V_1 to V_2 is thus:

$$\text{work done} = \sum_{V_1}^{V_2} p \, \delta V$$

If we consider δV tending to zero then we can write

$$\text{work done} = \int_{V_1}^{V_2} p \, dV$$

Example

Determine the equation for the work done in compressing a gas from a volume V_1 to V_2 if the gas obeys Boyle's law, i.e. $pV = $ a constant.

For the gas we have $pV = k$ and hence $p = k/V$. The work done is thus:

$$\text{work done} = \int_{V_1}^{V_2} p \, dV = \int_{V_1}^{V_2} \frac{k}{V} \, dV = [\ln V + C]_{V_1}^{V_2}$$

Hence the work done is $\ln V_2 - \ln V_1$.

Revision

16 Determine the equation for the work done in compressing a gas from a volume V_1 to V_2 if the gas obeys the law $pV^{5/3} = $ a constant.

22.4 Root-mean square values

In Section 17.2 the mean (average) value of a waveform was identified as being given by:

mean = (area under the graph)/(range considered)

For a current i which varies with time t, the mean value between the times t_1 and t_2 is thus:

$$\text{mean} = \frac{1}{t_2 - t_1} \int_{t_1}^{t_2} i \, \mathrm{d}t$$

The power dissipated by an alternating current i when passing through a resistance R is $i^2 R$. The mean power dissipated over a time interval from $t = 0$ to $t = T$ will thus be:

$$\text{mean power} = \frac{1}{T-0} \int_0^T i^2 R \, \mathrm{d}t = \frac{R}{T} \int_0^T i^2 \, \mathrm{d}t$$

If we had a direct current I generating the same power then we would have:

$$I^2 R = \frac{R}{T} \int_0^T i^2 \, \mathrm{d}t$$

and:

$$I = \sqrt{\frac{1}{T} \int_0^T i^2 \, \mathrm{d}t}$$

This current I is known as the *root-mean-square* current (see Section 17.3). There are other situations in engineering and science where we are concerned with determining root-mean-square quantities. The procedure is thus to determine the mean value of the squared function over the required interval and then take the square root.

Example

Determine the root-mean-square current value of the alternating current $i = I \sin \omega t$ over the time interval $t = 0$ to $t = 2\pi/\omega$, i.e. one cycle.

Using the above equation:

$$I = \sqrt{\frac{1}{T} \int_0^T i^2 \, \mathrm{d}t} = \sqrt{\frac{\omega}{2\pi} \int_0^{2\pi/\omega} I^2 \sin^2 \omega t \, \mathrm{d}t}$$

$$= \sqrt{\frac{I^2 \omega}{2\pi} \int_0^{2\pi/\omega} \tfrac{1}{2}(1 - \cos 2\omega t) \, \mathrm{d}t} = \sqrt{\frac{I^2 \omega}{4\pi} \left[t - \frac{1}{2\omega} \sin 2\omega t \right]_0^{2\pi/\omega}}$$

$$= \frac{I}{\sqrt{2}}$$

Figure 22.7   *Example*

Example

Determine the root-mean-square value of the waveform shown in Figure 22.7 over a period of 0 to 2 s.

From $t = 0$ to $t = 1$ s the waveform is described by $y = 2$. From $t = 1$ s to $t = 2$ s the waveform is described by $y = 0$. Thus the root-mean-square value is given by:

$$y_{\text{rms}} = \sqrt{\tfrac{1}{2}\left(\int_0^1 4\,dt + \int_1^2 0\,dt\right)} = \sqrt{\tfrac{1}{2}[4t]_0^1} = \sqrt{2}$$

Revision

17 Determine the root-mean-square values of the following functions between the specified limits:

(a) $y = \sin x + 1$ from $x = 0$ to $x = 2\pi$, (b) $y = \sin 2x$ from $x = 0$ to $x = \pi$

18 Determine the root-mean-square value of a half-wave rectified sinusoidal voltage. Between the times $t = 0$ and $t = \pi/\omega$ the equation is $v = V \sin \omega t$ and between $t = \pi/\omega$ and $t = 2\pi/\omega$ we have $v = 0$.

22.5 Current

Electrical *current*, *i* is the rate of movement of charge q with time t. Thus if δq moves in a time δt then the average current over that time is $\delta q/\delta t$. In the limit as δt tends to zero, the current i at an instant is given by:

$$i = \frac{dq}{dt}$$

If we require to know the charge moved through a circuit when a current flows then we require the reverse of the above. Since integration is the reverse of differentiation then:

$$q = \int i\,dt$$

Example

Derive an equation for the current through a circuit element if the charge q, in coulombs, entering the element is related to the time t, in seconds, by the equation $q = 4t$.

Current is the rate of movement of charge. Thus:

$$i = \frac{dq}{dt} = 4 \text{ A}$$

Example

Determine the charge that has flowed through a circuit between times of $t = 1$ s and $t = 2$ s when the current i, in amperes, is given by $2\,e^{-0.1t}$.

Using the equation given above, then:

$$q = \int i\,dt = \int_1^2 2\,e^{-0.1t}\,dt = \left[2 \times \left(-\frac{1}{0.1}\right)e^{-0.1t} + C\right]_1^2$$

$$= -20\,e^{-0.2} + 20\,e^{-0.1} = -16.37 + 18.10 = 1.73\ \text{C}$$

Revision

19 Derive an equation for the current through a circuit element if the charge q, in coulombs, entering the element is related to the time t, in seconds, by $q = 10t^2$.

20 Derive an equation for the current through a circuit element if the charge q, in coulombs, entering the element is related to the time t, in seconds, by $q = 4\,e^{-3t}$.

21 Derive an equation for the current through a circuit element if the charge q, in coulombs, entering the element is related to the time t, in seconds, by $q = 2t + e^{-5t}$.

22 Derive an equation for the current through a circuit element if the charge q, in coulombs, entering the element is related to the time t, in seconds, by $q = 2\cos 5t$.

23 Determine the charge that has flowed through a circuit between times $t = 0$ s and $t = 1$ s when the current i, in amperes, is given by $3 + e^{-0.1t}$.

24 Determine the charge that has flowed through a circuit between times $t = 1$ s and $t = 2$ s when the current i, in amperes, is given by $2t$.

22.6 Capacitors

The circuit element called a capacitor is essentially just a pair of parallel conducting plates separated by an insulator. When a voltage is applied across a capacitor then one of the plates becomes negatively charged and the other positively charged as a result of charge flowing on to one of the plates and leaving the other plate (Figure 22.8). The amount of charge gained by one plate is the same as that lost by the other plate. It is found that the amount of charge q on a capacitor plate is directly proportional to the potential difference v between the plates. Hence we can write $q \propto v$ and so

$$q = Cv$$

Applied voltage

Figure 22.8 *Capacitor*

where C, the constant of proportionality, is termed the *capacitance*.

Consider a small element of charge δq moving onto one plate of the capacitor and off the other. Then the resulting change in potential difference δV is given by:

$$\delta q = C\, \delta v$$

If we divide both sides of this equation by the time interval δt during which the change is taking place, then:

$$\frac{\delta q}{\delta t} = C\frac{\delta v}{\delta t}$$

If we consider the changes when δt tends to a zero value, then we can write:

$$\frac{dq}{dt} = C\frac{dv}{dt}$$

Since current i is the rate of movement of charge, we can write the above equation as:

$$i = \frac{dq}{dt} = C\frac{dv}{dt}$$

Thus the current in the circuit at some instant of time is the product of the capacitance and the rate at which the potential difference across the capacitor is changing with time.

Example

The voltage v, in volts, across a capacitor is continuously adjusted so that it varies with time t, in seconds, according to the equation $v = 2t$. Derive an equation indicating how the current varies with time for a capacitance of 1 mF.

With $v = 2t$ then we have:

$$i = C\frac{dv}{dt} = 0.001 \times 2 = 0.002 \text{ A}$$

The current is thus maintained constant at 2 mA.

Example

The voltage v, in volts, across a capacitor varies with time t, in seconds, according to the equation $v = 10 \sin 3t$. How will the current vary with time if the capacitance is 0.5 mF?

With $v = 10 \sin 3t$ then:

$$i = C\frac{dv}{dt} = 0.005 \times 30 \cos 3t = 0.15 \cos 3t \text{ A}$$

The current is thus given by a cosine curve.

Revision

25 The voltage v, in volts, across a capacitor varies with time t, in seconds, according to the equation $v = 10t$. How will the current vary with time if the capacitance is 0.5 mF?

26 The voltage v, in volts, across a capacitor varies with time t, in seconds, according to the equation $v = 4 \sin 10t$. How will the current vary with time if the capacitance is 1 mF?

27 The voltage v, in volts, across a capacitor varies with time t, in seconds, according to the equation $v = 4 \cos 10t$. How will the current vary with time if the capacitance is 1 mF?

28 The voltage v, in volts, across a capacitor varies with time t, in seconds, as shown in Figure 22.9. How will the current vary with time if the capacitance is 1 mF?

Figure 22.9 *Revision problem 28*

22.7 Inductors

An inductor is essentially just a coil of wire. When there is a current passing through a wire then a magnetic field is produced. We talk of there being magnetic flux generated by the current. The magnetic flux is proportional to the current. When the current through the coil changes then the magnetic flux generated by the current changes. The magnetic flux passing through the coil thus changes. Hence, as a result of electromagnetic induction, an e.m.f. is induced in the coil. This induced e.m.f. is in such a direction as to oppose the change producing it. Thus if it was produced by an increasing current the direction of the e.m.f. will be such as to slow down the increase. If it was produced by a decreasing current then the direction of the e.m.f. is such as to slow the decrease in current. The induced e.m.f. is proportional to the rate of change of magnetic flux. Hence, since the amount of magnetic flux is proportional to the current we have the induced e.m.f. proportional to the rate of change of current. We can thus write:

$$\text{e.m.f.} \propto -\frac{di}{dt}$$

The minus sign is because the e.m.f. opposes the change in current responsible for it. This equation can then be written as:

$$\text{e.m.f.} = -L\frac{di}{dt}$$

L, the constant of proportionality, is called the *inductance*.

When we have a current through a resistance in a circuit then the potential difference across the resistor is given by $V = RI$. However, if we are concerned with a pure inductance, i.e. one which has only inductance and no resistance, then since $R = 0$ there can be no potential difference across the inductor due to its resistance. To maintain a current through the inductor then the source must supply a potential difference v across the inductor which is just sufficient to cancel out the induced e.m.f. Thus

$$v = L\frac{di}{dt}$$

Then we have $(v + \text{induced e.m.f.}) = 0$. The potential difference applied across the inductor is thus proportional to the rate of change of current through it.

Example

The current i, in amperes, through an inductor of inductance 2 H varies with time t in seconds, being related to the time by the equation $i = 2t$. Derive a relationship for the voltage across the inductor.

Using the equation $v = L \, di/dt$:

$$v = L\frac{di}{dt} = 2 \times 2 = 4 \text{ V}$$

The voltage is thus constant at 4 V.

Example

The current i, in amperes, through an inductor of inductance 0.2 H varies with time t in seconds, being related to the time by the equation $i = 2 \sin 10t$. Derive a relationship for the voltage across the inductor.

Using the equation $v = L \, di/dt$,

$$v = L\frac{di}{dt} = 0.2 \times 2 \times 10 \cos 10t = 4 \cos 10t \text{ V}$$

The voltage is given by a cosine graph, the current by a sine graph. The voltage and current are thus out of phase.

Example

The voltage v, in volts, applied to an inductor of inductance 0.1 H varies with time t, in seconds, being related to the time by the equation $v = 20 \, e^{-10t}$. Derive a relationship for how the current varies with time if the current is zero when $t = 0$.

We need to reverse the equation $v = L \, di/dt$ for this problem. Hence:

$$i = \frac{1}{L} \int v \, dt = \frac{1}{0.1} \int 20 \, e^{-10t} \, dt = -20 \, e^{-10t} + C$$

When $t = 0$ then $i = 0$. Thus $0 = -20 + C$. Hence $C = 20$. Thus the equation is:

$$i = = -20 \, e^{-10t} + 20 = 20(1 - e^{-10t}) \text{ A}$$

Revision

29 The current i, in amperes, through an inductor of inductance 2 H varies with time t in seconds, being related to the time by the equation $i = 5t$. Derive a relationship for the voltage across the inductor.

30 The current i, in amperes, through an inductor of inductance 0.1 H varies with time t, in seconds, according to $i = 2(1 - e^{-100t})$. Determine how the potential difference across the inductor varies with time.

31 The current i, in amperes, through an inductor of inductance 0.01 H varies with time t, in seconds, and is described by $i = 10 \, e^{-2t}$. Determine how the potential difference across the inductor varies with time.

32 The voltage v, in volts, across an inductor of inductance 1 H varies with time t, in seconds, and is described by the equation $v = 9t^2$. Determine how the current through the inductor varies with time if the current is zero at $t = 0$.

Problems

1 Determine the velocity and acceleration after a time t for an object which has a displacement x from a fixed point which changes with time according to the equation $x = 2t^2 - 3t + 4$.

2 Determine the velocity and acceleration after a time t for an object which has a displacement x from a fixed point which changes with time according to $x = 10 \cos 2t$.

3 Determine the velocity and acceleration after a time t for an object which has a displacement x from a fixed point which changes with time according to the equation $x = 1 - e^{-2t}$.

4 Determine the velocity and acceleration after a time t for an object which has a displacement x from a fixed point which changes with time according to the equation $x = 3t^3 - t$.

5 Determine the velocity and acceleration after a time t for an object which has a displacement x from a fixed point which varies with time according to the equation $x = 1 + 6t - 2t^2$.

6 Determine the velocity and the acceleration after 2 s for an object for which the displacement x, in metres, is related to the time t, in seconds, by $x = 20 + 10t - 2t^2$.

7 Determine the velocity and the acceleration after 2 s for an object for which the displacement x, in metres, is related to the time t, in seconds, by $x = 20t - 4t^2$.

8 A particle has a displacement x, in metres, which varies with time t, in seconds, according to $x = 10 + 20t - t^2$. At what time will the velocity be zero?

9 The velocity v, in m/s, of a moving object depends on the time t, in seconds, and is given by the equation $v = 2t + 1$. Determine the acceleration.

10 A particle has a displacement x, in metres, which varies with time t, in seconds, according to $x = 20t + 12t^2 - 2t^3$. At what time will the acceleration be zero?

11 The velocity v, in m/s, of a moving object varies with time t, in seconds, and is given by the equation $v = 36 - 4t$. If at time $t = 1$ s the displacement is 20 m derive an equation for how the displacement varies with time.

12 The velocity v, in m/s, of a moving object varies with time t, in seconds, and is given by the equation $v = 4t + 6t^2$. If at time $t = 0$ the displacement is 20 m what will be the displacement at 4 s?

13 The velocity v, in m/s, of a moving object varies with time t, in seconds, and is given by the equation $v = 4t + 1$. If at time $t = 0$ the displacement is 0 what will be the displacement at 1 s?

14 The velocity v, in m/s, of a moving object varies with time t, in seconds, and is given by the equation $v = 2 e^{-t}$. If at time $t = 0$ the displacement is 0 what will be the displacement at 4 s?

15 The velocity v, in m/s, of a moving object varies with time t, in seconds, and is given by the equation $v = 48 - 3t$. If at time $t = 0$ the displacement is 0 what will be the time when the object is back at its starting point?

16 The acceleration a, in m/s^2, of a moving object varies with time t, in seconds, and is given by the equation $a = 4t^2$. Determine the velocity–time relationship given that the velocity is 5 m/s at $t = 1$ s.

17 The acceleration a, in m/s^2, of a moving object varies with time t, in seconds, and is given by the equation $a = 6t + 2$. Determine the velocity–time relationship given that the velocity is 10 m/s at $t = 1$ s.

18 The acceleration a, in m/s^2, of a moving object varies with time t, in seconds, and is given by the equation $a = 2\sqrt{t}$. Determine the velocity–time relationship given that the velocity is 10 m/s at $t = 4$ s.

19 Determine the work done in stretching a spring when a force F is applied and causes a displacement change in its point of application, i.e. an extension, from 0 to x if $F = k\sqrt{x}$, where k is a constant.

20 A horizontal force F, in newtons, acts on an object and causes the object to move from its starting position at $s = 0$ through a distance of 2 m in the direction of the force. If $F = 3s^2 + 1$, determine the work done.

21 An object is projected upwards against gravity in air. It experiences a resistive force of $2s$, where s is the distance vertically upwards from the point of projection. Determine the work done against the resistance in the object moving vertically upwards from its start position to height h.

22 A horizontal force F, in newtons, acts on an object and causes the object to move from its starting position at $s = 0$ through a distance of 4 m in the direction of the force. If $F = 4s + 3$, determine the work done.

23 The force F between two charged particles is given by $F = k/x^2$, where x is the distance between them and k a constant. Determine the work done in fetching a charged particle from infinity, where $F = 0$, up to a distance r from another charged particle.

24 Determine the equation for the work done in compressing a gas from a volume V_1 to V_2 if the gas obeys the law $pV^{1.4} = $ a constant.

25 Determine the root-mean-square values of the following functions between the specified limits:

(a) $y = 5 \sin 2x$ from $x = 0$ to $x = \pi$, (b) $y = 2 \sin x$ from $x = 0$ to $x = 2\pi$.

Figure 22.10 *Problem 26*

26 Determine the root-mean-square value of the periodic triangular voltage shown in Figure 22.10 over one complete period T.

27 Determine the mean value of a voltage v between the time $t = 0$ and $t = \pi/100$ if $v = 10 \sin 250t$.

28 Derive an equation for the current through a circuit element if the charge q, in coulombs, entering the element is related to the time t, in seconds, by $q = 2t^2$.

29 Derive an equation for the current through a circuit element if the charge q, in coulombs, entering the element is related to the time t, in seconds, by $q = 5 e^{-2t}$.

30 Derive an equation for the current through a circuit element if the charge q, in coulombs, entering the element is related to the time t, in seconds, by $q = 2t + 5 e^{-2t}$.

31 Derive an equation for the current through a circuit element if the charge q, in coulombs, entering the element is related to the time t, in seconds, by $q = 5 \sin 2t$.

32 Determine the charge that has flowed through a circuit between times of $t = 1$ s and $t = 2$ s when the current i, in amperes, varies with time and is given by the equation $0.2\, e^{-0.5t}$.

33 Determine the charge that has flowed through a circuit between times of $t = 0$ and $t = 2$ s when the current i, in amperes, is given by the equation $i = 0.2t$.

34 The voltage v across a capacitor varies with time t. Derive equations for how the current varies with time if (a) $v = a$, with a being a constant, (b) $v = at$, with a being a constant, (c) $v = V \cos \omega t$.

35 The voltage v, in volts, across a capacitor varies with time t, in seconds, according to the equation $v = 2 \sin 100t$. How will the current vary with time if the capacitance is 1 mF?

36 The voltage v, in volts, across a capacitor varies with time t, in seconds, according to the equation $v = 4t^2$. How will the current vary with time if the capacitance is 2 mF?

37 The current i, in amperes, through an inductor of inductance 0.1 H varies with time t in seconds, being related to the time by the equation $i = 0.2t$. Derive a relationship for the voltage across the inductor.

38 The current i, in amperes, through an inductance of 0.01 H varies with time t, in seconds, according to the equation $i = 10\, e^{-3t}$. Determine how the potential difference across the inductor varies with time.

39 The current i, in amperes, through an inductance of 0.2 H varies with time t, in seconds, according to the equation $i = 4\, e^{-20t} - 2\, e^{-40t}$. Determine how the potential difference across the inductor varies with time.

40 The current applied to an inductor increases at a constant rate for 1 s and then becomes constant. Determine the waveform of the voltage across the inductor.

41 The voltage v, in volts, across an inductor of inductance 1 H varies with time t, in seconds, and is described by the equation $v = 4t$. Determine how the current through the inductor varies with time if the current is zero at $t = 0$.

Answers

Chapter 1   *Revision*

1 (a) 1.3, (b) 0.01, (c) 10, (d) 0.15, (e) 0.7
2 (a) 2^7, (b) a^7, (c) x^9, (d) a^5b^{10}, (e) 2^6x^6, (f) 2^7, (g) x^3, (h) x^4
3 (a) $2^4 \times 2^{-2} = 2^2$, (b) $a^5 \times a^{-2} \times a^{-4} = a^{-1}$, (c) $(2x)^2 \times (3x)^3 \times (2x)^{-4}$
4 (a) $a^{7/2}$, (b) $x^{3/2}$, (c) $x^{1/2}$, (d) $2^{-1/2}$, (e) $a^{10/6}$
5 (a) 1.2×10^2, (b) 5.6×10^3, (c) 6.5×10^8, (d) 4.5×10^{-3}, (e) 1.2×10^{-8}
6 (a) 243, (b) 0.03125, (c) 1.621, (d) 1.442, (e) 2.83, (f) 4.50, (g) 0.885, (h) 0.61
7 (a) 120 km, (b) 5.1 mm, (c) 120 mA, (d) 1.2 kV, (e) 4.7 MΩ
8 (a) 100 V, (b) 2.5×10^{-3} A = 2.5 mA, (c) 4200 Ω = 4.2 kΩ

Problems

1 (a) 2, (b) 2, (c) 1, (d) 4
2 (a) 120, (b) 54.0, (c) 0.0151, (d) 0.000 693
3 (a) 9.0, (b) 0.009, (c) 1.51, (d) 4.5, (e) 0.21, (f) 0.165, (g) 0.14, (h) 114.8
4 (a) 3^8, (b) a^5, (c) a^3, (d) a^2, (e) $2^2 3^2 = 36$, (f) x^{-6}, (g) $a^{1/2}$, (h) $x^{-3/2}$, (i) $a^{5/2}$, (j) $x^{5/2}$, (k) 1/16, (l) $2^{7/2}$, (m) ab^2, (n) $(a + 1)/a^2$, (o) $a/4b^4$
5 (a) 128, (b) 0.0625, (c) 2.408, (d) 3.085
6 (a) 2×10^3, (b) 2×10^2, (c) 2.0×10^{-5}
7 (a) 5.2 kV, (b) 1.5 kW, (c) 160 mA, (d) 120 μA, (e) 10 mm, (f) 120 Mm, (g) 4.0 mm, (h) 1.5 MΩ
8 (a) 0.071 m = 71 mm, (b) 0.025 m/s^2 = 25 mm/s^2, (c) 0.5 s
9 (a) 240 N, (b) 0.2 N, (c) 2500 kg = 2.5 Mg, (d) 1.3 m/s^2

Chapter 2   *Revision*

1 (a) 250 mm^2, (b) 600 mm^2, (c) 450 mm^2
2 (a) 235 mm^2, (b) 261 mm^2, (c) 80.7 mm^2, (d) 112 mm^2
3 35 m^3
4 101 mm^2
5 4.2×19^6 mm^3
6 (a) 96 cm^3, (b) 25 cm^3, (c) 107 cm^3
7 4.2×10^4 mm^3
8 (a) 1.2×10^5 mm^3, 1.61×10^4 mm^2, (b) 93.3 m^3, 142 m^2
9 3.9×10^5 m^3
10 3.6×10^5 m^2
11 About 9.4 square units
12 About 9.4 square units
13 About 9.4 square units
14 About 139 m
15 About 20.7 J
16 17.8 cm^3

Problems

1 (a) 54.4 mm^2, (b) 31.4 mm^2, (c) 50.5 mm^2, (d) 57.4 mm^2, (e) 47 mm^2, (f) 45.8 mm^2, (g) 4000 mm^2, (h) 2.1 × 10^4 mm^2

2 2.8 × 10^5 cm^3

3 1.3 × 10^5 mm

4 1800 m^3

5 (a) 1.0 × 10^5 mm^3, (b) 1.2 × 10^5 mm^3, (c) 6.3 × 10^3 mm^3

6 7.5 × 10^3 cm^2

7 (a) 2.83 × 10^7 mm^3, 7.54 × 10^5 mm^2, (b) 5.3 m^3, 20.5 m^2, (c) 8.1 × 10^8 mm^3, 5.4 × 10^6 mm^2

8 1.1 × 10^5 mm^3

9 15.4 m^2

10 1.74 × 10^7 mm^3

11 312 cm^3

12 About 71 square units

13 216 cm^2

14 19.37 m^2

15 249 square units

16 173 m

17 71 m

18 23.9 J

19 12.1 m^3

20 134.4 m^3

Chapter 3   *Revision*

1 (a) $x = 5$, (b) $x = 2$, (c) $x = 1$, (d) $x = 24$, (e) $x = 6$, (f) $x = 2$, (g) $x = 2$, (h) $x = 7$, (i) $x = -7/2$, (j) $x = 5/6$, (k) $x = 16$, (l) $x = 4$

2 (a) 5 Ω, (b) 0.5 N/m, (c) 10 N, (d) 8.7 Ω

3 (a) $6 + 2x$, (b) $1 - x$, (c) $6x + 22$, (d) $13x + 16$, (e) $2x^2 + 13x + 21$, (f) $x^2 - 6x + 9$, (g) $4x^2 + 4x + 1$, (h) $x^2 - 1$, (i) $2x^2 + x$, (j) $x^2 + 5x + 5$

4 (a) $\dfrac{2y + 5x}{xy}$, (b) $\dfrac{2y - x^2}{x^2 y}$, (c) $\dfrac{3x + 3}{(x + 5)(x - 1)}$, (d) $\dfrac{x^2 - 4x - 1}{(x + 1)(x - 3)}$

5 (a) $a = \dfrac{v - u}{t}$, (b) $r_1 = \dfrac{F_2 r_2}{F_1}$, (c) $x = \dfrac{FL}{AE}$, (d) $L = \dfrac{T^2 g}{4\pi^2}$, (e) $X = \sqrt{Z^2 - R^2}$, (f) $m = \dfrac{pV}{RT}$, (g) $v = \sqrt{\dfrac{2E}{m}}$, (h) $a = \dfrac{2(s - ut)}{t^2}$, (i) $d = \sqrt{\dfrac{6Z}{b}}$

6 0.000 04 per °C

7 4000 N

8 0.15 m^3

9 N/m^2 (or Pa)

10 N m

11 J kg^{-1} K^{-1}

12 Ω m

13 kg m^2 s^{-2}

Problems

1 (a) $x = 1/3$, (b) $x = 16$, (c) $x = 3/8$, (d) $x = 4/15$, (e) $x = 11$, (f) $x = 7$, (g) $x = 7$, (h) $x = 8$, (i) $x = \frac{1}{2}$, (j) $x = 2$, (k) $x = 16$, (l) $x = 2$, (m) $x = 4$, (n) $x = 5$

2 (a) $4x + 5$, (b) $1 - 2x$, (c) $7x - 8$, (d) $8x - 1$, (e) $2x^2 + 8x + 1$, (f) $x^2 + x - 6$, (g) $x^3 + 2x^2 + x + 2$, (h) $x^2 - 6x + 9$, (i) $4x^2 - 4x + 1$, (j) $5x^2 - 6x + 9$, (k) $x^2 - 9$, (l) $4x^2 - 1$

3 (a) $\dfrac{5y - 2x}{xy}$, (b) $\dfrac{7x + 11}{(x + 1)(2x + 3)}$, (c) $\dfrac{3x^2 + 9x}{(x + 5)(x - 1)}$, (d) $\dfrac{-x^2 + 2x - 3}{(x + 1)(x - 1)}$

4 79.8 mm

5 5 s

6 1.02 m

7 J/kg

8 kg/m^3

9 N m

10 (a) $h = \dfrac{p_1 - p_2}{\rho g}$, (b) $R = \dfrac{\rho L}{A}$, (c) $r = \dfrac{V - E}{I}$, (d) $T = Ia + mr(a + g)$,

(e) $F = \dfrac{2m}{t^2}(s - ut)$, (f) $x = \dfrac{mgL}{EA}$, (g) $R = \dfrac{EI}{M}$, (h) $m = \dfrac{2E}{v^2}$,

(i) $r = \dfrac{3V}{\pi h}$, (j) $R = \dfrac{P}{I^2}$, (k) $a_2 = -\dfrac{m_1 a_1}{m_2}$, (l) $v = \sqrt{\dfrac{Fr}{m}}$,

(m) $d = \dfrac{\varepsilon_0 \varepsilon_r A}{C}$

11 $T_2 = \dfrac{V_2 T_1}{V_1}$, 600 K

12 $w = \dfrac{2R - W}{L}$, 4 kN

Chapter 4

Revision

1 (a) $x = 1$, (b) $x = 2$, (c) $x = 3$, (d) $x = 0.86$, (e) $x = 1$, (f) $x = 5/2$, (g) $x = -0.83$, (h) $x = 3$, (i) $x = -6$, (j) $x = 4.4$, (k) $x = 1.9$, (l) $x = 6.94$, (m) $x = 12.2$

2 (a) 0.75 A, (b) 4.35×10^{-3} °C^{-1}, (c) 2.15 Ω

3 (a) $x = 1, y = 1$, (b) $x = 2, y = 1$, (c) $x = 1, y = 3$, (d) $x = 4, y = -1$, (e) $x = -1, y = 2$, (f) $x = 5, y = -2$, (g) $x = 1, y = 5$, (h) $x = \frac{1}{2}, y = 1$, (i) $x = 2, y = 1$, (j) $x = 2, y = 4$

4 $R_1 = 3.85$ kN, $R_2 = 1.65$ kN

5 $I_1 = -0.71$ A, $I_2 = 1.14$ A

6 $F_1 = 1$ N, $F_2 = 1$ N

7 $u = 2$ m/s, $a = 3$ m/s^2

8 As given in revision problems 3, 4, 5, 6 and 7

Problems

1 (a) $x = 1$, (b) $x = 3.53$, (c) $x = 2$, (d) $x = -0.70$, (e) $x = 840$, (f) $x = -42$, (g) $x = -0.25$, (h) $x = 1.26$, (i) $x = -6$

2 $R = 10 \Omega$

3 $F = 1$

4 (a) $x = 2, y = -1$, (b) $x = 3, y = 1$, (c) $x = -1, y = 4$, (d) $x = 2, y = 2$, (e) $x = 3, y = -1$, (f) $x = 4, y = -1$, (g) $x = 1, y = 4$, (h) $x = 3, y = 5$, (i) $x = 2, y = 6$, (j) $x = 1, y = 3$

5 $a = 2$ N, $b = \frac{1}{2}$
6 $I_1 = 1$ A, $I_2 = 0.5$ A
7 $F_1 = 2$ N, $F_2 = 3$ N
8 $u = 2$ m/s, $a = 0.5$ m/s^2
9 $u = 1$ m/s, $a = 2$ m/s^2
10 $R_1 = R_2 = 40$ N

Chapter 5   *Revision*

1 Linear (a), (b), (g); quadratic (c), (d), (f), (h)
2 (a) $(x + 2)(x + 1)$, (b) $(x + 5)(x + 2)$, (c) $(2x + 1)(x - 3)$,
    (d) $(3x + 2)(5x + 3)$, (e) $(3x + 2)^2$, (f) $(2x + 3)(2x - 3)$, (g) $x(5x + 3)$,
    (h) $(2x + 5)(3x - 2)$
3 (a) $-2, +1$, (b) $-4, +2$, (c) $+1, +4$, (d) $-2, +2$, (e) $-2, -2$, (f) $+1, +1$,
    (g) $0, -7$, (h) $-1, + 1$, (i) $-3, +1$, (j) $+2, +3$
4 (a) $x^2 - 5x + 4 = 0$, (b) $x^2 - x - 2 = 0$, (c) $x^2 - 6x = 0$,
    (d) $x^2 + 5x + 6 = 0$, (e) $x^2 - 2x = 0$, (f) $x^2 + 3x + 2 = 0$,
    (g) $x^2 - 4x + 3 = 0$
5 (a) $-2, -6$, (b) $-3.65, 1.65$, (c) $-1.14, 1.46$, (d) $-1.16, 3.16$,
    (e) $-3.13, 0.64$, (f) $0.22, -1.55$, (g) $1.46, -5.46$, (h) $3.41, 0.59$
6 As given in revision problem 5
7 (a) $-0.56, 3.56$, (b) $-2, -6$, (c) $-4.08, 0.08$, (d) $-1.16, 3.16$,
    (e) $-4.54, 1.54$
8 18.8 m or 1.6 m
9 5.41 cm and 8.41 cm
10 431.7 rad/s
11 1.84 m and 18.16 m
12 6 s
13 4.42 cm
14 12 cm
15 (a) Imaginary, (b) two real distinct roots, (c) imaginary, (d) two real
    coincident roots, (e) two real distinct roots

Problems

1 (a) $-2, +1$, (b) $-1, -3$, (c) $+1, +3$, (d) $-1, -1$, (e) $-3, + 3$, (f) $-5, + 5$,
    (g) $0, -3$, (h) $-5/2, -1$, (i) $-1/3, -1/2$, (j) $-5/2, -2/3$, (k) $5, -3/2$,
    (l) $7/3, 4/3$
2 (a) $x^2 - 1 = 0$, (b) $x^2 - 4x + 4 = 0$, (c) $x^2 + 4x + 3 = 0$, (d) $x^2 - x = 0$
3 (a) $-2, 4$, (b) $-2.3, 1.3$, (c) $0.33, 1$, (d) $-1.5, 1$, (e) $-3.45, 1.45$,
    (f) $-0.88, 0.68$, (g) $-2.5, 1.5$, (h) $0.31, 1.29$, (i) $-1.69, 0.44$
4 12.8 cm by 7.8 cm
5 10 cm by 3 cm
6 2.73 s and 7.47 s (on the way up and the way down)
7 31.9°C and 326°C
8 18.94 m or 1.06 m
9 2.61 cm
10 0.385 A or -10.39 A
11 16 mm
12 4.6 mm, 17.4 mm
13 8.41 cm, 5.41 cm

14 14 and 19 or −14 and −19
15 4 cm, 7 cm
16 10 cm by 10 cm
17 9 or −10
18 5 cm
19 12 and 13 or −12 and −13
20 14 m, 19 m
21 (a) Two distinct real roots, (b) two distinct real roots, (c) imaginary roots, (d) imaginary roots, (e) two real coincident roots

Chapter 6   *Revision*
1 (a) 64 m, (b) 15 m/s, (c) 8 m, (d) 4 s, (e) 12 m, (f) 2 s, (g) 5 m/s
2 30 m/s
3 400 m
4 20 s
5 (a) $s = (v^2 − u^2)/2a$, (b) $a = (v^2 − u^2)/2s$, (c) $u = \sqrt{(v^2 − 2as)}$
6 (a) 7.5 s, (b) 112.5 m
7 300 m
8 36 s
9 10.1 m, 5.2 m/s
10 0.51 s
11 31.3 m/s
12 $V = \sqrt{(PR)}$
13 15 Ω
14 5 V
15 (a) 2 V, (b) 0.04 W
16 (a) 30 Ω, (b) 2.73 Ω
17 (a) 6.32 Ω, (b) 0.0911 A
18 6 Ω
19 (a) 0.25 A, (b) 1 V, 3 V
20 (a) 1 A, 0.33 A, (b) 5.33 W
21 (a) 22 Ω, 0.55 A, (b) 6 Ω, 2 A, (c) 12.8 Ω, 0.94 A, (d) 50 Ω, 0.24 A
22 (a) 20 Ω, (b) 0.6 A, 0.2 A, 0.4 A, (c) 7.2 W
23 (a) 22 Ω, (b) 5.45 V, 6.55 V, (c) 6.55 W
24 (a) 0.71 A, (b) −0.44 A

Problems
1 3 m/s
2 0.2 m/s^2
3 1650 m
4 12 m/s
5 500 m
6 20 s
7 (a) $a = (v − u)/t$, (b) $t = (v − u)t$, (c) $u = v − at$
8 (a) $a = 2(s − ut)/t^2$, (b) $a = (s − \frac{1}{2}at^2)/t$
9 9.8 m/s
10 1.6 s
11 9.9 m/s

12 4.1 m, 0.92 s

13 2 s, 5 s

14 22.5 m

15 180 m

16 0.50 m

17 (a) 200 V, (b) 2000 Ω

18 0.05 A

19 (a) 5 V, (b) 0.25 W

20 (a) 20 V, (b) 0.4 W

21 0.2 A, 4 V, 6 V, 10 V

22 (a) 1.43 W, 0.41 W, 1.02 W, (b) 5 W, 2 W, 7 W

23 6.88 mA

24 2.03 V, 9.97 V

25 3.43 Ω, 2 A, 1 A, 0.5 A

26 1.875 A

27 2.22 A

28 (a) 15 Ω, (b) 6.8 Ω

29 (a) 0.2 A, (b) 2.2 A

30 (a) 1.5 A, (b) 1.2 A, (c) 0.29 A

Chapter 7

Revision

1 (a) 2 then add 2, (b) 5 and then double, (c) 1 and multiply by -1

2 (a) 6, 8, 10, 12, 14, (b) 3, 7, 10, 13, 16

3 25

4 (a) 2, 8, 32, 128, 512, (b) 4, -2, 1, $-\frac{1}{2}$, ¼

5 -243

6 (a) 56.5, (b) 190, (c) 220

7 25

8 (a) 453.3, (b) 23.98, (c) 1023

9 24

10 (a) $1 + 7x + 21x^2 + 35x^3 + 35x^4 + 21x^5 + 7x^6 + x^7$, (b) $16 + 32x + 24x^2 + 8x^3 + x^4$, (c) $32x^5 + 240x^4y + 720x^3y^2 + 1080x^2y^3 + 810xy^4 + 243y^5$, (d) $1 - 3x + \frac{15}{4}x^2 + \frac{5}{2}x^3 + \frac{15}{16}x^4 + \frac{3}{16}x^5 + \frac{1}{64}x^6$

11 (a) $1 + x + \frac{3}{2}x^2 + \frac{5}{2}x^3 + ..., x < 1$, (b) $1 + x - x^2 + \frac{5}{3}x^3 + ..., 3x < 1$

12 $1 + 0.01 - 0.01^2 + ...$

13 $1 - 0.2 + 0.04 - 0.008 + 0.0016 + ...$

14 $1 + 6x + 24x^3 - 80x^3 + ..., -\frac{1}{2} < x < \frac{1}{2}$, (b) $\frac{1}{16} + \frac{x}{32} + \frac{3}{256}x^2 + \frac{1}{256}x^3$, $-4 < x < 4$

15 $+6\%$

16 Reduction by 6%

17 $1 + 9x/2$

Problems

1 (a) $a = 2, d = 1$, arithmetic, (b) $a = -3, r = -2$, geometric, (c) $a = 5, d = 1$, arithmetic, (d) $a = 1, r = 4$, geometric, (e) $a = -1, r = -1$, geometric

2 (a) 29, (b) 46, (c) 41

3 (a) 2, 5, 8, 11, 14, (b) $-4, -2, 0, 2, 4$

4 (a) 1458, (b) 243, (c) $-1/32$

5 (a) 24, 12, 6, 3, 1.5, (b) 10, 1, 0.1, 0.01, 0.001
6 £3800
7 140, 98, 68.6, 48.0 cm
8 1.78
9 (a) 630, (b) 610, (c) 202
10 11
11 12 or 16
12 37
13 (a) 85, (b) 128, (c) 305.25
14 (a) 27/5, (b) 24, (c) 5/0.9
15 31.5
16 £872.1
17 (a) $243 + 405x + 270x^2 + 90x^3 + 15x^4 + x^5$, (b) $128 + 448x + 672x^2 + 560x^3 + 280x^4 + 84x^5 + 14x^6 + x^7$, (c) $1 + 8x + 24x^2 + 32x^3 + 16x^4$, (d) $64 - 192x + 240x^2 - 160x^3 + 60x^4 - 12x^5 + x^6$, (e) $y^3 - 6y^2x + 12yx^2 - 8y^3$, (f) $x^4 + 8x^3y + 24x^2y^2 + 32xy^3 + 16y^4$, (g) $1 + 2x + 3x^2 - 4x^3 + ...$, $|x| < 1$, (h) $\sqrt{2}(1 + x/4 - x^2/32 + x^3/384 - ...)$, $|x| < 1$, (i) $1 + x + 3x^2/2 + 5x^3/2 + ...$, $|x| < \frac{1}{2}$, (j) $1 + x - x^2 + 5x^3/3 + ...$, $|x| < 1/3$, (k) $1 - 2x + 3x^2 - 4x^3 + ...$, $|x| < 1$, (l) $1/3 + 2x/9 + 4x^2/27 + ...$, $|x| < 3/2$
18 1%
19 (a) +1%, (b) −3%
20 (a) 1.008 02, (b) 1.020 41, (c) 0.868 13
21 (a) $1 + 7x/2$, (b) $1 + 2x$, (c) $1 - x$

Chapter 8 *Revision*

1 (a) The negative sign, (b) N_0, (c) λ small
2 (a) The power is positive, (b) L_0, (c) α high means a high expansion
3 (a) 0, (b) 3 A
4 (a) 2, infinite, (b) 10, 0, (c) 0, 2, (d) 2, 0, (e) −4, 0,(f) 0, 0.5, (g) 0, 4, (h) 10, 0, (i) 0, 0.2
5 (a) 403.43, (b) 3.48×10^{-3}, (c) 1.49, (d) 0.741, (e) 1.99×10^5, (f) 3.70×10^{-3}
6 5, 13.59, 36.94, 100.43, infinite; 5, 1.84, 0.68, 0.25, 0; 0, 3.16, 4.32, 4.75, 5
7 10, 73.89, 546.00, 4034.29, infinite; 10, 1.35, 0.18, 0.02, 0; 8.65, 9.82, 9.98, 10
8 2, 3.30, 5.44, 8.96, infinite; 2, 1.21, 0.74, 0.45, 0; 0, 0.78, 1.26, 1.55, 2
9 10, 11.05, 12.21, 13.50, infinite; 10, 9.05, 8.19, 7.41, 0; 0, 0.95, 1.81, 2.59, 10
10 (a) 18.10 V, (b) 7.36 V
11 9.96×10^4 Pa
12 (a) 0, (b) $0.86E/R$ A
13 (a) 0, (b) $0.86E$
14 0.904 84
15 $1 + 5x + 17x^2/2 + 37x^3/6 + ...$
16 (a) 2^{8x}, (b) 3^{-2x}, (c) e^8, (d) e^{8t}, (e) e^{-2t}, (f) e^{-12t}, (g) $1 + 2 e^{2t} + e^{4t}$, (h) e^{-3t}, (i) e^{-2t}, (j) e^{-8t}, (k) $5 e^{3t}$, (l) $0.4 e^{4t}$

Problems

1 100 V, decreasing
2 (a) Negative sign, (b) C_0, (c) slower
3 (a) 2, infinite, (b) 12, infinite, (c) 3, 0, (d) 10, 0, (e) 2, 0, (f) 0, 2,
    (g) 0, 12, (h) 0, 2
4 (a) 54.60, (b) 0.18, (c) 1.49, (d) 0.67, (e) 19.48, (f) 2.87, (g) −19.78,
    (h) 9.50, (i) 0.19, (j) 2.43
5 10, 12.21, 14.92, 18.22, infinite; 10, 8.19, 6.70, 5.49, 0; 0, 1.81, 3.30,
    4.51, 10; 10, 11.05, 12.21, 13.50, infinite; 10, 9.05, 8.19, 7.41, 0; 0,
    0.95, 1.81, 2.59, 10
6 0.95 μC
7 (a) 0, (b) 1.57 A, (c) 2.53 A, (d) 3.11 A, (e) 4 A
8 5.13 V
9 5637 Ω
10 (a) 1.26 A, (b) 1.72 A, (c) 1.90 A
11 0.03 g
12 (a) 8.61×10^4 Pa, (b) 7.41×10^4 Pa
13 (a) 198.0 V, (b) 190.2 V
14 (a) $-E$, (b) $-0.61E$
15 (a) 0, (b) $0.63E$
16 (a) 200°C, (b) 134.1°C, (c) 0°C
17 (a) 100%, (b) 95.1%, (c) 60.7%, (d) 0%
18 $+$ gives growth, $-$ decay.
19 (a) 4^{3x}, (b) 5^{2x}, (c) e^{9x}, (d) e^{-4x}, (e) e^{4t}, (f) $1 + 2\,e^{2t} + e^{4t}$,
    (g) $4 + 6\,e^{4t} + 9\,e^{8t}$, (h) $4\,e^{-4t}$, (i) $3\,e^{-2t}$, (j) $0.25\,e^{-6t}$
20 1.051 23
21 (a) $-1 - x + x^2/2 + 5x^3/6 + 11x^4/24 + ...,$
    (b) $1 + x + x^2 + 2x^3/3 + 2x^4/3 + ...$

Chapter 9 *Revision*

1 (a) $x = \lg_{10} 7.8$, (b) $x = \lg_{10} 0.97$, (c) $x = \ln 1.54$, (d) $x = \ln 0.99$,
    (e) $x = \lg_2 4.5$, (f) $x = \lg_5 1.5$, (g) $x = \lg_{10} 4.5$, (h) $x = \ln 1.8$
2 (a) $7.5 = 10^x$, (b) $0.45 = 10^x$, (c) $9.9 = e^x$, (d) $0.88 = e^x$, (e) $20 = 2^x$,
    (f) $12 = 5^x$, (g) $3 = e^x$
3 (a) 4, (b) 2, (c) −2, (d) 10 000, (e) 9, (f) 16
4 (a) 0.892, (b) −0.0132, (c) 0.432, (d) −0.0101, (e) 0.875, (f) −0.347,
    (g) 2.293, (h) −0.128, (i) 1.61, (j) 1.11, (k) 8.067, (l) 3.380, (m) 0.580,
    (n) 0.273, (o) 11 013
5 (a) 4.800, (b) −0.981, (c) 0.062, (d) −5.203, (e) −11.525, (f) 1.779,
    (g) 2, (h) −6
6 (a) −0.438, (b) 0.270, (c) 0256, (d) 0.0477
7 (a) 0.926, (b) 1.639, (c) 0.931, (d) −2, (e) −8.64, (f) 2.35, (g) 0.693,
    (h) 2.386
8 $\ln 2/\lambda$
9 1.31×10^{-4} /°C

Problems

1 (a) $x = \lg_{10} 2.4$, (b) $x = \lg_{10} 2.5$, (c) $x = \lg_{10} 0.68$, (d) $x = \ln 1.3$,

(e) $x = \ln 55.6$, (f) $x = \ln 5.8$, (g) $x = \ln 0.35$
2 (a) $34 = 10^x$, (b) $0.54 = 10^x$, (c) $345 = 10^x$, (d) $37 = e^x$, (e) $0.86 = e^x$, (f) $397 = e^x$
3 (a) 0.380, (b) 0.340, (c) −0.167, (d) 0.262, (e) 4.018, (f) 1.758, (g) −1.050, (h) 1.531, (i) −0.268, (j) 2.538, (k) 3.611, (l) −0.151, (m) 5.984
4 (a) −1.478, (b) 2.870, (c) 1.967, (d) −0.972, (e) 16.014, (f) 6.300
5 (a) 0.858, (b) 205 936, (c) 0.242, (d) 0.507
6 (a) 1.532, (b) 1.151, (c) 0.511, (d) 1.783
7 0.527
8 (a) 4.19, (b) 1.76, (c) 0.44, (d) 1.15, (e) −1.56, (f) 0.347, (g) −1.19
9 7.16×10^4
10 9.16×10^{-6} s
11 5.36×10^{-4} s^{-1}
12 36.4

Chapter 10 *Revision*
1 (a) 1.13 rad, (b) 2.18 rad, (c) 4.28 rad, (d) 0.35 rad, (e) 2.08 rad, (f) 5.45 rad
2 (a) 17.2°, (b) 108.9°, (c) 200.5°, (d) 30°, (e) 77.1°, (f) 36°
3 (a) 3/5, (b) 4/5, (c) 3/4, (d) 5/3, (e) 5/4, (f) 4/3
4 (a) 0.342, (b) 2.924
5 (a) 0.707, (b) 1.414
6 (a) 0.577, (b) 1.732
7 (a) 0.819, (b) 0.966, (c) 1.732, (d) 1.743, (e) 5.759, (f) 2.145, (g) 0.783, (h) 0.454, (i) 0.309, (j) 1.277, (k) 1.065, (l) 2.070
8 (a) 58.2°, (b) 69.5°, (c) 56.3°, (d) 64.2°, (e) 12.1°, (f) 31.8°, (g) 39.8°, (h) 39.7°
9 (a) 2.88, (b) 5.74, (c) 2.75, (d) 22.64, (e) 9.46, (f) 5.09
10 16.8 m
11 14.5°
12 11.1 m, 4.9 m, 7.0 m
13 6.8°
14 270.8 m, 44.6° W of N
15 66.4 m
16 (a) $12\sqrt{3}$ m, (b) $8/\sqrt{3}$ m, (c) $5\sqrt{3}$ m
17 (a) 10 m, (b) $6/\sqrt{2} = 3\sqrt{2}$ m, (c) 5 m
18 As given in the problem
19 (a) 0.819, (b) −0.819, (c) −0.176, (d) −0.342, (e) −0.766, (f) 5.672, (g) −0.940, (h) 0.500, (i) - 0.364, (j) −0.259, (k) 0.598, (l) −0.999, (m) 1.778, (n) −0.612
20 ½
21 $-\sqrt{2}/2$
22 (a) 59.5°, 239.5°, (b) 53.1°, 126.9°, (c) 134.4°, 225.6°, (d) 210°, 330°, (e) 36.9°, 323.1°
23 (a) 406.0 m, (b) 177.7 m, (c) 36.2°, (d) 52.4°, (e) 51.3 mm, (f) 51.5 mm, (g) 83.8 mm, (h) 46°
24 5.49 m, 4.71 m

25 28.9 m
26 (a) 63.2° or 116.8°, 91.8° or 38.2°, (b) 56.8° or 123.2°, 81.2° or 14.8°
27 (a) 12.5, (b) 44.4°, (c) 62.9, (d) 16.9, (e) 79.7°, (f) 15.6, (g) 33.1 mm, (h) 63.4 mm
28 100.3°
29 101°, 79°
31 59.0°, 70.5°, 50.5°0
31 (a) 353.6 mm^2, (b) 2.30 m^2, (c) 3288.9 mm^2, (d) 5.91 m^2, (e) 738.8 mm^2, (f) 1285.6 mm^2, (g) 344.1 mm^2
32 1184.8 m^2
33 (a) 6 m^2, (b) 24 m^2, (c) 34.98 m^2, (d) 17.3 cm^2, (e) 26.8 cm^2, (f) 14.7 cm^2, (g) 9.8 cm^2
34 4.79 cm^2

Problems
 1 (a) 0.391, (b) 0.965, (c) 0.354, (d) 1.788, (e) 1.942, (f) 0.649, (g) 0.479, (h) 0.353, (i) 0.627, (j) 1.332, (k) 2.010, (l) 0.355, (m) 0.643, (n) −0.866, (o) −1.192, (p) −0.174, (q) −0.966, (r) 1.192, (s) −0.985, (t) 0.342, (u) −1.000, (v) −0.643, (w) 0.985, (x) −0.988, (y) 1.898, (z) −3.133
 2 (a) 51.3°, (b) 81.4°, (c) 62.2°
 3 (a) 2.48, (b) 6.97, (c) 1.41, (d) 19.20, (e) 1.37, (f) 1.23
 4 27.5 m
 5 2.1 m
 6 6.4 m
 7 23.0 km
 8 4.4 m
 9 7.7 m
10 824.3 m
11 5.02 m
12 5.91 m
13 As given in the problem
14 (a) 20.5°, 159.5°, (b) 123.4°, 236.6°, (c) 53.1°, 306.9°, (d) 5.7°, 185.7°, (e) 123.7°, 303.7°., (f) 208.7° , 331.3°, (g) 141.1°, 321.1°, (h) 113.6°, 293.6°, (i) 203.6°, 336.4°, (j) 41.8°, 138.2°
15 (a) 0.5, (b) infinity, (c) 0, (d) 0.5, (e) −1, (f) −0.87, (g) 0.58
16 (a) 42°, 41.4 mm, 29.9 mm, (b) 45°, 77°, 206.8 mm, (c) 75°, 4.29 cm, 9.81 cm, (d) 110°, 9.24 mm, 46.08 mm, (e) 99.3°, 30.7°, 38.85 mm, (f) 70°, 136.2 mm, 146.2 mm, (g) 56°, 53.3 mm, 53.3 mm, (h) 65°, 138.2 mm, 141.9 mm, (i) 50.9°, 49.1°, 126.6 mm, (j) 53.1°, 96.9°, 99.3 mm or 126.9°, 23.1°, 39.2 mm
17 6.35 m, 5.37 m
18 2.01 m
19 44.9°
20 (a) $A = 4.8°$, $B = 153.9°$, $C = 21.3°$, (b) $a = 85.8$, $B = 50.1°$, $C = 99.6°$, (c) $A = 31.5°$, $B = 48.9°$, $C = 99.6°$, (d) $A = 120°$, $B = 38.2°$, $C = 21.8°$, (e) $A = 34°$, $B = 101.5°$, $C = 44.5°$, (f) $a = 9.67$, $B = 45.5°$, $C = 54.5°$, (g) $B = 47.3°$, $C = 66.7°$, $a = 49.7$ mm, (h) $A = 34.1°$, $B = 44.4°$, $C = 101.5°$, (i) $A = 76.9°$, $B = 29.1°$, $c = 98.7$ mm

21 118.2 km
22 128.4 mm
23 31.2 mm
24 (a) 199.2 mm^2, (b) 919.3 mm^2, (c) 3.94 m^2, (d) 19.7 m^2, (e) 13.42 cm^2, (f) 1.73 m^2, (g) 30 cm^2
25 33.7°, 146.3°
26 34.8 cm^2

Chapter 11

Revision
1 (a) cos θ, (b) tan θ, (c) sin θ, (d) sin$^2\theta$
2 As given in the problem
3 As given in the problem
4 As given in the problem
5 As given in the problem
6 As given in the problem
7 As given in the problem
8 −sin A
9 As given in the problem
10 30°, 90°

Problems
1 As given in the problem
2 As given in the problem
3 As given in the problem
4 (a) −cos A, (b) cos A

Chapter 12

Revision
1 See Figure A.1
2 5 at an angle of 36.9° to **a**
3 8.3 at an angle of 19.9° north of east
4 (a) 6.08 N at 34.7° to the horizontal, (b) 6.71 N at 26.6° west of north, (c) 7.61 N at 28.5° to 5 N force, (d) 10.2 N at 13.0° to 8 N force, (e) 11.6 N at 12° to 7 N force, (f) 17.9 N at 18.9° to 10 N force, (g) 13.5 N at 45.7° to 12 N force
5 3.17 kN at 9.4° to 1.2 kN force
6 21 N, 16 N
7 (a) 11.18 km/h at 26.6° east of north, (b) 4.25 km/h at 48.3° east of north, (c) 6.24 km/h at 76.1° east of north
8 60° to the bank
9 17.4 km/h in southerly direction
10 About 134 kN at 23°
11 About 4.7 N at 49°
12 (a) 4.47 at 26.6° south of west, (b) 4.4 at 37° north of east, (c) 10.75 at 9.0° north of east, (d) 7.61 at 36.5° east of north
13 65 m/s due south
14 15.9 m/s at 57.8° east of south
15 32°

Figure A.1 *Chapter 12 Revision problem 1*

16 15.9 m/s at 32.2° south of east
17 (a) 7.7 N, 6.4 N, (b) 5.1 kN, 14.1 kN, (c) 11.3 N, 4.1 N, (d) 5.2 kN, 29.5 kN
18 28.2 N, 10.3 N
19 (a) 0.78 N at 39.9° south of west, (b) 4.17 N at 4.5° north of east, (c) 6.85 N at 40.7° south of east
20 28.3 m/s, 28.3 m/s
21 375.9 m/s, 136.8 m/s
22 7.07 m, 7.07 m

Problems
1 (a) 3.61 at 56.3° north of east, (b) 6.45° at 26° west of north, (c) 2.95 at 73° north of east, (d) 3.61 at 56.3° south of west, (e) 2.83 at 3.5° north of east, (f) 5.60 at 30.4° south of east
2 (a) 4.47 kN at 26.5° north of west, (b) 22.3 N at 63.4° from horizontal, (c) 83.2 N at 22.5° from 40 N force, (d) 27.5 N at 16.2° from vertical, (e) 13.5 N at 45.7° from 12 N force, (f) 17.4 kN at 83.4° from 12 kN force
3 14.0 N
4 7.3 N, 16.3 N
5 34.3°
6 44.7 kN at 26.6° to the 40 kN force
7 58° and 48° with respect to the 90 N force
8 22.8 kN at 70.3° to the horizontal
9 (a) 5 km/h at 36.9° east of north, (b) 10.8 m/s at 21.8° to the vertical, (c) 27.5 m/s at 16.2° to 20 m/s velocity, (d) 14.7 km/h at 16.3° west of north, (e) 10 m/s north
10 11.2 m/s at 63.4° to the horizontal
11 239.9 km/h at 17.1° south of east
12 8.6 km at 54.5° north of east
13 9.27 km at 62.8° west of south
14 7.2 N at 9° east of north
15 21.4 km/h at 119.7° west of north
16 5 m/s at 36.9° west of north
17 5.0 N, 8.7 N
18 5.3 N at 33.8° west of north
19 12.3 kN, 8.6 kN
20 34.3 N at 27.8° west of north
21 4.1 N at 6° north of west

Chapter 13 *Revision*
1 See Figure A.2
2 See Figure A.3
3 See Figure A.4
4 (a) See Figure 13.13, (b) see Figure A.5, (c) see Figure 13.11, (d) see Figure A.6(a), (e) see Figure A.6(b)
5 (a), (b), (c), (d), (e) See Figure A.7
6 See Figure A.8

Figure A.2 *Chapter 13 Revision problem 1*

7 (a) $x = 1$, $y = 2$, (b) $x = 2$, $y = -1$, (c) $x = -2.2$, $y = 4.8$, $x = 0.69$, $y = 0.48$, (d) $x = -0.4$, $y = 0.6$, $x = 0.6$, $y = 1.4$
8 (a) 5, (b) -2, -1, (c) 3.4, 0.6, (d) -1.8, 0, 2.8, (e) 1, (f) -2, 1, 3
9 (a) 0, (b) $+2$, -2, (c) -1, -1, (d) -2, -1, (e) no real roots
10 Shifted vertically by $3 - 1 = 2$ units
11 (a) 1, (b) 1/5, (c) 1/4, (d) 3/5, (e) -2
12 (a) 5, 3, (b) 1, -3, (c) 2, 1/2, (d) -2, 2
13 (a) $y = 2x + 3$, (b) $y = 4x - 3$, (c) $y = -2x + 1$, (d) $y = 5x$
14 3, 4
15 (a) 4, (b) 5, (c) 13
16 (a) 4 square units, (b) 2.5 square units, (c) 1/3 square units, (d) 23/3 square units
17 (a) 8 square units, (b) 8.7 square units, (c) 20 square units
18 51 square units
19 1235 m

Figure A.3 *Chapter 13 Revision problem 2*

Figure A.4 *Chapter 13 Revision problem 3*

Figure A.5 *Chapter 13 Revision problem 4(b)*

Figure A.6 *Chapter 13 Revision problem 4(d) and 4(e)*

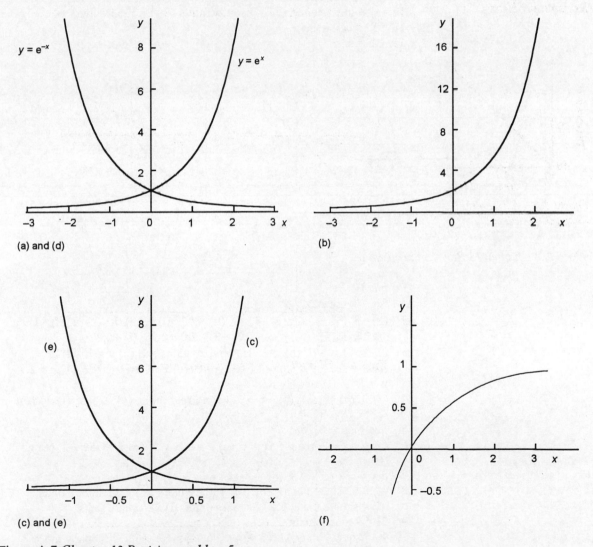

Figure A.7 *Chapter 13 Revision problem 5*

Figure A.8 *Chapter 13 Revision problem 6*

Problems

1   See Figure A.9
2   (a) −6, (b) +1, +2, (c) −1, −3, (d) −2.2, +0.7, (e) 0, −3, +1, (f) −1.95, +0.33, +1.52, (g) −1.5, +0.3, +3.7, (h) +0.95, (i) +0.87
3   (a) $x = 1$, $y = 1$, (b) $x = −1$, $y = 1$, $x = 3$, $y = 9$, (c) $x = 1$, $y = 2$, (d) $x = 1.63$, $y = 8$, $x = −1.63$, $y = 8$, (e) $x = −1.79$, $y = 3.21$, $x = 2.79$, $y = 7.79$, (f) $x = −1$
4   (a) 0, (b) +3, −3, (c) −4, + 1, (d) no real roots, (e) −2, −2, (f) no real roots, (g) −3, +2
5   See Figure A.10
6   (a) $y = x + 2$, (b) $y = 5x − 1$, (c) $y = −3x + 2$, (d) $y = −5x − 2$, (e) $y = 3$, (f) $y = 2x − 5$
7   (a) 7, (b) −2, (c) 3, (d) 8, (e) 6
8   (a) 8.63 square units, (b) 1.01 square units, (c) 2.83 square units, (d) 4 square units, (e) 12 square units, (f) 28 square units
9   21.66 square units
10 40.5 square units with 5 strips, 46.7 square units with many more
11 71.25 square units
12 1.49 square units

Figure A.9 *Chapter 13 Problem 1*

(a)

(b)

(c)

(d)

(e)

(f)

Figure A.10 *Chapter 13 Problem 5*

Chapter 14

Figure A.11 *Chapter 14 Revision problem 5*

Figure A.13 *Chapter 14 Revision problem 8*

Revision

1 A, C changing velocity, B constant velocity, D at rest
2 (a) 2 m/s, (b) 0, (c) 1.5 m/s, (d) 1.1 m/s
3 (a) 19.6 m/s, (b) 29.4 m/s, (c) 39.2 m/s
4 (a) 16.8 m/s, (b) −8.3 m/s, (c) −15.1 m/s, (d) −5.6 m/s
5 See Figure A.11.
6 See Figure A.12. (a) 0.9 m/s^2, (b) −1.8 m/s^2, (c) 3150 m
7 (a) 1 m/s^2, (b) 60 m
8 See Figure A.13. 12 m/s^2, 98.3 m
9 See Figure A.14

Figure A.12 *Chapter 14 Revision problem 6*

Figure A.14 *Chapter 14*
Revision problem 9

Problems

1 (a) 34 mm/s, (b) 50 mm/s, (c) 66 mm/s
2 (a) 8 m/s, (b) 11 m/s, (c) 16 m/s
3 (a) 16 mm/s, (b) 24 mm/s, (c) 32 mm/s
4 (a) 4 mm/s, (b) 6 mm/s, (c) 32 mm/s
5 (a) 18 mm/s, (b) 26 mm/s, (c) 8 mm/s
6 -1.0 m/s^2, 11 m
7 (a) 10 m/s^2, (b) 10 m/s^2, (c) 11.25 m
8 (a) 0.69 m/s^2, (b) -1.39 m/s^2, (c) 903 m
9 1.2 m/s^2, 12.6 m
10 54.8 s
11 1.25 m/s^2, 256 m
12 -4 m/s^2, 54 m
13 (a) Distance–time graph non-zero gradient straight line, velocity–time graph straight line with zero gradient, (b) distance–time graph curved, velocity–time graph non-zero gradient straight line, (c) as (a), (d) as (b).

Chapter 15

Revision

1 (a) $V = 0.07\theta + 19$, (b) $n = 0.15V - 3$, (c) $L = -2.91\theta + 2540$, (d) $\sigma = -0.025\theta + 85.2$
2 (a) $y = 3x^2 - 10$, (b) $y = x^2 + 2$, (c) $y = -2x^2 + 5$
3 (a) $y = \frac{1}{x} + 1$, (b) $y = \frac{2}{x} - 1$, (c) $y = -\frac{1}{x} + 2$
4 (a) $y = 3x^2 - x$, (b) $y = x^2 + 2x$, (c) $y = -x^2 + x$
5 (a) $y = x^3$, (b) $y = 2x^{-2}$, (c) $y = -2x^{1/2}$
6 (a) $y = 0.5\,e^{0.3x}$, (b) $y = 3\,e^{-0.2x}$, (c) $y = -2\,e^{0.1x}$
7 As problems 5 and 6
8 (a) log-linear, 1 cycle, (b) log-log, 1 cycle $\times$ 1 cycle, (c) log-log, 1 cycle $\times$ 2 cycles

Problems

1 (a) $E = 0.45L + 5$, (b) $R = 4.2L$, (c) $v = 0.1t + 4$, (d) $\sigma = -0.098\theta + 160$
2 (a) $y = 2x^2 - 1$, (b) $y = -x^2 + 3$, (c) $y = \frac{3}{x} + 1$, (d) $y = -\frac{2}{x} - 4$,
   (e) $y = 2x^2 + x$, (f) $y = -x^2 - x$, (g) $y = 0.5x^{1/2}$, (h) $y = -0.6x^3$,
   (i) $y = 100\,e^{-0.1x}$, (j) $y = 10\,e^{0.3x}$
3 (a) T against $\sqrt{L}$, gradient $2\pi/\sqrt{g}$, intercept 0,
   (b) s/t against t, gradient $a/2$, intercept u,
   (c) e/θ against θ, gradient b, intercept a,
   (d) R against θ, gradient $R_0\alpha$, intercept R_0,
   (e) $\ln i$ against t, gradient $1/RC$, intercept $\ln I$,
   (f) p against $1/V$, gradient k, intercept 0,
   (g) y against L^4, gradient $w/8EI$, intercept 0, or $\lg y$ against $\lg L$, gradient 4, intercept $\lg (w/8EI)$.
4 $R = \dfrac{1200}{V} + 50$
5 $R = \dfrac{0.16}{d^2}$
6 $T = 2.0L^{1/2}$

Figure A.15   *Chapter 16*
Revision problem 1

7   $V = 16p^{-1}$
8   $T = 500p^{0.28}$
9   $C = 0.001n^3 + 30$
10 $s = 0.1v^2 + 0.5v$
11 $I = 0.001V^2$
12 $E = 4.1f - 4.3$
13 2×2
14 2
15 $T = 0.2R^{3/2}$ (this is known as one of Kepler's laws)
16 $v = 10.2\ e^{-0.1t}$
17 $\theta = 800\ e^{-0.2t}$
18 $Q = 2.6h^{2.5}$
19 $A = 400\ e^{-0.02t}$
20 $T = 50\ e^{0.3\theta}$

Chapter 16

Revision

1 See Figure A.15
2 (a) (4.47, $\angle 26.6°$) or (4.47, $\angle 0.46$), (b) (2.83, $\angle -45°$) or (2.83, $\angle -\pi/4$), or (2.83, $\angle 315°$) or (2.83, $\angle 7\pi/4$), (c) (4.47, $\angle 116.6°$) or (4.47, $\angle 2.03$), (d) (2.83, $\angle 225°$) or (2.83, $\angle 5\pi/4$), or (2.83, $\angle -135°$) or (2.83, $\angle -3\pi/4$), (e) (2, $\angle 0°$) or (2, $\angle 0$), (f) (2, $\angle 90°$) or (2, $\angle \pi/2$), (g) (4, $\angle 180°$) or (4, $\angle \pi$), (h) (4, $\angle 270°$) or (4, $\angle 3\pi/2$)
3 (a) (2.8, 2.8), (b) (1.5, 2.6), (c) (−1.7, 1.0), (d) (1.0, −2.8), (e) (1.4, 1.4), (f) (3.5, 2.0), (g) (−3.0, 0)
4 (a) As Figure 16.6 with radius 3, (b) as Figure 16.7 with angle $\pi/6$ rad, (c) as Figure 16.8 with all radii doubled, (d) see Figure A.16

Figure A.16   *Chapter 16*
Revision problem 4(d)

Problems

1 (a) (3, $\angle 90°$) or (3, $\angle \pi/2$), (b) (2, $\angle 180°$) or (2, $\angle \pi$), (c) (2.83, $\angle 135°$) or (2.83, $\angle 3\pi/4$), (d) (5, $\angle 53.1°$) or (5, $\angle 0.93$), (e) (5, $\angle 126.9°$) or (5, $\angle 2.22$), (f) (4.47, $\angle 333.4°$) or (4.47, $\angle 5.82$), or (4.47, $\angle -26.6°$) or (4.47, $\angle -0.46$), (g) (3, $\angle 0°$) or (3, $\angle 0$)
2 (a) (1.3, 1.5), (b) (−2.8, 1.0), (c) (−0.5, −3.0), (d) (3.8, −1.4), (e) (1.0, 1.7), (f) (0.5, 0.9), (g) (−3.0, 0), (h) (0, 4.0)
3 (a) As in Figure 16.6 with radius 4, (b) as in Figure 16.7 with angle $\pi/4$, (c) see Figure A.17(a), (d) see Figure A.17(b)
4 See Figure A.18
5 As in the problem.

(a)

(b)

Figure A.17   *Chapter 16*
Problem 3

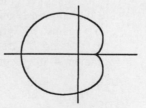

Figure A.18   *Chapter 16 Problem 4*

Chapter 17

(a)

(b)

(c)

Figure A.19   *Chapter 17 Revision problem 15*

Revision

1 (a) 1 V, (b) 0, (c) 2 A, (d) 1 A, (e) 0.9 V, (f) 1 V
2 As given in the problem
3 14 mA
4 8.2 V
5 As given in the problem
6 7.5 mA, 19 mA, 2.53
7 (a) $v = 10 \sin 2\pi 50t$ V, (b) (i) 5.88 V, (ii) 9.51 V, (iii) −5.88 V
8 (a) $i = 50 \sin 2\pi \times 2 \times 10^3 t$ mA, (b) (i) −47.6 mA, (ii) −29.4 mA, (iii) 47.6 mA
9 (a) 0, (b) 10 V, (c) 1.99 V
10 0, 0.20, 0.39, 0.56, 0.72, 0.84, 0.93, 0.99, 1.0
11 5, 2.5, 0, −2.5, −5, −2.5, 0, 2.5, 5
12 (a) 0.87 V, (b) −0.87 V
13 (a) 100 mA, (b) 100 Hz, (c) 0.25 rad or 14.3° lagging
14 (a) 12 V, (b) 50 Hz, (c) 0.5 rad or 28.6° leading
15 See Figure A.19
16 $22.4 \sin (\omega t + 1.11)$ mA
17 $6.81 \sin (\omega t + 0.147)$ V
18 $126.2\sqrt{2} \sin (\omega t − 0.071)$ V
19 $8.72\sqrt{2} \sin (\omega t − 0.639)$ A
20 $2.95 \sin (\omega t − 0.50)$ V
21 $6.4 \sin (\omega t + 0.90)$ A
22 $85.0 \sin (\omega t + 0.728)$ mA

Problems

1   (a) 1.13 A, (b) 0
2   (a) 5 V, (b) 0
3   As given in the problem
4   As given in the problem
5   2.83 V
6   4 A
7   (a) 6.50 V, (b) 7.14 V, (c) 1.10
8   (a) 15.0 mA, (b) 17.1 mA, (c) 1.14
9   As given in the problem
10 As given in the problem
11 5 V, 7.07 V, 1.4
12 1 V
13 1.4 V
14 7.1 mA
15 141.4 mA
16 10 V
17 (a) 0, 1.68, 1.82, 0.28, −1.51, −1.92; 0, 1.98, 3.89, 5.65, 7.17, 8.41; 0, 4.55, −17.20, −1.40, 4.95, −2.72
    (b) 0, 8.66, 10.00, 8.66, 0; 10.00, 5.00, 0, −5.00, −10.00; 8.66, 8.66, 5.00, 0, −5.00
18 4.79 A
19 1.99 V
20 (a) 500 rad/s, (b) 79.6 Hz, (c) 17.1 V

(a)

(b)

Figure A.20   *Chapter 17*
Problem 21

21 See Figure A.20
22 (a) 20 mA, (b) 250 Hz, (c) 0.5 rad or 28.6° lagging
23 (a) 5 V, (b) 500 Hz, (c) 0.2 rad or 11.5° leading
24 $4.47 \sin (\omega t + 0.464)$ V
25 $6.24 \sin (\omega t - 0.281)$ V
26 $6.94 \sin (\omega t + 0.935)$ V
27 $43.6 \sin (\omega t + 0.648)$ mA
28 $50 \sin (\omega t + 0.927)$ mA
29 $5.66 \sin (\omega t + \pi/4)$ V
30 $25 \sin (\omega t - 0.644)$ V
31 $14.7 \sin (\omega t - 0.5)$ mA
32 $17.8 \sin (\omega t + 0.691)$ mA
33 $4.47 \sin (\omega t + 0.464)$ V
34 $8.77 \sin (\omega t - 0.524)$ V
35 (a) $12.81 \sin (\omega t - 0.151)$ mA, (b) $12.81 \sin (\omega t + 1.198)$ mA
36 $10 \sin (\omega t + \pi/3)$ A
37 $3.1 \sin (\omega t - 0.33)$ A

Chapter 18

Revision
1 (a) $6x$, 0, 6, (b) $2x + 1$, 1, 5, (c) 3, 3, 3, (d) $-1/2x^2$, infinity, $-1/2$,
   (e) 0, 0, 0, (f) 3, 3, 3, (g) 1/4, 1/4, 1/4
2 As given in the problem
3 (a) $6x^2$, (b) $6x^5$, (c) $-4x$, (d) $(3/2)x^{1/2}$, (e) $\frac{1}{2}x^{-3/4}$, (f) $30x^9$, (g) $-(3/5)x^{-6/5}$,
   (h) $0.4x^3$
4 10
5 24
6 $-1/2$
7 62.8 mm²/mm
8 (a) $2x$, (b) $8x^3 + 6x$, (c) $12x^2 - 4x + 3$, (d) $12x^2 - 6x + 4$,
   (e) $10x^4 + 9x^2 - 2$, (f) $20x^4$, (g) $-2x^{-2} - 10x^{-3}$, (h) $4x + 3x^{-2}$
9 (a) $\cos x$, (b) $5 \cos 5x$, (c) $-4 \sin 4x$, (d) $-3 \cos 3x$, (e) $6 \cos 3x$,
   (f) $-40 \sin 4x$
10 $v = -50 \sin 50t$
11 $i = 200 \cos 100t$
12 (a) $2 e^{2x}$, (b) $-3 e^{-3x}$, (c) $\frac{1}{2} e^{x/2}$, (d) $6 e^{3x}$, (e) $2 e^{2x}$, (f) $1 + 6 e^{2x}$,
   (g) $2 e^x + 6 e^{2x}$
13 $-\lambda N_0 e^{-\lambda t} = -\lambda N$
14 $1/x$
15 (a) 2, (b) 4, (c) $6x + 4$, (d) $-\sin x$, (e) $4 e^{2x}$
16 (a) $-\dfrac{CV}{RC} e^{-t/RC}$, (b) $\dfrac{CV}{R^2 C^2} e^{-t/RC}$

Problems
1 (a) $-4/x^2$, (b) $-2/x^3$
2 (a) $12x^5$, (b) $-10/x$, (c) $(7/3)x^{4/3}$, (d) $0.6x^2$, (e) $x^{-1/2}$, (f) $-3x^{-5/2}$, (g) $8x + 1$,
   (h) $6x^2 + 6x$, (i) $9x^2 - 10x + 3$, (j) $-15x^{-4}$, (k) $6x^2 - 2x^{-3}$,
   (l) $-10x^{-3} + 2x^{-2}$, (m) $4 \cos 4x$, (n) $-2 \sin 2x$, (o) e^x, (p) $2 e^{2x}$,
   (q) $-(2/3) e^{-2x/3}$, (r) $1/x$, (s) $(3/2)x^{-1/2}$, (t) $\cos x - 2 \sin 2x$

3 6

4 11

5 *at*

6 (a) 3, (b) –2.12, (c) 0

7 218.4

8 –0.30

9 20 mm^2/mm

10 $I\omega$

11 CV

12 $12 - 10t$

13 $-p_0c\,e^{-ch}$

14 434.8 A/s

15 –1091 V/s

16 4.8 mV/°C

17 (3, –40), (–1, 12)

18 –112 m^3/min

19 (a) 6, (b) 4 e^{2x}, (c) –4 cos 2x, (d) 6

Chapter 19 *Revision*

1 (a) $5x^4 + 3x^2 + 6x$, (b) $e^{4x} + 4x\,e^{4x}$, (c) $-2\sin x \sin 2x + \cos x \cos 2x$,
(d) $e^{-2x} - 2(x + 5)e^{2x}$, (e) $3x^2(3x^2 + 2x + 1) + x^3(6x + 2)$

2 (a) $-(2x^2 + 2x + 4)/(x^2 - 2)^2$, (b) $-(3x^2 + 6x)/(x^3 + 3x^2)^2$,
(c) $(x - 3)/(1 + x)^3$, (d) $(1 - \cos x)/\sin^2 x$

3 (a) $6x(2x + 1)^2$, (b) $x/\surd(x^2 + 1)$, (c) $3x^2 \cos x^3$, (d) 5 cos (5x + 2),
(e) $-4(6x^3 - x)^{-5}(18x^2 - 1)$, (f) $2(x + \sin x)(1 + \cos x)$

Problems

1 (a) $2x \sin x + x^3 \cos x$, (b) $2x(3x + 5) + 3x^2$, (c) $e^{4x} + 4x\,e^{4x}$,
(d) $14/(x + 5)^2$, (e) $4/(1 - 3x)^2$, (f) 4 cos (4x – 5), (g) $-5 \sin x \cos^4 x$,
(h) $6(x^2 - 6x + 1)^2(x - 3)$, (i) $4(x + 3)^3$, (j) $8(x^4 + 2x^3 + 3)^7(4x^3 + 6x^2)$,
(k) $3x^2/2(x^3 + 5)^{1/2}$, (l) –cot x cosec x, (m) $6x(x \cos 2x + \sin 2x)$,
(n) 3 sec^2 3x, (o) –2 cosec^2 2x, (p) $(3x + 2)/\surd(3x^2 + 4x - 1)$,
(q) $e^{3x}(4 \cos 4x + 3 \sin 4x)$, (r) 10 $e^{2x + 3}$

Chapter 20 *Revision*

1 (a) x = 3, minimum, (b) x = –1 maximum, x = 2 minimum, (c) x = 0
minimum, x = 2 maximum, (d) x = –3 maximum, x = 2 minimum,
(e) –3/2 minimum, (f) 1/4 maximum, (g) π/2 maximum, 3π/2 minimum,
5π/2 maximum, etc.

2 $x = \frac{1}{2}(a - b)$

3 48 m^2

4 $L/2$

5 20, 20

6 4 s

7 r = 0.34 m, h = 2.15 m

8 15 cm, 15 cm

9 2.89 cm × 2.89 cm × 5.77 cm

10 $I = E/2r$

11 33.4 V

12 $-a/2b$

Problems

1 (a) $x = 1$, minimum, (b) $x = -3$ minimum, (c) $x = 0$, minimum,
 (d) $x = 1/\sqrt{3}$ minimum, $x = -1/\sqrt{3}$ maximum, (e) $x = 0$, maximum,
 $x = -2/3$ minimum, (f) $x = 3$ minimum, $x = -2$ maximim,
 (g) $x = -2$ minimum, (h) $x = 2$ minimum, $x = -2$ maximum,
 (i) $x = 0$ inflexion, (j) $x = 1$ minimum, (k) $x = 1$ minimum,
 (l) $x = 1$ minimum, $x = -1$ maximum, (m) $x = 1/2$ minimum,
 $x = -2$ maximum, (n) $x = -2$ maximum, $x = 4$ minimum

2 40

3 6, 12

4 $L = 100$ m

5 $t = 3/2$ s

6 39.2 mm

7 $r = 0.86$ m, $h = 0.86$ m

8 50 m

9 As given in the problem

10 500 000

11 $r = 5.42$ cm, $h = 10.84$ cm

12 2.8 A

Chapter 21   *Revision*

1 (a) $2x^2 + C$, (b) $2x^3 + C$, (c) $\frac{1}{2}x^2 + 2x + C$, (d) $-\cos x + C$, (e) $e^x + C$

2 (a) $\frac{1}{4}x^4 + C$, (b) $\frac{1}{2}x^4 + C$, (c) $-\frac{1}{6}x^{-6} + C$, (d) $2x^{1/2} + C$, (e) $\frac{3}{4}x^{4/3} + C$,
 (f) $-\frac{1}{3}\cos 3x + C$, (g) $\frac{1}{2}\sin 2x + C$, (h) $\frac{1}{4}e^{4x} + C$, (i) $-\frac{1}{5}e^{-5x} + C$

3 (a) $\frac{5}{2}x^2 + 3x + C$, (b) $x^4 + x^3 + 2x + C$, (c) $-\cos x + \frac{1}{2}\sin 2x + C$,
 (d) $2x - \frac{1}{3}e^{-3x} + C$, (e) $\ln x + \frac{3}{2}x^2 + C$, (f) $\frac{3}{2}x^2 - \frac{1}{4}\cos 4x + C$

4 (a) $y = x^2 + 3x - 7$, (b) $y = 3x^3 + x^2 + 2x + 1$, (c) $y = 2x^2 - 9x + 5$,
 (d) $y = 2x + \frac{1}{3}x^{-1} + \frac{2}{3}$

5 $x = 10t + t^2 + 2$

6 $x = t^2 + 5t$

7 (a) 20 sq units, (b) 7 sq units, (c) 12 sq units, (d) 8 sq units,
 (e) 1 sq unit, (f) 0.71 sq units, (g) 0.86 sq units, (h) 2.25 sq units,
 (i) 21.3 sq units, (j) 7.5 sq units

8 8 force × distance units

9 109.5 force × distance units

Problems

1 (a) $\frac{1}{6}x^6 + C$, (b) $-\frac{1}{2}x^2 + C$, (c) $\frac{2}{3}x^{3/2} + C$, (d) $\frac{1}{2}x^{1/2} + C$, (e) $x^2 + C$,
 (f) $-\frac{1}{3}e^{-3x} + C$, (g) $2 e^{x/2} + C$, (h) $\sin x + C$, (i) $-\frac{1}{4}\cos 4x + C$,
 (j) $-\frac{1}{2}\cos 2x + C$, (k) $\frac{1}{2}\sin 2x + C$, (l) $x^2 + \frac{1}{3}x^3 + C$,
 (m) $\frac{1}{4}x^4 + x^2 + 4x + C$, (n) $3x - \frac{1}{2}e^{-2x} + C$,
 (o) $-\frac{1}{2}\cos 2x + \frac{1}{3}\sin 3x + C$

2 (a) $y = x^3 + 2$, (b) $y = 2x^2 - 2x + 3$, (c) $y = 2x^3 + 5x^2 - 5x - 1$,

(d) $y = 2x - 1$

3   $q = -\frac{3}{100} \cos 100t$ mC

4   $x = 5t^2 + t$

5   $\theta = \frac{1}{2}cI^2$

6   (a) 64/3, (b) 1, (c) 8, (d) 0.366, (e) 1.099, (f) −2,25, (g) 50/3

7   (a) 52/6 sq units, (b)1/4 sq units, (c) 8 sq units, (d) 66 sq units, (e) 12 sq units, (f) 32 sq units, (g) 0.71 sq units, (h) 52 sq units, (i) 4 square units, (j) 25.5 square units

Chapter 22

Revision

1   (a) $v = 10 - 2t$, (b) $a = -2$, (c) $v = 6$ m/s, $a = -2$ m/s^2

2   (a) $v = \frac{1}{2}t^{-1/2}$, (b) $a = -\frac{1}{4}t^{-3/2}$

3   2 m/s upwards

4   12 m/s, −4 m/s^2

5   $v = \frac{1}{2}\cos 2t - \frac{1}{3}\sin 3t$, $a = -\frac{1}{4}\sin 2t - \frac{1}{9}\cos 3t$

6   (a) $v = 6t$, (b) $a = 6$, (c) $v = 6$ m/s, $a = 6$ m/s^2

7   27 m

8   36 m

9   38 m

10   −29/3 m

11   $v = t^2 + t + 4$

12   (a) $v = 3t^2 + 4$, (b) $x = t^2 + 4t + 10$

13   $\frac{1}{3}kx^3$

14   30 J

15   18 J

16   $-\frac{3}{2}k\left(V_2^{-2/3} - V_1^{-2/3}\right)$

17   (a) 1.23, (b) 0.707

18   $V/2$

19   $i = 20t$ A

20   $i = -12\,e^{-3t}$ A

21   $i = 2 - 5\,e^{-5t}$ A

22   $i = -10 \sin 5t$ A

23   3.0 C

24   3 C

25   5 mA, constant

26   0.04 cos 10t A

27   −0.04 sin 10t A

28   See Figure A.21

29   $v = 10$ V

30   $v = 20\,e^{-100t}$ V

31   $v = -0.2\,e^{-2t}$ V

32   $i = 3t^3$

Problems

1   $v = 4t - 3$, $a = 4$

2   $v = -5 \sin 2t$, $a = -2.5 \cos 2t$

3   $v = 2\,e^{-2t}$, $a = -4\,e^{-2t}$

4   $v = 9t^2 - 1$, $a = 18t$

Figure A.21   *Chapter 22 Revision problem 28*

5 $v = 6 - 12t$, $a = -12$

6 2 m/s, -4 m/s^2

7 4 m/s, -8 m/s^2

8 10 s

9 2 m/s^2

10 2 s

11 $x = 36t - 2t^2 + 20$

12 180 m

13 3 m

14 1.96 m

15 32 s

16 $v = \frac{4}{3}t^3 + \frac{11}{3}$

17 $v = 3t^2 + 2t + 10$

18 $v = 4t^{3/2} + 32$

19 $\frac{2}{3}kx^{3/2}$

20 10 J

21 h^2

22 44 J

23 $-\dfrac{k}{r}$

24 $-0.4k\left(V_2^{-0.4} - V_1^{-0.4}\right)$

25 (a) 3.54, (b) 1.41

26 $V/\sqrt{3}$

27 1.27

28 $i = 4t$ A

29 $i = -10\,e^{-2t}$ A

30 $i = 2 - 10\,e^{-2t}$ A

31 $i = 10 \cos 2t$ A

32 0.024 C

33 0.4 C

34 (a) 0, (b) Ca, (c) $-CV\omega \sin \omega t$

35 $i = 0.2 \cos 100t$ A

36 $i = 16t$ mA

37 $v = 0.02$ V

38 $v = -0.3\,e^{-3t}$ V

39 $v = -16\,e^{-20t} + 16\,e^{-40t}$ V

40 See Figure A.22

41 $i = 2t^2$

V

0   1

Time in ms

Figure A.22 *Chapter 22*
Problem 40

Index

Alternating waveforms
    average value, 227
    defined, 227
    form factor, 232
    mean value, 227
    phasors, 233
    root-mean-square value, 230,
      302
Approximations, 102
Areas
    circle, 13
    cone, 18
    cylinder, 15
    compound shapes, 13
    irregular, 21, 193
    parallelogram, 12
    prisms, 16
    rectangle, 12
    sphere, 15
    trapezium, 12
    triangle, 12, 150

Brackets, 34

Capacitors, 304
Cosine rule, 148
Current, 303

Definite integral, 289
Degrees, 130
Differentiation
    chain rule, 272
    constant multiplier rule, 258
    defined, 252
    derivatives list, 266
    from first principles, 253
    function of a function, 272
    of a constant, 258
    of ax^n, 255
    of exponentials, 263
    of logarithmic functions, 264
    of trigonometric functions, 260
    product rule, 269

    quotient rule, 270
    second derivative, 265
    sum rule, 258

Equations
    graphs, 180
    linear, 45
    manipulating, 30
    polynomials, 57
    quadratic, 57, 189
    simultaneous, 46, 187
    solving graphically, 187
    straight line, 191, 205
Exponential
    decay, 108, 109
    defined, 107
    e, 109, 114
    graphs, 186
    growth, 107, 109
    manipulating, 115
    solving equations, 125

Factors, 58
Force–distance graphs, 24, 300
Form factor, 232
Formulae, manipulation, 37
Frusta
    prisms, 17
    pyramids, 19

Graphs
    area under, 21, 193, 288, 290
    Cartesian, 178
    gradients, 190, 252
    log graph paper, 212
    maxima and minima, 274
    of equations, 180
    of exponentials, 186
    of sums of functions, 186
    point of inflexion, 275
    scales, 179
    trigonometric functions, 182
    turning points, 174

Indefinite integral, 282
Indices
    defined, 5
    fractional, 7
    negative, 6
    rules for handling, 5
    standard form, 7
Inductors, 306
Integration
    as area under graph, 288
    common functions, 284
    of sum, 286
    particular solution, 287
    reverse of differentiation, 282

Kirchhoff's laws, 83

Linear laws, finding, 205
Logarithm
    common, 120
    defined, 120
    Napierian, 120
    natural, 120
    properties, 123

Maxima and minima, 274
Mid-ordinate rule, 22, 25, 193, 228

Ohm's law, 77

Periodic functions, 182
Phasors
    adding, 239
    defined, 233
    phase difference, 236
    representation, 235
    subtracting, 244
Polar
    coordinates, 221
    graphs, 223
Polynomials, 57
Pythagoras theorem, 157
Root-mean-square value, 230, 302

Quadratic equations
    completing the square, 61
    defined, 57
    factors, 58
    formula, 64

roots, 58, 68, 189
solving graphically, 188

Radians, 130
Resistance
    Ohm's law, 77
    parallel, 79
    series, 78
    series–parallel, 81
Roots, 59

Significant figures, 1
Sequence
    arithmetic, 93
    defined, 92
    geometric, 93
Series
    arithmetic, 95
    binomial, 98
    defined, 95
    geometric, 96
Sine rule, 145
SI units
    defined, 8
    prefixes, 9
Standard index form, 7
Straight-line motion
    distance–time graphs, 198
    equations, 71
    simultaneous equations, 74
    using calculus, 294
    velocity–time graphs, 200

Transposition, 32
Trapezoidal rule, 22, 24, 193
Trigonometric ratios
    angles of any magnitude, 140
    defined, 131
    differentiation, 260
    for common angles, 138
    for double angles, 160
    for sums of angles, 159
    graphs, 182
    relationships, 133, 157

Units
    manipulation, 40
    representation, 30
    SI, 8

Vectors
  adding, 163
  components, 171
  parallelogram rule, 164
  polygon, 168
  relative, 170
  representation, 162
  resolution, 171
  subtraction, 169
  triangle rule, 163

Volumes
  cone, 18
  cylinder, 15
  irregular solids, 24
  prisms, 15
  pyramids, 18
  sphere, 15

Work, 24, 300